Textbooks in Telecommunication Engineering

Series Editor

Tarek S. El-Bawab
Professor and Dean
School of Engineering
American University of Nigeria, Yola, Nigeria

The Textbooks in Telecommunications Series:

Telecommunications have evolved to embrace almost all aspects of our everyday life, including education, research, health care, business, banking, entertainment, space, remote sensing, meteorology, defense, homeland security, and social media, among others. With such progress in Telecom, it became evident that specialized telecommunication engineering education programs are necessary to accelerate the pace of advancement in this field. These programs will focus on network science and engineering; have curricula, labs, and textbooks of their own; and should prepare future engineers and researchers for several emerging challenges.

The IEEE Communications Society's Telecommunication Engineering Education (TEE) movement, led by Tarek S. El-Bawab, resulted in recognition of this field by the Accreditation Board for Engineering and Technology (ABET), November 1, 2014. The Springer's Series Textbooks in Telecommunication Engineering capitalizes on this milestone, and aims at designing, developing, and promoting high-quality textbooks to fulfill the teaching and research needs of this discipline, and those of related university curricula. The goal is to do so at both the undergraduate and graduate levels, and globally. The new series will supplement today's literature with modern and innovative telecommunication engineering textbooks and will make inroads in areas of network science and engineering where textbooks have been largely missing. The series aims at producing high-quality volumes featuring interactive content; innovative presentation media; classroom materials for students and professors; and dedicated websites.

Book proposals are solicited in all topics of telecommunication engineering including, but not limited to: network architecture and protocols; traffic engineering; telecommunication signaling and control; network availability, reliability, protection, and restoration; network management; network security; network design, measurements, and modeling; broadband access; MSO/cable networks; VoIP and IPTV; transmission media and systems; switching and routing (from legacy to next-generation paradigms); telecommunication software; wireless communication systems; wireless, cellular and personal networks; satellite and space communications and networks; optical communications and networks; free-space optical communications; cognitive communications and networks; green communications and networks; heterogeneous networks; dynamic networks; storage networks; ad hoc and sensor networks; social networks; software defined networks; interactive and multimedia communications and networks; network applications and services; e-health; e-business; big data; Internet of things; telecom economics and business; telecom regulation and standardization; and telecommunication labs of all kinds. Proposals of interest should suggest textbooks that can be used to design university courses, either in full or in part. They should focus on recent advances in the field while capturing legacy principles that are necessary for students to understand the bases of the discipline and appreciate its evolution trends. Books in this series will provide high-quality illustrations, examples, problems and case studies.

For further information, please contact: Dr. Tarek S. El-Bawab, Series Editor, Professor and Dean of Engineering, American University of Nigeria, telbawab@ieee.org; or Mary James, Senior Editor, Springer, mary.james@springer.com

More information about this series at http://www.springer.com/series/13835

Dario Sabella

Multi-access Edge Computing: Software Development at the Network Edge

 Springer

Dario Sabella
Intel (Germany)
Munich, Germany

ISSN 2524-4345   ISSN 2524-4353 (electronic)
Textbooks in Telecommunication Engineering
ISBN 978-3-030-79620-4   ISBN 978-3-030-79618-1   (eBook)
https://doi.org/10.1007/978-3-030-79618-1

This Springer imprint is published by the registered company Springer Nature Switzerland AG
The registered company address is: Gewerbestrasse 11, 6330 Cham, Switzerland

Foreword

The cloud has become in recent years one of the most important ingredients in any computing or networking recipe. But, as in all technology fields since the invention of fire, its use is at the same time fueling its evolution by discovering unresolved issues and finding ways to pushing its edge forward. As the attentive reader has surely noted, this "pushing the edge" is an intended pun to set the scene for the major topic of this book: edge computing and how it can be applied to develop enhanced applications.

What we can call the classic cloud architecture is highly centralized, relying on the network to facilitate the access of mobile applications to the central datacenters where processing takes place. This approach has several limitations, and we can name a few. The most immediate is related to latency, critical in any environment requiring a tight control loop, such as industrial processes, driving or surgery. But we should not forget privacy implications, and the requirement for personal information to not abandon a concrete realm, and security aspects, when dealing with data not to be exposed to environments beyond the trusted domain. It is worth adding a final consideration on the ossification this highly centralized approach implies, tying services not only to a software developer but also to an infrastructure provider.

Edge computing constitutes a technology proposal trying to address the above limitations, leveraging on the growing convergence between cloud and network infrastructures and providing the foundations for an evolution to a converged, transparent information processing environment able to support distributed, mobile applications addressing a much wider range of use cases, and becoming a key enabler for bringing next-generation networking (5G and beyond) promises into reality.

Such a global, pervasive environment requires the use of common standards and best practices to guarantee application availability to users and provide a stable reference framework to developers. This book provides a complete view of MEC, the most mature and comprehensive approach to edge computing. It starts with the foundations and principles of edge computing, introduces the MEC approach and analyzes its relationship with essential technologies such as virtualization and the mobile network internals. And it goes well beyond a pure theoretical approach, introducing readers to the practical matters of MEC APIs, application development and service lifecycle management. Finally, the author is one of the "usual suspects" on the edge computing arena worldwide, with a deep knowledge of all the standardization and implementation back- and foreground of the technology.

This book is not only a useful base for initiation to edge computing, including a consistent practical approach to MEC but also a valuable reference for application developers, infrastructure providers and advanced users to this next generation of pervasive networking and computing, with 5G and cloud-native on its banners.

Madrid, Spain
Diego Lopez
Senior Technology Expert at
Telefonica I+D, Chair of ETSI NFV

Preface

The book will cover the main aspects of Edge Computing, from the description of the technology to the standards and industry associations working in the field. The book is conceived as a textbook for developers, researchers, engineers, and it is particularly suitable for undergraduate/graduate lectures and courses on Edge Computing, with main emphasis on the ETSI MEC standard (Multi-access Edge Computing) and other SDOs related to 5G systems, in line with the recent efforts on harmonizing standards for edge computing (i.e. leveraging ETSI ISG MEC and 3GPP specifications).

In line with the spirit of Springer's Series Textbooks in Telecommunication Engineering, the book is aiming to provide a modern and innovative telecommunication engineering textbook, especially in a field so important as MEC/Edge Computing, where still literature doesn't provide sufficient support for Ph.D. lectures and courses. Moreover, the book aims at producing a high-quality volume, integrated by useful open source code references as classroom materials for engineers, developers, students and professors.

The book will include relevant topics in the following areas:

- Principles of Edge Computing, Fog and Cloud Computing
- MEC APIs, on MEC platforms and services
- MEC in NFV deployments and in 5G systems
- Application mobility aspects from radio and computation perspectives
- Compliance, testing and proofs-of-concept
- OpenAPI and MEC software resources

The book is written mainly for software developers, engineers and researchers, and also for MS students, Ph.D. and postdoc students, as a textbook for courses for telecommunication engineers, computer science, IT and mobile applications. Finally, due to the innovative nature of the topic, also senior researchers, team leaders and professors may adopt this book, as an advanced reference for their work.

Munich, Germany   Dario Sabella

Acknowledgements

This book is a tremendous effort of many years. I wrote most of the content, but definitely, its finalization and whole publication would not have been possible without the help and support of many friends and colleagues. Many thanks to all Contributors and Reviewers[1] (in alphabetical order[2]):

Alex Reznik, Amr Mokhtar, Angelo Corsaro, Anish Rawat, Antonio Virdis, Babu Pillai, Bob Gazda, Chang Hong Shan, Claudio Cicconetti, Daniele Brevi, Danny Moses, Debashish Purkayastha, Edoardo Bonetto, Elian Kraja, Giada Landi, Giovanni Nardini, Giovanni Stea, Greg Allison, Jane Shen, Jim Blakley, Jing Zhu, Joey Chou, Julien Enoch, Kishen Maloor, Luca Cominardi, Manuel Femia, Marco Picone, Masaki Suzuki, Maurizio Floridia, Michele Carignani, Miltiadis Filippou, Neal Oliver, Olivier Hecart, Pietro Piscione, Purvi Thakkar, Riccardo Scopigno, Rolf Schuster, Sindhura Gaddam, Stefano Mariani, Uwe Rauschenbach, Walter Featherstone, Yizhi Yao.

A special thanks to Diego Lopez (from Telefonica) for his nice foreword to the book. Also, I'm very grateful for all the fruitful exchanges of technical opinions with many colleagues and partners in ETSI MEC, including all experts in the industry, the academia and the whole edge ecosystem.

Finally, the biggest acknowledgment to my wife, for her infinite support...

...and my two children for their patience, as their father is always busy!

---

[1]Details of the various contributions are provided at the end of the book.

[2]Contributions ordered per chapter are listed and detailed at the end of the book, in Annex E.

Contents

Introduction

Edge Computing is currently one of the hottest topics in communication networks, since it represents a key technology for offering better performances and new innovative services in 5G systems.

As indicated in the figure below, most of the new services related to the introduction of 5G networks are expected to be generated by a variety of devices and heterogeneous use cases, e.g. from connected vehicles and all automotive-related use cases (the so-called C-V2X services) to Virtual Reality (VR), Augmented Reality (AR), IoT (Internet-of-Things) applications, etc…

Moreover, edge computing (leveraging network virtualization and cloud technologies) is giving more degree of flexibility for the deployment of server applications, with great benefits not only for the end user (in terms of end-to-end performances) but also for the other players in the ecosystem, like network providers, mobile/fixed operators and service/content providers (in terms of operational and management costs).

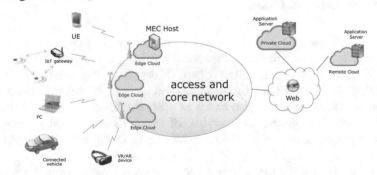

A comprehensive understanding of edge computing requires not only a knowledge of network infrastructure and related communication technologies, but also higher-level aspects including application development practice. Many different stakeholders are thus in principle interested to have an overview on edge computing, as the topic is involving an entire ecosystem of companies, from operators to network infrastructure providers, to various IT companies, and application and content providers.

The following bulleted list is highlighting the main specific features that the reader will find in this book, differentiating it from the rest of literature:

- An accurate and comprehensive description of Edge Computing technologies, with particular reference with the standard introduced by ETSI MEC (Multi-access Edge Computing), as a reference for edge developers, researchers, students and professionals. The textbook contains also practical exercises and classroom materials for students and professors, as a tool for a faster and concrete learning.
- The extensive work conducted in this book will clarify technological aspects around edge computing from many points of view, in order to constitute a useful reference material for both telecommunication and IT experts, coming from all the stakeholders in the eco-system. In fact, the acquisition of the needed knowledge on Edge Computing from IT stakeholders cannot avoid having some background on telecommunication, and vice versa. So, this textbook aims at providing a perfect balance of the two aspects.
- This book is capitalizing on the author's long lasting and extensive background in the industry, from mobile network operator to data-centric technology provider, his multi-years' experience on MEC and 5G networks, and his engagement with edge computing companies, software developers and startups. The book will provide real-world examples and open-source based reference material for researchers, students and professionals, but also mobile application developers, and system integrators.

Outline of this Textbook

The various chapters of this book cover all main aspects of edge computing, and thus it is divided in four parts: the first one is providing a more general overview of edge computing. In **Part 1**, the Chap. 1 describes the principles of Edge Computing, Fog and Cloud Computing, starting from a background on Cloud Computing, and considering also basics of MEC (Multi-access Edge Computing); instead, Chap. 2 is giving an overview of standards and industry associations around Edge Computing (e.g. 5GAA, AECC, OFC, OEC) including also research projects.

Part 2 is about Edge Computing standards, where Chaps. 3 and 4 are treating more in details the work done by ETSI on MEC, which is a leading international standard for edge computing. In particular, the former covers the specifications on edge platforms, while the latter on edge services. More detail on some selected APIs (e.g. Radio Network Information) is provided by Chap. 5, due to the special nature and wide applicability of these specifications for developers. The description will be mainly based on the completed MEC Phase 1 and Phase 2 standardization work, as the Phase 3 work just started in 2021. Future editions of this book may include a comprehensive Phase 3 work description.

Part 3 is covering Edge Computing deployments. Chapter 6 treats the aspects related to MEC in virtualized environments (with emphasis on NFV—Network Functions Virtualization). Chapter 7 cover the importance of deploying Edge Computing in 5G networks, including a mapping of MEC in 3GPP System Architecture, and a description of the recent efforts to produce harmonized standards for edge computing leveraging ETSI MEC and 3GPP specifications. In Chap. 8, federation and mobility aspects are discussed, both from radio and computation perspectives. The MEC Federation is clearly a Phase 3 work (at the time of publishing this book, a normative work related to this topic just started).

Finally, **Part 4** of the book is more about Edge Computing Software Development, where Chap. 9 is describing some relevant Software frameworks, and Chap. 10 represents an overview of various exemplary initiatives in the area of MEC, from OpenAPI representations, to MEC Sandbox, compliance testing activities, hackathon, PoCs and MEC trials. In fact, market adoption is as well essential to complement the pure knowledge of the technology.

Tips and Tricks: Didactical Features Included in this Book

Through all the book, several textboxes with nice graphics will provide an easily understandable support to the reader, in order to better find references, software resources, or to help practicing on MEC topics with examples/exercises and/or quiz/questions. In particular:

- The following textbox will be adding useful reference for the reader, for further documentation and deepening on the subject matter.

 Useful Reference/source code
Title of the reference

Download: Web Link to the reference

- The textbox below on Terminology will be recalling key concepts, and guide the reader in the view of a better understanding and alignment on the extensive terminology and literature sources.

Terminology: Definition
Download: Web Link to the definition

- At the end of each book part, the reader can find an useful quiz, with key questions on the topics described in the part (solutions will be provided).

Quiz—Part 1

- Moreover, in alternative to quiz, at the end of the other chapters, the reader can find some exercises or examples, that will help especially the student to practice on MEC topics learned. NOTE: they are structured not as traditional exercises, but more as "case studies", thus accompanied by their suggested solution, in order to help the reader in the learning process.

Exercises/examples—Chapter 3

About the Author

Dario Sabella works with Intel as Senior Manager Standards and Research, driving new technologies and edge cloud innovation for advanced systems, involved in ecosystem engagement and coordinating internal alignment on edge computing across standards and industry groups. In February 2021, he has been elected as Chairman of ETSI MEC (Multi-access Edge Computing), while from 2019, he was serving as Vice-Chairman, previously Lead of Industry Groups, and from 2015, Vice-Chair of IEG WG. Since 2017, he is also a delegate of 5GAA (5G Automotive Association). Before 2017, he worked in TIM (Telecom Italia group), as responsible for various research, experimental and operational activities on OFDMA technologies (WiMAX, LTE, 5G). Author of several publications (40+) and patents (30+) in the field of wireless communications, energy efficiency and edge computing, he is an IEEE senior member and has also organized several international workshops and conferences. From January 2021, he is also the Innovation Manager of the EU-funded 6G flagship Hexa-X project.

Part I
Edge Computing Overview

Chapter 1
Principles of Edge Computing, Fog and Cloud Computing

In this chapter, we will describe the principles of edge computing, starting from a background on cloud computing, and considering also fog computing and MEC basics, as defined by ETSI [MEC-web].

1.1 Background on Cloud Computing

A cloud consists essentially of a set of technologies to abstract away the details of hardware infrastructure to present a homogeneous set of abstracted resources to software, shared by the users through Internet technologies. This is one of the most promising paradigms of the present era, thanks to the exploitation of recent advances in software, networking, storage and processor technologies. The cloud model permits to offer scalable and elastic services, and able to dynamically acquire and use computing resources to support variable workloads.

According to the official NIST definition [NIST-CC], "cloud computing is a model for enabling ubiquitous, convenient, on-demand network access to a shared pool of configurable computing resources (e.g. networks, servers, storage, applications and services) that can be rapidly provisioned and released with minimal management effort or service provider interaction". Figure 1.1 shows that cloud computing is composed of five essential characteristics, three service models and four deployment types.

According to this classification, there is no one single type of cloud; instead, the actual realization of a cloud (and its specific characteristics) depends on the specific deployment model and service model considered. In particular, when speaking of the service models, it is important to provide a mapping to the application stack used in the different cases (Infrastructure as a Service (IaaS), Platform as a Service (PaaS) and Software as a Service (SaaS)).

Figure 1.2 shows the different layers of the stack, namely:

© Springer Nature Switzerland AG 2021
D. Sabella, *Multi-access Edge Computing: Software Development at the Network Edge*, Textbooks in Telecommunication Engineering, https://doi.org/10.1007/978-3-030-79618-1_1

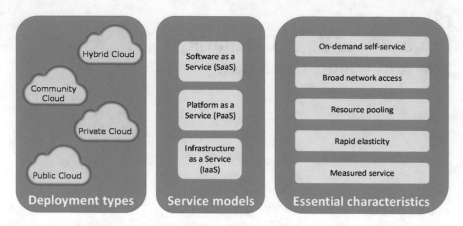

Fig. 1.1 The NIST cloud definition framework

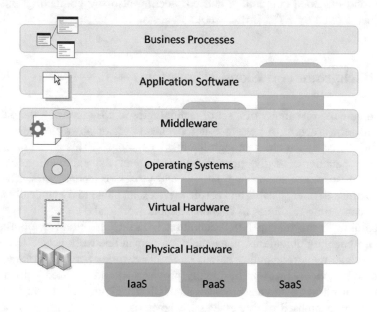

Fig. 1.2 Association between application stack and cloud service models

- Physical Hardware: all tangible infrastructure, e.g. servers, storage, networks connecting servers, as well as buildings housing the data centers, etc.
- Virtual Hardware (when present[1]): hardware components mapped into virtual counterparts, by a "hypervisor". By means of virtualization, users perceive the system as composed of virtual computing, storage and networking resources.

---

[1]In fact, another possibility is to have Bare-metal based container stacks.

Containerisation is also a possible variant, that will be described more in detail in Chap. 6.

- Operating System: this is the basic software installed on top of virtualized or physical infrastructure, and providing functions to higher levels. Examples of Operating System (OS) are Microsoft Windows Server, Linux or Apple OS X Server.
- Middleware: a software installed on an OS, providing a suitable environment for the execution of applications and handling of data storage. Examples: Java Virtual Machine, IBM Websphere, MySQL
- Application Software: custom applications interfacing with users, and providing the tools for the realization of certain tasks or activities. Example: email, FTP, web browsing, etc.
- Business Processes: these are complex processes or sets of activities, e.g. managed by a company, for example, order processing, budget approval processes, charging, etc.

In particular, as we can see in the above figure, Infrastructure as a Service (IaaS) is the lowest level of cloud service model, and it makes (physical or virtual) hardware accessible to customers (e.g. servers, network resources or storage). On the other hand, with Platform as a Service (PaaS) an execution environment is offered to customers so that they can deploy their specific applications or components, and possibly expose to third parties. Finally, Software as a Service (SaaS) is offering directly an application layer to the users, e.g. through a Draphical User Interface (GUI) or by providing a set of customized Application Programming Interfaces (APIs).

The NIST published also a Cloud Computing Reference Architecture (CCRA) [NIST-ccra], which identifies the major actors, their activities and functions in cloud computing. The diagram in Fig. 1.3 depicts a generic high-level architecture and is intended to facilitate the understanding of the requirements, uses, characteristics and standards of cloud computing.

This conceptual reference model from NIST introduces also five major actors:

- *Cloud Provider*: a person, organization or entity responsible for making a service available to interested parties.
- *Cloud Consumer*: a person or organization that maintains a business relationship with, and uses service from, Cloud Providers.
- *Cloud Auditor*: a party that can conduct an independent assessment of cloud services, information system operations, performance and security of the cloud implementation.
- *Cloud Broker*: an entity that manages the use, performance and delivery of cloud services and negotiates relationships between Cloud Providers and Cloud Consumers.
- *Cloud Carrier*: an intermediary that provides connectivity and transport of cloud services from Cloud Providers to Cloud Consumers.

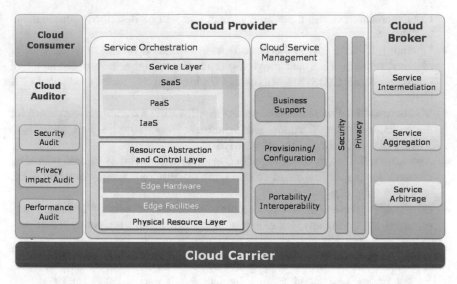

Fig. 1.3 Cloud computing reference architecture (CCRA) [NIST-ccra][2]

These roles are important also from a business perspective, essentially because each actor is an entity (a person or an organization) that participates in a transaction or process and/or performs tasks in cloud computing. This aspect will be clearer when talking about MEC Federation and the recent GSMA work on TEC (Telco Edge Cloud), described in Chap. 8.

1.2 From Remote Cloud to Edge Cloud

The introduction of edge computing is extending the general cloud computing paradigm to the edge of the network. In fact, traditionally every cloud service is offered by exploiting computing resources, that in principle can be physically delocalized (and remotely connected to the end user). With edge computing, user traffic is terminated in close proximity to the user, by providing a suitable environment characterized by better latency, high throughput and end-to-end performance.

Exercise 1.1 at the end of the chapter illustrates how degraded network quality translates into slower response times for latency-sensitive applications. But end-to-end latency improvement is not enough to explain the better network quality provided by edge computing. In fact, the higher latency and packet error rate which

---

[2]Reprinted courtesy of the National Institute of Standards and Technology, U.S. Department of Commerce.

result from a longer network path, degrade the network performance as perceived by higher layers. In other words, not only small delay but also a better channel error rate is needed to improve network performances.

This phenomenon is well illustrated by the Mathis formula for TCP throughput [Mathis97], which is drawing a relation between the Quality of Experience (QoE) in IP networks with Network Latency (expressed by the Round Trip Time, *RTT*) and Packet Loss (P_{loss}), with a certain MSS (maximum segment size):

$$Throughput \leq \frac{MSS}{RTT\sqrt{P_{loss}}} \qquad (1.1)$$

Validated by TCP transfer measurement between Internet sites, the equation is widely accepted as a good model of the real world, and is used to predict the bandwidth (BW) of Congestion Avoidance-based TCP implementations under many conditions. This simple formula is also proving that Bandwidth increase and Quality of Service (QoS) Functionalities are not sufficient to improve User Experience, even in properly designed and managed networks, as this is an inherent limitation of today's networks. Instead, QoE platforms are needed, and in that perspective edge

---

 Example 1

• Fiber Bit Error Rates (BER) are typically 10E-13.
• Some optical gear treats a link as down at a BER of 10E-6 (one bad bit in 1 million of bits).
• Assuming a stream of 1460 byte packets, that's one bad packet approximately every 85 packets.

---

computing is a potentially disruptive enabler to optimize operators' networks.

The following examples show some typical values for the throughput and

---

 Example 2

• Consider that a packet nominally contains 10,000 bits (1250 Bytes), or 10E4 bits.
• We should be looking for links that are experiencing BER loss that exceeds 10E-10.
• As a consequence, we need a packet loss figure of 10E-10 (BER) / 10E-4 (the packet size in bits) = 10E-6 (packet loss rate).
• Also we should note that LAN interfaces should typically have a loss rate that's several orders of magnitude less than a WAN link or wireless link.

---

latency in fixed communication networks with fiber connection:

The packet loss can be also derived from design figures, like the required bit error rate (BER), as clarified in the following example:

By reverting the formula, we can also calculate the maximum packet loss for a certain needed BW.

$$p \leq \left(\frac{MSS}{BW \cdot RTT} \right)^2 \qquad (1.2)$$

Thus, given a required Network Latency (expressed by the Round Trip Time, *RTT*) and bandwidth (BW), we can calculate (for a certain packet length) the needed maximum packet loss (P_{loss}).

With the Mathis formula, we have explained that network performances perceived by higher layers are influenced by latency and channel error rate. These two aspects can be improved thanks to the introduction of edge computing, thus not only through proximity to the end user (improving latency) but also through the adoption of QoS platforms and mechanisms (implemented as traffic flow processing at the edge) aiming at the packet loss improvement. An example of these mechanisms is provided by the TCP Throughput Guidance discussed at IETF [MTG17]. In order to effectively produce traffic flow improvement in the access link (which is more subjected to errors), some local information is needed about the radio channel.

As a consequence, we may summarize that the two key technology advantages provided by edge computing from a communication network point of view are:

(1) Proximity to the end users, shortening the end-to-end latency of packets (at user plane, UP);
(2) Access to local radio information (at control plane, CP), which is helping the implementation of mechanisms for improving the channel error rate.

In addition to that, of course, other indirect benefits are provided by edge computing, ranging from the possibility to run local processing tasks exploiting huge amount of local data (e.g. collected by local sensors and devices), to the opportunity to provide distributed computation capability that may offer task offloading from end devices (often with limited processing capabilities).

Furthermore, having the possibility to process locally measured data locally will provide by definition better security and possibility to guarantee privacy, e.g. by anonymizing personal information at the edge, while transmitting only aggregated and clustered data to the central cloud.

Before talking about fog computing (in the following section), it is worth mentioning the concept of cloudlets, introduced by Satyanarayan et al. in [Satya96, Satya09], as edge cloud platform constituting the intermediate layer between end-device and a centralized datacenter. The objective of cloudlet is to extend the remote datacenter cloud services in close proximity to the end users. For this reason, from an architectural point, of view, cloudlets are applicable to the definition of edge computing. Anyway, in Chap. 2 more attention will be reserved for them, thanks to the relevant work done by OEC (Open Edge Computing) initiative [OEC-web].

1.2.1   Fog Computing

In the last years, advancements in the research field have seen many studies related to fog computing, fog networking, mist computing, etc., but not associated to any specific standards or well-recognized products. Often, most of this terminology is even mixed up with edge computing, making very difficult a complete under-standing of the boundaries between different technologies and solutions. This phenomenon is also exacerbated by the absence of fog standards or well-recognized product segments in the market, which would permit to clearly identify all the different variants and solutions found in the scientific literature.

As a consequence, this redundant (and not always properly adopted) terminol-ogy is often leading to confusion, so it can be hard to understand the relationship between one solution and another, or whether different technologies are competing or complementary. On the other hand, as we will see in the next section, edge computing is instead clearly recognized as a key technology for 5G systems [TS23.501], and also defined by an international standard, i.e. ETSI MEC [MEC-WP].

Nevertheless, it is anyway important to shed some light on the "fog" termi-nology, also because of the emerging diffusion of the Internet-of-Things (IoT) and the progressive application of this paradigm to many market segments.

According to Wikipedia [WIKI-fog], fog computing (or fog networking), also known as fogging, is an "architecture that uses one or more collaborative end-user clients or near-user edge devices to carry out a substantial amount of storage (rather than stored primarily in cloud data centers), communication (rather than routed over the internet backbone), control, configuration, measurement and management (rather than controlled primarily by network gateways such as those in the LTE core network)".

But fog computing is also a concept introduced by CISCO in 2012, as "an extension of cloud computing paradigm from the core to the edge of the network. It enables computing at the edge of the network, closer to IoT and/or the end-user devices" [Cisco-fog].

More recently, also the NIST (National Institute of Standards and Technology) initiated in 2017 an effort (Special Publication 800–191 (Draft) [NIST-fog]) that defines Fog computing as "a horizontal, physical or virtual resource paradigm that resides between smart end-devices and traditional cloud computing or data center".

From these definitions, it is still hard to clearly identify the key aspects of fog computing. A more detailed work is instead done by the OpenFog Consortium (OFC), an association of major tech companies aimed at standardizing and pro-moting fog computing. According to OpenFog Consortium [OFC-fog], fog com-puting is "a horizontal, system-level architecture that distributes computing, storage, control and networking functions closer to the users along a cloud-to-thing continuum". The OpenFog Reference Architecture (RA) is obviously not an international standard but is a joint effort of many relevant companies in the market (also belonging to different vertical segments), and for that reason can be considered as a relevant reference for the future evolution of fog products.

More in detail, the RA document produced by OFC provides a few key elements and characteristics (also depicted in the figures below) that permit to clearly identify what fog computing is:

- fog computing is characterized by a multi-tier deployment. Nodes consider both VM and other virtualization technologies.

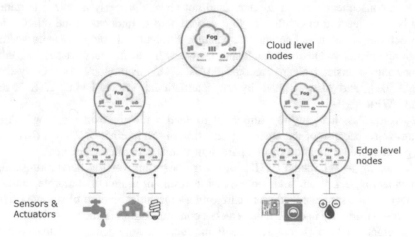

- Moreover, unlike cloudlet and MEC, fog is tightly linked to the existence of a cloud, i.e., it cannot operate in a standalone mode.

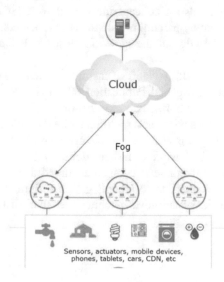

- Fog nodes are self-aware, peer-aware and community-aware. Fog is about distributed Intelligence, infrastructure, management and control plane. Intelligence is part of the infrastructure. (Note: the architectural elements of a node will vary based on its role and position within an N-tier fog deployment).

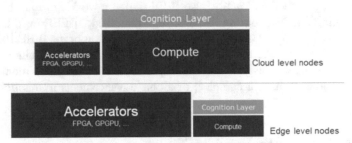

- Fog nodes may be linked to form a mesh to provide load balancing, resilience, fault tolerance, data sharing and minimization of cloud communication.

  - Architecturally, this requires that fog nodes have the ability to communicate laterally (peer to peer or east to west) as well as up and down (north to south) within the fog hierarchy. The node must also be able to discover, trust, and utilize the services of another node in order to sustain RAS (reliability, availability and serviceability).

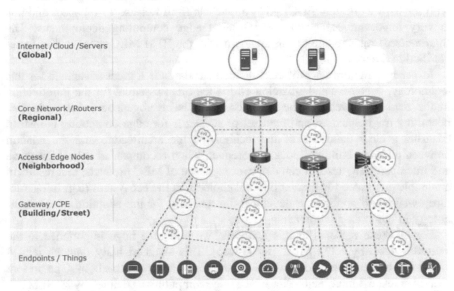

All these elements, as defined by OFC RA, permit also to better clarify the difference between cloud, fog and edge computing.

In addition to these efforts done by OFC (as industry group), the ISO (International Standardization Organization) started a technical report [ISO-edge] with the purpose to describe all activities in the field of edge and fog computing, clarify the terminology and identify the standardization gaps in order to take future actions in that perspective. Finally, also IEEE started a working group on "FOG - Fog Computing and Networking Architecture Framework" [IEEE-fog], leveraging the work carried out by OFC and leading to the publication of the IEEE 1934–2018 ("IEEE Standard for Adoption of OpenFog Reference Architecture for Fog Computing"). For this reason, it will be important for the reader to monitor these activities, in order to have a complete of all the advancements in this field.

1.3   Edge Computing and MEC

Now that we have clarified the relation between cloud, fog and edge, it's time to talk more in detail about edge computing. As a matter of fact, even if there are also other initiatives in the field of edge computing (that will be described in the next chapter), we can clearly state that ETSI MEC is the leading international standard available in the space of edge computing. At the same time, it is worth clarifying that recently also 3GPP started to work seriously on edge computing, in particular with the SA6 work item EDGEAPP, applicable from Release 17 on specs and this is very important especially for 5G-based edge computing deployments. The (harmonized) relationship of this standard with the ETSI MEC specs is thus better clarified in Chap. 7.

In general, the presence of consolidated standards is a great advantage for this technology, because standardization efforts are often essential for the introduction in the market, especially when the ecosystem is heterogeneous, and could risk becoming fragmented. A common set of standards for edge computing is instead ensuring a clear definition of the technology, the architecture and the relation between different building blocks. Moreover, when accompanied by conformance and interoperability tests, it can define compliance of MEC products, that results in a suitable guarantee for the various stakeholders in the ecosystem (e.g. infrastructure owners and application developers), especially for the adoption of the technology in complex environments.

For all these reasons, in this textbook, we will give huge importance to the activities done by ETSI MEC [MEC-web]. This standard body (and its related work) will be described more in detail in the next chapters of this book. The present section is instead important to introduce edge computing as defined by ETSI MEC:

	Terminology:
	ETSI definition of MEC
	Multi-access Edge Computing offers to application developers and content providers cloud-computing capabilities and an IT service environment at the edge of the network

Web Link to the definition:
https://www.etsi.org/technologies-clusters/technologies/multi-access-edge-computing

Let's analyze more in detail the meaning of this definition:

- ETSI is talking about "multi-access": this means that the definition of MEC is access agnostic (thus applicable in principle to 4G/5G, Wi-Fi or fixed access)
- Edge computing is a technology offered to application developers and content providers, which are the end users of the specifications; this means that, even if MEC is defining all the aspects needed to implement platforms (thus, useful for technology and infrastructure providers), the actual "core business" of the ETSI standard consists in the publication of APIs, which can be used by application developers and content providers
- MEC is providing "cloud-computing capabilities and an IT service environment"; thus, MEC is essentially enriching cloud technologies, by adding a suitable IT service environment that can enable application providers to develop new applications and exploit MEC services exposed from the platform. As we will see, MEC applications can also produce services that can be exposed to the platform and consumed by other MEC applications in the system.
- The concept of edge of the network is intentionally not explained in detail (i.e. how deep in the "edge" should be deployed MEC hosts? In the edge of mobile network? In C-RAN aggregation points? Co-located with core network?); the main reason of this generic definition is given by the need to include many deployment options: in fact, operators and infrastructure providers need to have a certain degree of flexibility, and exploit virtualization technologies to instantiate MEC apps and platforms conveniently, and based on the different needs (e.g. costs, traffic, presence of users, tariffs, etc....). All the deployment options should be in principle allowed via the definition of MEC.

In the next chapter, we will provide an overview of the ETSI MEC standards and other main MEC-related industry groups and associations. The details of MEC technology (divided by topics) will follow in the subsequent chapters of this textbook.

	Exercises/examples—Chapter 1

Exercise 1.1 Goal of the exercise: Understand how degraded network quality translates into slower response times for latency-sensitive applications.

Consider the following figure, showing network quality measured over a certain time period between 3 homes located within the same neighborhood in a city

	5%	10%	50%	90%	95%
Home A & B	18.5	19.2	26.4	77.8	133.6
Home B & C	36.4	37.2	44.9	87.2	98.0
Home C & A	38.8	39.3	44.9	75.1	92.6

(b) Latency distribution (milliseconds)

	5%	10%	50%	90%	95%
Home A & B	0.5	0.6	1.9	2.3	2.3
Home B & C	0.5	0.7	0.8	0.9	0.9
Home C & A	0.5	0.5	0.8	0.9	0.9

(a) Network connections

(c) Upload Bandwidth distribution (Mbps)

Fig. 1.4 Example of network quality indicators between homes

(1 km^2 area). In many practical situations [VMHO15], end-to-end latency and bandwidth can be poor despite physical proximity. This degrades further when different ISPs are involved.

Home A and B are connected to ISP1 and home C to ISP2 (Fig. 1.4).

The numbers provided in the above tables are based on real observations [VMHO15] over a 1-week period between the three homes that are located within a one square mile area and can lead easily (for all the three cases) to several tens of milliseconds of latency (see last column, related to 95% percentile). Thus, we can already observe that a relative "physical" proximity (from a geographical point of view) doesn't necessarily mean also a corresponding "logical" proximity (from an actual end-to-end latency perspective). This differentiation should be always kept in mind when considering zoning and locality concepts introduced by ETSI MEC (i.e. when talking about consumption of MEC services locally or in other MEC servers nearby).

Now, if we compare the three cases, we can notice that the path between homes A & B (connected to the same service provider ISP1) provides better throughput and latency compared to the other two cases (path among homes connected to different services providers, ISP1 and ISP2). In particular, assuming asymptotic values corresponding to the 95% percentile, the path between homes A & B provides a throughput of 2.3 Mbps and a latency of 133.6 ms, while the path between home C (connected to ISP2) and homes A or B (connected to ISP1) provides a throughput of 0.9 Mbps and a latency between 92.6 and 98 ms. This comparison gives also a rough idea of the possible network performance gain between a traditional client-server communication (with a remote app instantiated on the ISP1 cloud, as indicated in the figure below) and an edge computing deployment, with a server application located at ISP2 premises (indicated in the figure below by a MEC app, running one MEC host), thus closer to the user (the UE client application at home C) (Fig. 1.5).

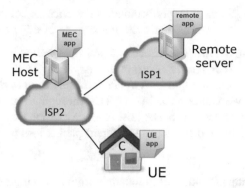

Fig. 1.5 Comparison between traditional client-server communication and edge computing deployment

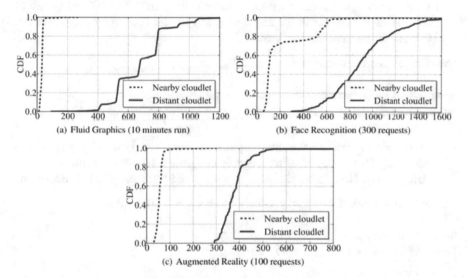

Fig. 1.6 CDF of Response Time (milliseconds) [VMHO15]

Additionally, actual response time (perceived by the end user) depends on the specific kind of application considered. The figures below show the CDF (Cumulative distribution function) of three different workloads (fluid graphics, face recognition and augmented reality) comparing again a communication between a UE client located at home C and a server application (located at home C, nearby, or at home A, distant) (Fig. 1.6).

In all cases, a nearby MEC application yields significantly lower response times. For example, the median response time of Face Recognition is 104 ms when it is associated to the nearby MEC app, but it is 882 ms when deploying the server app on a distant cloud. Since this performance difference is solely due to different network conditions, it confirms again the importance of edge computing.

Exercise 1.2 Goal of the exercise: *collect performance measurements for a simple client-server TCP packet transfer for a specific TCP implementation and compare the obtained throughput with the bound expression of the Mathis formula.*

In this exercise, we validate the Mathis TCP throughput model, by means of a simple experiment between two hosts: host A sends packets for a duration of 30 s to host B on a link configured to emulate different packet loss probabilities:

- Use `netem` (https://wiki.linuxfoundation.org/networking/netem), a Network Emulation functionality for testing protocols capable to emulate the properties of wide-area networks, including variable delay, loss, duplication and packets re-ordering.
- set up the network interface with a constant latency (RTT) of 100 ms (given by 50 ms at each host, ass in the example below) and a loss probability from 1 to 10% (on all packets leaving host A).

Example of `netem` configuration (with P_{loss} of 2%):

```
host_B# tc qdisc add dev eth0 root netem latency 50ms
host_A# tc qdisc add dev eth0 root netem latency 50ms loss 2%
```

- use `iperf` to measure the throughput from host A to host B by transmitting data during 30 s (`-t`) and using a certain congestion window algorithm (e.g. `-Z reno`, for the New Reno algorithm, or `-Z cubic` for the CUBIC algorithm)

Example of a single `iperf` experiment (with CUBIC algorithm):

```
host_B# iperf -s
host_A# iperf -c host_B -t 30 -Z cubic
```

- all measures can be collected by varying the P_{loss} from 1 to 10% in different `iperf` experiments, and putting the Tput values into a chart, like the following one:

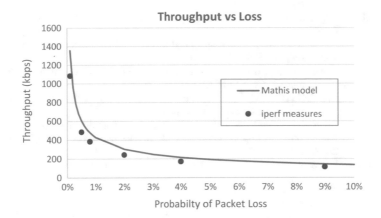

- This set of measures can be compared with the Mathis model, with RTT = 100 ms, and considering that the default TCP Maximum Segment Size is 536 bytes (according to RFC 879, page 1, Sect. 1).

References

[MEC-web]   ETSI   MEC,   https://www.etsi.org/technologies-clusters/technologies/multi-access-edge-computing

[NIST-CC]   Peter Mell and Tim Grance, (NIST, National Institute of Standards and Technology's): "The NIST Definition of Cloud Computing", Sept. 2011, https://csrc.nist.gov/publications/detail/sp/800-145/final

[NIST-ccra]   Cloud Computing Reference Architecture, Rec. National Institute of Standards and Technology, 2011. Available at: https://doi.org/10.6028/NIST.SP.500-292

[Mathis97]   Mathis et al (1997) The Macroscopic Behavior of the TCP Congestion Avoidance Algorithm. ACM SIGCOMM Computer Communication Review 27(3):67–82

[MTG17]   A. Jain et al. "Mobile Throughput Guidance Inband Signaling Protocol", March 2017. Available at https://tools.ietf.org/html/draft-flinck-mobile-throughput-guidance-04

[Satya96]   M. Satyanarayanan, "Fundamental Challenges in Mobile Computing", in Proc. of 15th Annual ACM Symposium on Principles of Distributed Computing, Philadelphia, May 1996.

[Satya09]   M. Satyanarayanan, P. Bahl, R. Caceres, and N. Davies, "The Case for VM-Based Cloudlets in Mobile Computing", IEEE Pervasive Computing, Vol.8, No.4, Oct.-Dec. 2009, pp. 2–11

[OEC-web]   Open Edge Computing initiative, http://openedgecomputing.org

[TS23.501]   "3GPP TS 23.501: System Architecture for the 5G System", June 2018, Link: http://www.3gpp.org/ftp/Specs/archive/23_series/23.501/23501-f20.zip

[MEC-WP]   ETSI White Paper: "Mobile Edge Computing: A key technology towards 5G", Sept 2015; https://www.etsi.org/images/files/ETSIWhitePapers/etsi_wp11_mec_a_key_technology_towards_5g.pdf

[WIKI-fog]   Wikipedia: "Fog   Computing",   Link: https://en.wikipedia.org/wiki/Fog_
                computing
[Cisco-fog]   https://www.cisco.com/c/dam/en_us/solutions/trends/iot/docs/computing-
                overview.pdf
[NIST-fog]   NIST, "Fog Computing Conceptual Model", March 2018; Link: https://doi.org/
                10.6028/NIST.SP.500-325
[OFC-fog]   https://www.openfogconsortium.org/wp-content/uploads/OpenFog_Reference_
                Architecture_2_09_17-FINAL.pdf
[ISO-edge]   ISO/IEC JTC 1/SC 41 - Internet of Things and related technologies; Link: https://
                www.iso.org/committee/6483279.html
[IEEE-fog]   IEEE Std 1934–2018 - IEEE Standard for Adoption of OpenFog Reference
                Architecture for Fog Computing, Link: https://standards.ieee.org/standard/1934-
                2018.html
[VMHO15]   Kiryong Ha et al.: Adaptive VM Handoff Across Cloudlets. June 2015

Chapter 2
MEC: Standards and Industry Associations Around Edge Computing

As already explained in Chap. 1, MEC (Multi-access Edge Computing), as defined by ETSI MEC ISG (Industry Standard Group), offers to application developers and content providers cloud computing capabilities and an IT service environment at the edge of the network [MEC-web]. In this chapter, we will clarify the relation between MEC standards on edge computing, 5G systems (as defined by 3GPP) as well as other access technologies, and the relationship with industry associations related to edge computing from different vertical market segment perspectives.

2.1 MEC, Edge Computing, 5G and Verticals

We've already clarified that, currently, ETSI MEC [MEC-web] is a leading international standard available in the space of edge computing. In fact, while ETSI started to define edge computing in late 2014, more recently, also 3GPP is seriously working on it, in order to introduce the support in 5G networks. On the other hand, of course, MEC is not only supposed to be deployed in mobile networks: in fact, MEC deployments can include Wi-Fi and also fixed networks. For this reason, the ETSI MEC specifications (that by the way are also harmonized with 3GPP [ETSIWP28, ETSIWP32]) are generic enough and provide a good international reference standard for edge computing.

The MEC main specifications will be analyzed more in detail soon in the next chapters of this book. In the meantime, due to the market relevance of MEC in 5G deployments [SabRezFraz], some preliminary and important questions should be answered first, in order to have a complete picture and global understanding of this technology with respect to the 5G ecosystem. In particular: what is the relation between MEC and 5G, from a standardization perspective? And what is the applicability of edge computing (and MEC standard) to the different vertical

© Springer Nature Switzerland AG 2021
D. Sabella, *Multi-access Edge Computing: Software Development at the Network Edge*, Textbooks in Telecommunication Engineering, https://doi.org/10.1007/978-3-030-79618-1_2

segments in future 5G systems? Moreover, how the different verticals are standardized for 5G systems, leveraging ETSI MEC activities?

In order to answer these fundamental questions, first of all, it is essential to keep in mind that edge computing is commonly recognized as a key technology for 5G networks [ETSIWP11, 5GPPP], and currently already part of the 5G system architecture [TS23.501]. This should not confuse the reader (who could be induced to think that there are two standard bodies defining the same technology). In fact, from a general perspective, the two SDOs (Standards Developing Organizations) have different scopes and points of view, and their work should be seen as complementary (due to different focuses) although also synergic (for the reasons that we will explain better in Chap. 7). In fact:

- 3GPP is typically more focused on (radio and core) network technologies, and only recently (from Rel.15 on) is inserting edge computing in the overall picture of 5G systems, by describing it in a general way, and defining further (in SA6) the support from Rel.17 on;
- ETSI ISG MEC is a standard group explicitly conceived to define edge computing technology, and it is thus defining more in detail how edge computing is realized from an IT perspective, in an access agnostic manner; ETSI MEC is thus specifying the set of cloud technologies to be implemented by MEC providers and application/content providers, in order to realize MEC platforms and applications.

This basic distinction of scopes is very important because, on one hand, the two standard bodies don't necessarily overlap. On the other hand, when it comes to 5G systems, the two bodies are clearly referring to the same technology and are committed to provide a set of harmonized standards [ETSIWP36] for edge computing. The aim of the industry in fact is to ensure that the two activities are coherent and compatible, in order to guarantee a smooth deployment of MEC technologies in 5G systems. These aspects will be analyzed more in detail in Chap. 7. For the time being, it is sufficient to clarify the basic alignment from a standardization perspective, even if of course, at the present moment some implementation aspects (e.g. especially related to management and orchestration) are still under definition in 3GPP and ETSI MEC, also due to different standardization roadmaps and priorities (which is a reasonable aspect, to be always kept in mind by the reader).

MEC technology is applicable to many heterogeneous 5G use-cases and service scenarios (briefly described in Sect. 2.3), so in principle, MEC can cover many different vertical market segments, like automotive, as well as other industrial automation. Each of these verticals is generating very specialized technical requirements toward 5G networks (and also with respect to the MEC standard). For this reason, many specialized industry associations are emerging in the international scene, with the aim to study in detail each vertical market segment and to create a wide consensus among all stakeholders for deriving (and push to standard bodies) the respective requirements. An example for the automotive sector is provided by 5GAA (5G Automotive Association), which is an association *"created to connect telecom industry and vehicle manufacturers and work closely together to develop end-to-end solutions for future mobility and transportation services*" (Christoph Voigt, AUDI AG, 5GAA Board Chairman [5GAA-web]).

The main essence of 5GAA work is to put together different stakeholders in the automotive domain:

- Telecommunication companies (like operators, network technology providers, chip makers, etc.), that are providing Connectivity and Networking Systems, Devices and Technologies and
- Automotive players (like car makers, OEM suppliers, system integrators, etc.), that are working on Vehicle Platform, Hardware and Software Solutions.

As the reader can easily see, the main challenge here is of course to harmonize the view of these two huge (and different) categories of stakeholders and to make sure that the consortium can produce a set of comprehensive technical requirements and technologic solutions able to realize intelligent transportation, mobility systems and smart cities. In addition to that, since automotive ecosystem is quite complex and heterogeneous, the need of standardized solutions is very relevant, especially to guarantee interoperability among different players.

Nevertheless, since 5GAA is an industry group (and not an SDO), it is not publishing "standards". For that reason, 5GAA needs to have a standardization impact on the relevant bodies which are affected by the identified solutions. Consequently, the scope of 5GAA is to build a relevant consensus and produce suitable solutions that are pushed as standard contributions to the respective SDOs. The figure below is depicting a possible landing zone for 5GAA toward 3GPP standards (as regards 5G network technologies, i.e. access and core network-related aspects) and ETSI MEC standards (as regards edge computing technologies, i.e. cloud-related aspects) (Fig. 2.1).

The synergy described in the above figure is bi-directional since SDOs (like 3GPP and ETSI MEC) need to receive a correct guidance from experts of the vertical segment (like car makers, which usually in their turn don't have much telecommunication expertise, as they are supposed not to build but to build this technology). For that reason, ETSI MEC (that is covering many use-cases) is establishing collaborations with several industry organizations (specialize in the respective industry domains), in order to ensure that the MEC standard will be able to accommodate all the different needs coming from the different vertical market segments. Some examples of industry groups are VRARA (VR/AR association), ICC (Industrial Internet Consortium), 5GAA, SCF (Small Cell Forum), GSMA, BBF (Broadband Forum).

We will better describe ETSI ISG (Industry Specification Group) MEC ISG is conducting not only a more traditional work of producing standards but also a wide set of ecosystem engagement activities, including the coordination with industry groups impacted by MEC technologies. The latter activities are traditionally categorized not properly as traditional "standard" activities but recently have gained attention (also by other ISGs) since they are considered equally important as specifications production, for the actual market adoption of MEC.

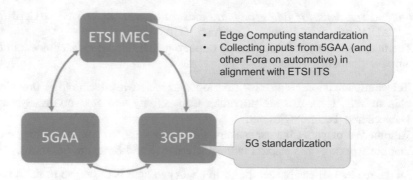

Fig. 2.1 Example of synergy between ETSI MEC, 3GPP standard and industry group (5GAA, 5G Automotive Association) [15MEC002p2]

2.2   MEC Service Scenarios

Before describing in detail the ETSI ISG MEC standardization body and its organization, it's important to have an overview of the MEC service scenarios, in order to better understand the applicability of edge computing to the various and heterogeneous use-cases.

For that purpose, it's worth having a look at the initial work done in November 2015 by ETSI ISG MEC, who published a set of high-level service scenarios [IEG004] applicable to MEC (and only briefly described in this section).

These service scenarios should be seen simply as meaningful examples of how MEC can support innovative services and applications that would create a better quality of experience for the end user. More detailed use-cases can also be found in the ETSI GS MEC-002 specification [15MEC002p2], which is described in Chap. 3 . Anyway, it's worth noting that MEC could enable even a larger number of new kinds of applications and services for multiple sectors (e.g. consumer, enterprise, health).

Intelligent Video Acceleration service scenario

In this scenario, a Radio Analytics application, running on a MEC server, is providing the video server with a near real-time information related to the estimated available throughput (at the radio interface). This information can assist TCP traffic flow, in order to improve the end user's Quality of Experience (QoE) (Fig. 2.2).

More in detail, the video server may use this information coming from the MEC server to assist TCP congestion control decisions, e.g.:

- selection of the initial window size,
- setup of the value of the congestion window during the congestion avoidance phase,
- and/or adjustment of the size of the congestion window when the conditions on the "radio link" deteriorate.

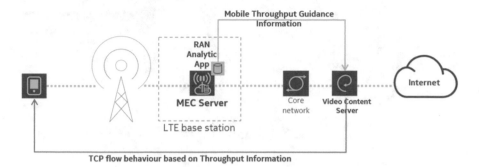

Fig. 2.2 Intelligent Video Acceleration service scenario [IEG004]. ETSI copyright license: ©
European Telecommunications Standards Institute 2015. Further use, modification, copy and/or
distribution are strictly prohibited

For interested readers, Chap. 5 will provide more deepening on radio analytics;
in particular, Exercise 5.1 will permit to better understand how network information
available from MEC server can support TCP traffic performance improvements.

Video Stream Analysis service scenario

This scenario is considering video-based monitoring systems with image recognitions algorithms (e.g. vehicle license plate recognition to monitor vehicles entering
and exiting an area of the city, car parks, for security purposes). Running video
analytics in a MEC host permits not only to reduce the cost of the cameras to be
deployed but also to save traffic transfer. In fact, performing the analysis locally
mitigates the need to transmit high data video streams through the core network to
remote cloud-based services. With data analytics on MEC, only the valuable data
(extracted from the video elaboration) can be transmitted to the application server.
The usage of MEC is thus more convenient and efficient with respect to both
placing video analytics on the cameras and in a remote server (Fig. 2.3).

Augmented Reality service scenario

In this scenario, a visitor of a certain point of interest (e.g. a museum, or art gallery,
etc.) is using the camera on his mobile device to capture the point of interest, while
the application is showing additional information related to what the visitor is
viewing (e.g. based on the local context). From an analytics point of view, this
scenario is similar to the previous one, since also in this case is more efficient to
process the video stream locally. Moreover, augmented information related to the
specific point of interest is also highly localized, thus hosting this information
locally on a MEC server is advantageous compared with hosting in a remote cloud.
The advantage of MEC is given by the exploitation of computation capabilities and
efficient storage of local information, providing high rate of data processing in a low
latency environment (Fig. 2.4).

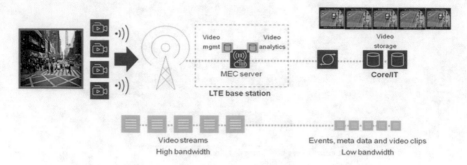

Fig. 2.3 Video stream analysis service scenario [IEG004] ETSI copyright license: © European Telecommunications Standards Institute 2015. Further use, modification, copy and/or distribution are strictly prohibited

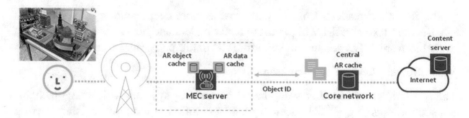

Fig. 2.4 Augmented Reality service scenario [18] ETSI copyright license: © European Telecommunications Standards Institute 2015. Further use, modification, copy and/or distribution are strictly prohibited

Assistance for intensive computation.

This service scenario is known also as task offloading, as a certain computationally intensive data processing is offloaded from a sensor (or generally a low-cost device), which needs to remain operational for a long period of time. Many examples of task offloading are possible, ranging from environmental sensors, to gaming and security applications. The usage of MEC for hosting intensive computation permits to increase the battery life of the device, with evident benefits in terms of cost of the sensor itself. Moreover, the data processing may require input from multiple sources, so again the usage of MEC permits to avoid the device to receive multiple information from different sources, as the data elaboration is all done by MEC (Fig. 2.5).

The convenience of the scenario highly depends on the specific application considered, which by the way should be designed in order to offload some computation task toward the MEC host. Moreover, the usage of MEC can be optimized based on different variables:

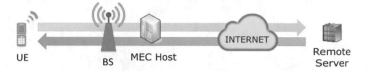

Fig. 2.5   Assistance for intensive computation (general principle)

- the amount of input data needed to perform the computation and the quality of the communication link between the sensor/device and the MEC host (performing the computation),
- the entity of computation capability needed to perform the task, and the actual computation power available at the MEC server (possibly by considering also its related energy efficiency),
- and also the amount of output data sent back from the MEC to the sensor, as a result of the computation task offloaded.

For interested readers, Exercise 2.1 will permit more deepening on task offloading and the related opportunity to exploit MEC.

Enterprise deployment of MEC

In this service scenario, located in an enterprise area, the integration of an IP-PBX with a MEC platform is envisaged, to provide seamless service between mobile network (e.g. an Operator small cell deployment) and the enterprise WLAN network. Employees can use their smartphones/tablets/laptops to connect to enterprise services (e.g. based on cloud-based platforms) by accessing (upon authorization) the same services whether connected through the cellular network or enterprise WLAN. The MEC platform can assist with network selection (e.g. by means of a MEC application performing access control management and integrated with the enterprise network) and classify different levels of services per end-user within the enterprise domain (Fig. 2.6).

Connected vehicles

This service scenario is introducing connected vehicles communication supported by MEC. The connectivity between cars, with road infrastructure and cellular network, is essential for the transmission of safety-critical messages (e.g. notification of road hazards, HD maps, messages to reduce traffic congestion, sensing the vehicle's behavior, etc.), and also to provide infotainment services (e.g. in-car video streaming delivery, etc.). In this scenario, MEC is used to store and process data at the edge (thus more efficiently than in a remote cloud and with lower latency), e.g. by instantiating MEC applications on MEC servers at the edge of the cellular network (e.g. deployed at the LTE base station site, small cell sites or aggregated site locations), to provide the roadside functionality. These applications can receive local messages directly from the applications in the vehicles and the roadside sensors, analyze them and then propagate them (with extremely low latency) to other cars in the area (Fig. 2.7).

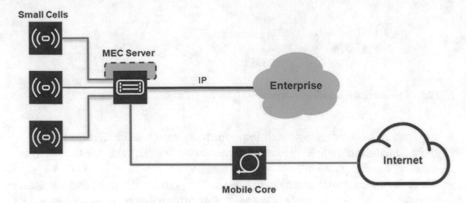

Fig. 2.6 MEC platform with breakout to Enterprise network [IEG004] ETSI copyright license: ©
European Telecommunications Standards Institute 2015. Further use, modification, copy and/or
distribution are strictly prohibited

Fig. 2.7 Connected vehicles
[IEG004]. ETSI copyright
license: © European
Telecommunications
Standards Institute 2015.
Further use, modification,
copy and/or distribution are
strictly prohibited

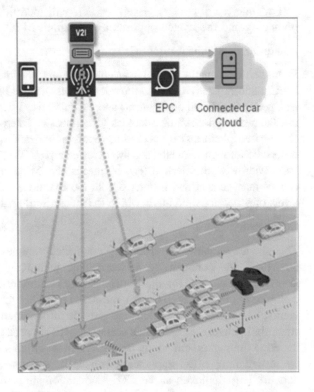

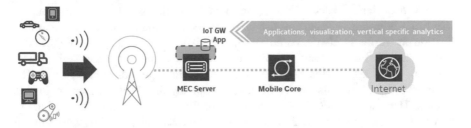

Fig. 2.8 IoT gateway service scenario [IEG004] ETSI copyright license: © European Telecommunications Standards Institute 2015. Further use, modification, copy and/or distribution are strictly prohibited

IoT gateway service scenario

This scenario is related to the Internet of Things, and the related presence of a multitude of heterogeneous sensors and devices, which are connected over different forms of connectivity (e.g. 3G, LTE, Wi-Fi or other spectra) and communicate with different protocols. The usage of MEC permits to aggregate various IoT device messages and provide analytics processing capability and a low latency response time. In addition, MEC Server could provide some additional compute and memory (e.g. utilized for aggregation and distribution services, analytics of the device messages, database logging, remote provisioning and access control to the end devices, etc.) (Fig. 2.8).

2.3 ETSI MEC Standard

The starting of ETSI work on edge computing can be placed at the end of 2014, with the launch of the MEC Industry Specification Group (ISG). Since the beginning, the goal was to create an open environment across multi-vendor cloud platforms located at the edge of the network, accessible by application/service providers and third parties, by providing better performances with respect to centralized cloud computing environments, both in terms of latency, and higher performances [ETSIWP11]. The proximity to end-users, together with the access to real-time network information, were the main key enablers provided by MEC technology.

Initially, the standardization work was focused on the introduction of MEC, intended as "Mobile Edge Computing", to promote and accelerate the advancement of edge-cloud computing in mobile networks. Only later on, in order to move toward a second phase of standardization, at the beginning of Term 2 the group decided to change the meaning of MEC acronym, by widening the scope of the work to "Multi-access Edge Computing", thus not only by considering mobile networks but also other accesses (e.g. fixed networks) and non-3GPP networks (e.g. Wi-Fi).

To better clarify, each Term (time period) of the standardization work in MEC ISG is lasting 2 years (by the way, the reader should notice that this organization in Terms is also a common characteristic of other ISGs in ETSI). But standardization Phases (and specifications releases) are not necessarily corresponding to the ISG Terms (which instead are the time periods approved by ETSI for an ISG to continue its work). In fact, we can say that MEC Phases are lasting 3 years. Thus, while MEC Phase 1 was obviously coinciding with Term 1 (2015–2016), the subsequent Term 2 (2017–2018) was essentially a migration between Phase 1 and Phase 2 (ended with the ISG Term 3, in 2020). The updated MEC acronym (from "mobile" to "multi-access" edge computing) was announced already from the beginning of Term 2 (see figure below), in order to allow the group to work on Phase 2 documents already from Term 2 on (Fig. 2.9).

The main drivers of these two phases of the MEC standardization work are:

- **PHASE 1—Mobile Edge Computing**

  - Focus on cellular networks; main benefits for mobile network operators: push data-intensive tasks toward the edge and locally process data in proximity to the users, thus reducing traffic bottlenecks in the core and backhaul networks; MEC is at the edge of access network, co-located with base stations or C-RAN aggregation points; introduction of RNI (radio network information) API [MEC012], in order to improve network performance

- **PHASE 2—Multi-access Edge Computing**

  - Focus not only on the cellular network but also on other access networks (Wi-Fi and fixed access); many other stakeholders are joining the group (e.g.

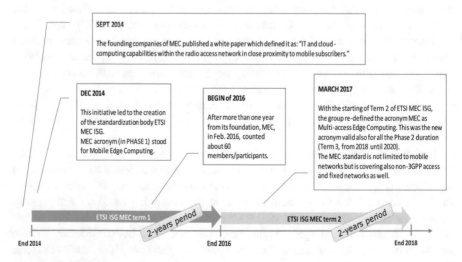

Fig. 2.9 Brief history of ETSI MEC, with few milestones until Term 2

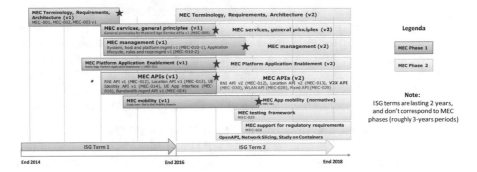

Fig. 2.10 Overview of ETSI MEC standardization work in Term 1 and Term 2

IT companies, cloud providers), as there is a common understanding that not only operators can benefit from the introduction of MEC technologies; the group is not only focusing on RNI API (conceived for cellular networks), but also similar work is also started both for Wi-Fi access points (WLAN API [MEC028]) and fixed networks (Fixed API [MEC029]); more intensive activities in collaboration with vertical segments, and related production of specific APIs (e.g. V2X API [MEC030]).

It should be noted that, even if initially ETSI MEC was driven by cellular network stakeholders, both phases of ISG MEC standardization since its infancy are characterized by the assumption that MEC is access agnostic. In fact, since the beginning, the intention was to define a decentralized cloud architecture independent from the specific underlying network, offering cloud computing abilities and an IT service environment at the edge of the network. So, from this perspective, the MEC architecture defined in Phase 1 is forward compatible with the subsequent work done in Phase 2 by widening the MEC applicability to other access networks (Fig. 2.10).

The above Figure is giving a high-level overview of the main standardization efforts in ETSI ISG MEC, by clearly distinguishing the specifications produced in the first two terms. All the published documents are also linked in the official ETSI MEC portal [MEC-web], while stable drafts (thus still not finalized specifications, but already at a quite mature stage) are usually put in the MEC open area [MEC-open], as a sort of preview for interested stakeholders who would have an earlier look at the standards before their finalization.

Phase 1 of MEC (textboxes colored in blue) was focused on the publication of the first set of specifications[1]; in particular, in that initial phase, the ISG worked on the preliminary basis of MEC: Terminology [MEC001p1], Technical Requirements [MEC001p1] and Framework and Reference Architecture [MEC001p1].

---

[1]According to ETSI document production policies, Global Specifications (GS) are normative documents, while Global Reports (GR) are informative documents.

This set of specifications is the foundation of MEC technology,[2] as MEC-001 is clarifying all the terms used by MEC, while MEC-002 is starting from detailed use-cases and deriving the technical requirements for the implementation of a MEC platform, and finally, MEC-003 is depicting the architecture of a MEC system (assuming a standalone deployment, thus without the need of an NFV-based network, and allowing both VNF- and PNF-based implementation versions).

This first set of specifications published (in Phase 1) drove the definition of:

- the MEC platform, specified by Application Enablement [MEC009], General principles for MEC Service APIs [MEC0011] and MEC Management, in two parts (System, host and platform management [MEC010-1] and Application lifecycle, rules and requirements management [MEC010-2]);
- The first version of MEC service APIs: Radio Network Information (RNI) API [MEC012], Location API [MEC013], UE Identity API [MEC014], Bandwidth management API [MEC015], UE app interface [MEC016].

Edge computing standards will be described in detail in the second part of this book: the MEC platform and Mp1 interface with MEC apps (specified by MEC-009, MEC-010 and MEC-011) are described in Chap. 3, while service APIs are instead described in Chap. 4; finally, practical examples of MEC service APIs are provided in Chap. 5.

In phase 1, mobility issues for the MEC system were also studied in a technical report on "End to End Mobility Aspects" [MEC018] that was the basis for the relative normative work started in phase 2. The structure of the ISG in phase 1 was including also a Working Group (WG) called IEG (Industry Enabling Group) and was specifically dedicated to ecosystem engagement activities. This WG produced a set of specifications related to "MEC service scenarios" [IEG004], "Proof-of-Concepts Framework" [IEG006] and "MEC Metrics Best Practice and Guidelines" [IEG008].

On the other hand, **Phase 2** specifications (textboxes in green) mainly continued the work done in the previous period (with publications of specifications versioned as 2.x.x documents[3]), and also added new APIs to the MEC services portfolio: WLAN API [MEC028], Fixed API [MEC029] and V2X API [MEC030].

In this phase of MEC, the work on mobility aspects started by the Global Report (GR) MEC-018 conducted to the normative work on "MEC Application Mobility Service API" [MEC021] that developed a specification for end-to-end ME application mobility support in a multi-access edge system. More deepening on MEC application mobility can be found in Chap. 8. Last but not the least, the ISG updated

---

[2]This is valid for every MEC phase.

[3]In phase 2, the version 2.x.x of all specifications needed to be aligned with the new definition of the MEC acronym (i.e. from "mobile" to "multi-access" edge computing), that required multiple updates to the MEC terminology and architecture documents, but also to MEC platform and service APIs (previously published as version 1.x.x).

the Terminology [MEC001p1] by publishing a new version of MEC 001, with all terms and definitions compliant with the new phase [MEC001p2].[4]

The need to address vertical market segments triggered also the publication of a "Study on MEC Support for V2X Use Cases" [MEC022], focused on identifying the necessary support provided by Multi-access Edge Computing for V2X applications. This technical report collects and analyzes the relevant V2X use-cases (by taking into account also the findings from external organizations like 5GAA). Moreover, as most of the GRs produced by MEC, it evaluates the gaps from the defined MEC features and functions and identifies the new requirements by recommending also the necessary normative work to close the identified gaps. The main outcome of this study was in fact the starting of the Work Item MEC-030, introducing a Vehicular-to-Everything (V2X) MEC service for automotive use-cases. This new MEC V2X API was conceived in order to facilitate V2X interoperability in a multi-vendor, multi-network and multi-access environment.

Moreover, in this phase, the ISG studied also the "MEC support for network slicing" [MEC0026], by collecting and analyzing the use-cases related to the deployment of MEC functions in combination with network slicing. This study is thus aiming at identifying the needed requirements for the MEC support of network slicing and discussing how the orchestration of resources and services from multiple domains could facilitate that. In this book, network slicing aspects are investigated more in detail in Chap. 7, entirely dedicated to MEC deployment in 5G systems.

For a better acceptance and adoption of MEC technology, the ISG specified also the MEC support for regulatory requirements [41], specifically in areas of Lawful Interception (LI) and Retained Data (RD) for the cases where underlying networks do not provide those.

MEC is based on virtualized infrastructure, and not only intended to be deployed in Virtual Machines (VMs). The Study on Containers [42] was thus focused on identifying the additional support that needs to be provided by MEC when MEC applications run as containers (note: as ETSI NFV is also working on containers; the MEC work is aligned with this group, where applicable).

Another important activity of MEC in phase 2 (although not indicated in the above figure) is the alignment with NFV (Network Function Virtualization). The interested reader can refer to Chap. 6, entirely dedicated to virtualized environments; what is important to point out at this stage is the fact that MEC architecture (as defined since Phase 1) is a standalone architecture, but conceived with the assumption to leverage the presence of an underlying virtualization infrastructure, and (for that purpose) it is reusing most of the NFV framework and definitions done by ETSI ISG NFV [30]. For this reason, in phase 1 the ISG MEC started a work item to study "MEC in NFV deployments" [31], thus with the aim of defining the MEC architecture variant, applicable when MEC is not standalone but deployed in

---

[4]For further details on the terminology difference between MEC Phase 1 and Phase 2, the reader is invited to read Annex A.

a network based on the ETSI NFV framework. This work flowed in enhancements of the MEC architecture document[17],[5] and it is essential also for a better understanding of the role of MEC in 5G systems (based on network virtualization).

In phase 2, ETSI ISG MEC conducted also a series of heterogeneous activities related to the ecosystem engagement. In particular,

- a work item MEC-023 aiming at describing ETSI MEC RESTful APIs using the OpenAPI specification (https://www.openapis.org/). The motivation is to enhance the accessibility of the APIs, particularly to the MEC development community. More details on the published MEC APIs representations [36] will be provided in Chap. 9.
- the definition of an official MEC Hackathon Framework (https://mecwiki.etsi. org/index.php?title=MEC_Hackathon_Framework). The idea is to engage SW and applications developers communities, provide a mean for them to practice MEC technologies usage and to stimulate the ecosystem of developers to adopt MEC for their applications. This work item will be described more in detail in Chap. 10.
- the definition of an official MDT (MEC Deployment Trial) Framework (https:// mecwiki.etsi.org/index.php?title=MEC_Deployment_Framework). The goal of MDTs is to demonstrate the viability of MEC in commercial trials/deployments and to provide feedback to the standardization work. This work item will be described more in detail in Chap. 10.
- The creation (in early 2019) of a WG DECODE on "Deployment and Ecosystem Development", with a focus on accelerating the market adoption and implementation of systems using MEC-defined framework and services exposed using MEC-standardized APIs. This WG includes all activities related to Proofs-of-Concept, MDTs (MCE Deployment Trials), Hackathons, OpenAPI representations, testing and conformance tests, and recently also the MEC Sandbox (described in Chap. 10).

In particular, as a natural prosecution of the work done in phase 1 with the IEG-006 specification (on MEC Metrics, Best Practice and Guidelines), in phase 2 the WG DECODE started a work item [40] for the definition of a "MEC testing framework", with the purpose to list all the functionalities and capabilities required by a MEC compliant implementation. This specification (better described in Chap. 10) is essential for MEC products compliance and interoperability, especially when multiple stakeholders in the value chain have to collaborate to produce an end-to-end solution that could work in operation.

A careful reader might realize that the activities of the DECODE WG are very heterogeneous, as they encompass Proofs-of-Concept, Deployment Trials, Hackathons, MEC Sandbox and also OpenAPI representations and MEC conformance tests. These activities are not part of the *traditional* standardization work

---

[5]Please see the differences between v1 and v2 of MEC003 (ref. [MEC003p1] [MEC003p2]), where the main updates coming out from the MEC017 study on "MEC in NFV" were added.

(intended as "production of standard specifications"), but equally important because focused on accelerating the market adoption and implementation of systems using MEC-defined framework and services exposed using MEC-standardized APIs. In fact, since the beginning, ETSI ISG MEC realized the importance of involving all categories of MEC stakeholders (e.g. both **infrastructure owners** and **application developers**), and for that purpose having even perfect standards is definitely not enough: early implementations of those standards (starting from proofs-of-concept to trials) can better convince infrastructure owners to move to more serious deployments; similarly, application developers need an understandable and more "usable" description of MEC APIs (e.g. via OpenAPI representations), and to try MEC for their applications (e.g. in a MEC Sandbox). Furthermore, both infrastructure owners and application developers need to test MEC products (in terms of platforms but also applications and services) and verify compliance with standards. These are all activities which are key, as well as MEC specs, for a successful introduction of MEC into the market. That's indeed the basic reason for the creation of a WG dedicated to "Deployment and Ecosystem Development" (Fig. 2.11).

Phase 3 of MEC standardization has started in 2021 with the ISG Term 4 (which will last again 2 years) and will produce a new set of specifications. Similarly, to what was done in the previous phases, MEC will continue and finalize the work done in the previous period, will update the reference architecture and also add new APIs to the MEC services portfolio. As a note, versions of MEC service APIs are aligned with the respective standardization phases; as a consequence, a similar trend is foreseen for the future Phase 3 of MEC, which is expected to start in 2019 (lasting again 2 years) and generate version 3.x.x documents.

As a general note, the remainder of this book will refer (by default) to the latest version of published specifications, both for alignment with the Phase 2 terminology and because the newer deliverables often contains bug fixes and an additional feature that are not present in the first publications.

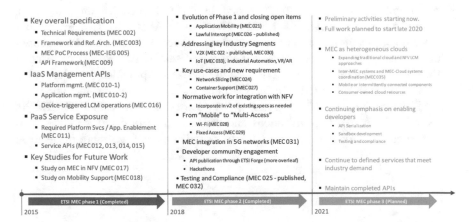

Fig. 2.11 Overview of ETSI MEC standardization work in Phase 1, Phase 2 (completed) and Phase 3 (planned). *Source* ETSI

2.4   Standardization Landscape on Edge Computing

Even if we already clarified that, currently, ETSI MEC is a leading international standard available in the space of edge computing, other standard bodies are working on complementary technologies, or just supporting MEC, or again studying specific topics (e.g. fog) or simply conducting a gap analysis to identify further standardization needs in this area. In summary, the main SDOs currently involved (more or less directly) in edge computing are thus:

- **ETSI ISG MEC**: the entire work of this group is focused on the definition of multi-access edge computing platforms and services, and on MEC ecosystem engagement activities; currently in phase 2, while a third phase of the ISG is planned (starting in 2021);
- **ETSI ISG NFV**: this group is working more in general on network functions virtualization; for that reason, MEC applications (as well as MEC platforms) can be considered just as a particular case of VNFs (virtual network functions);
- **3GPP**: traditionally, this SDO was mainly focusing on cellular network technologies, and only from Release 15 edge computing is described at high level. Specific working groups (within SA) have started working on MEC, for a support which is gradually defined with more details in the subsequent releases.
- **3GPP SA2** working group defined the 5G system architecture, where edge computing is supported from Release 15 of specifications [4, 5] and considered as a key part for 5G services delivery; for the sake of clarity, we have to point out that edge computing is not specified there in detail, as at that time (Rel. 15) a complete and detailed definition of the edge computing support was not a priority from a 3GPP standardization point of view (while implementation details are well covered by ETSI MEC, that acted as pioneering SDO in this field). More details about these specs will be given in Chap. 7 ("Edge Computing in 5G systems").
- **3GPP SA6** working group published in 2018 a specification [10] on a "Common API Framework for 3GPP Northbound APIs (CAPIF)" which considered architectural aspects and procedures related to common API functionalities such as registration, discovery and advertisement of the services. Among the other things, this SA6 work (related to Release 15) provided the relationship of CAPIF with the ETSI MEC API framework, by analyzing its support of CAPIF functionality. More recently, SA6 started a work called EDGEAPP, targeting Release 17 specifications and defining more in detail the edge computing support for 5G systems. This work will be analyzed more in detail in Chap. 7.

Useful References

More context on the CAPIF study can be found by the interested reader in the following online article:

• Erik Guttman, 3GPP TSG SA Chairman: "3GPP Initiates CommonAPI Framework Study", May 9, 2017, link:http://www.3gpp.org/news-events/3gpp-news/1854-common_api

• **3GPP SA5** working group started a technical report for Release 16 of the specifications (TR28.803, "Study on management aspects of edge computing"), with the goal to "study the use cases, and potential requirements and potential solutions for the issue regarding the deployment and management of edge computing, by taking into account the relevant works in ETSI ISG MEC" [5GAAWP]. More recently, SA5 started a study on enhancements of edge computing management and another one on charging aspects. These works (both targeting Release 17 specifications) will be analyzed more in detail in Chap. 7.

Useful References

More context on management aspects can be found by the interested reader in the following online article:

• Thomas Tovinger, 3GPP SA5 Chairman: "Man-agement, Or-chestration and Charging for 5G networks", March 22, 2017, link: http://www.3gpp.org/news-events/3gpp-news/1951-SA5_5G

• **ISO**: the JTC1 SC41 group started an Edge Computing Technical Report on information technology, IoT and related technologies. The scope of this Technical Report is to describe architectures, common concepts, terminologies, values, characteristics, concerns, challenges, use-cases and main technologies of Edge Computing for IoT systems applications. The Technical Report is also meant to assist in the identification of potential areas for standardization in Edge Computing for IoT.

• **IEEE**: in June 2018, the OpenFog Consortium's OpenFog Reference Architecture for fog computing was adopted as an official standard by the IEEE Standards Association (IEEE-SA). The new standard, known as IEEE 1934™, will rely on the reference architecture as a universal technical framework that enables the data-intensive requirements of the Internet-of-Things (IoT), 5G and Artificial Intelligence (AI) applications.

• **IETF**: many internet drafts can be associated to edge computing, at least indirectly because they enable some basic functionalities reused by ETSI MEC specifications. Some examples:

   – IETF RFC 6749: "The OAuth 2.0 Authorization Framework", available at https://tools.ietf.org/html/rfc6749

- IETF RFC 6455: "The WebSocket Protocol", available at https://tools.ietf.org/html/rfc6455
- IETF RFC 7231: "Hypertext Transfer Protocol (HTTP/1.1): Semantics and Content", available at https://tools.ietf.org/html/rfc7231
- IETF RFC 7159: "The JavaScript Object Notation (JSON) Data Interchange Format", available at https://tools.ietf.org/html/rfc7159

- **CCSA**: China Communications Standards Association (CCSA) is a non-profit legal person organization for carrying out standardization activities in the field of Information and Communications Technology (ICT) across China. CCSA is going to start working on MEC (their future work may leverage on ETSI MEC and/or contribute to that standard).[6]

2.5   Industry Groups on Edge Computing

As we described in Chap. 1, MEC technology is applicable to a number of heterogeneous use-cases and services, driven by different vertical market segments. For this reason, ETSI ISG MEC is establishing collaborations with different industry organizations, in order to collect specific requirements coming from these and ensure a proper production of specifications, that are suitable from the point of view of these vertical sectors.

A (not exhaustive) list of international associations and industry groups, relevant from an edge computing perspective, is the following:

- VR-IF (Virtual Reality Industry Forum), https://www.vr-if.org
- OFC (Open Fog Consortium), https://www.openfogconsortium.org
- 5GAA (5G Automotive Association), http://5gaa.org/
- SCF (Small Cell Forum), https://www.smallcellforum.org
- GSMA, https://www.gsma.com
- BBF (Broadband Forum), https://www.broadband-forum.org
- 5G-ACIA (5G Alliance for Connected Industries and Automation), https://www.5g-acia.org
- ECC (Edge Computing Consortium), http://en.ecconsortium.org/
- TIP (Telecom Infra Project), https://telecominfraproject.com/
- WBA (Wireless Broadband Alliance), https://wballiance.com/

The next sections describe just a few of these associations, as examples of industry groups with relevance and possible impact for MEC on various aspects.

---

[6]Internet blog: "Huawei creates new China-focused ecosystem for MEC development", Jul 4, 2017, https://www.telecomtv.com/content/mec/huawei-creates-new-china-focused-ecosystem-for-mec-development-15776/.

2.5.1   5GAA (5G Automotive Association)

The 5G Automotive Association (created in September 2016) brings together automotive, technology and telecommunications companies to work closely together to develop end-to-end connectivity solutions for future mobility and transportation services. The 5GAA Working Groups develop the frameworks, practical aspects, required standards and business cases for 5G and the future application of connected mobility solutions.

The association (counting more than 130 members globally) is organized with seven Working Groups, where in particular WG2 is dedicated to System Architecture and Solution Development (figure below) (Fig. 2.12).

In December 2017, 5GAA published a white paper "Toward fully connected vehicles: Edge computing for advanced automotive Communications" [5GAAWP]. This white paper provides an overview of automotive use-cases and shows how edge computing provides compute/storage/networking capabilities at the network edge, and how it can be considered a supporting technology for multiple services for connected AD vehicles. The 5G Automotive Association categorizes a comprehensive list of connected vehicle applications, categorized in four main groups of use-cases: Safety, Convenience, Advanced Driving Assistance and Vulnerable Road User (VRU). In this paper, 5GAA provides an analysis of uses cases relevant for MEC, as depicted in the Table 2.1.

An example of MEC relevance is given by the use-case on "Real-Time Situational Awareness & High Definition (Local) Maps". Edge computing deployment is ideally suited for this use-case, due to the real time and local nature of the data transferred to augment the situational awareness of road users.

Another relevant example is provided by the VRU (Vulnerable Road user) discovery use-case, where the VRU detection can be improved by exploiting local context and information available at the edge (for example, by using information made available by ETSI MEC APIs, e.g. Radio Network information and Location APIs [5GAAWP, IEC002], that can be exploited to improve the accuracy of the positioning information of all traffic participants).

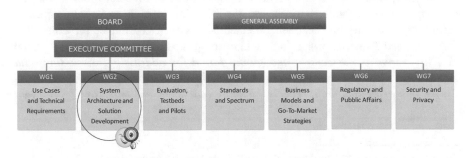

Fig. 2.12   5GAA structure and working groups. *Source* http://5gaa.org/

Table 2.1 5GAA use-cases with relevance for Edge Computing [IEG004]

Use-case	Description	Relevance for MEC
Intersection Movement Assist	Warn driver of collision risk through an intersection	High
Software Updates	Deliver and Manage Automotive Software Updates	Mid
Real-Time Situational Awareness & High Definition Maps	Alert driver of Host Vehicle (HV) moving forward of hazard (icy) road conditions in front	High
See-Through	Driver of Host Vehicle that signals an intention to pass a Remote Vehicle (RV) using the oncoming traffic lane is provided a video stream showing the view in front of the RV	High
Cooperative Lane Change (CLC) of Automated Vehicles	Driver of Host Vehicle (HV) signals an intention to change the lane with at least one Remote Vehicle (RV) in the target lane in the vicinity of the HV	High
Vulnerable Road User Discovery	Detects and Warns drivers of VRUs in the vicinity	High

More recently 5GAA has opened a work item,[7] on "MEC technology to support automotive services", called MEC4AUTO. This work item has the goal to show shows the benefits and capabilities of MEC for multiple automotive use-cases and services in a multi-OEM, multi-MNO and multi-vendor environments.

2.5.2 5G-ACIA (5G Alliance for Connected Industries and Automation)

Industrial automation is an important vertical segment for 5G systems (use-cases and requirements related to this sector are also treated in 3GPP [TR22.804]).

Recently, several companies representing relevant stakeholder groups from the OT (operational technology) industry, the ICT (information and communication technology) industry and academia, created 5G-ACIA (5G Alliance for Connected Industries and Automation). This new industry group has been established with the aim to: "*serve as the central and global forum for addressing, discussing, and evaluating relevant technical, regulatory, and business aspects with respect to 5G for the industrial domain. It reflects the entire ecosystem, encompassing all relevant*

---

[7]5GAA status update, ITU-T meeting, Oct 2020, Link: https://www.itu.int/en/ITU-T/extcoop/cits/Documents/Meeting-20200909-e-meeting/32_5GAA_Status-update.pdf.

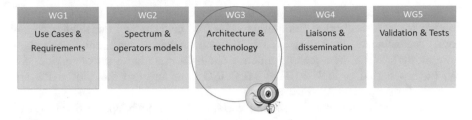

Fig. 2.13 5G-ACIA working groups. *Source* https://www.5g-acia.org

stakeholder groups from the OT (operational technology) industry, the ICT (information and communication technology) industry and academia".

5G-ACIA has five Working Groups and in particular WG3 "Architecture and Technology" (figure below) considers the overall architecture of future 5G-enabled industrial connectivity infrastructures, also including integration concepts and migration paths, the evaluation of key technologies emerging from 5G standardization bodies (Fig. 2.13).

According to a recent 5G-ACIA white paper [9], edge computing can represent an important technology enabler for industrial automation use-cases considered by this consortium.

> In manufacturing [...], 5G may have a disruptive impact as related building blocks, such as wireless connectivity, edge computing or network slicing, find their way into future smart factories. [...] One example are local data centers that support critical industrial applications by way of an edge computing approach. In this case, existing infrastructures must be modified to tackle the new challenges. For instance, industrial applications can be deployed locally within an edge data center to reduce latency.

5G-ACIA is not an SDO, and then needs to have a proper standardization landing zone (e.g. while 3GPP represents the mainstream for URLLC, similarly ETSI MEC is a possible target for their edge computing standardization goals). Dually, on the ETSI ISG MEC side, industrial automation is a very important vertical market segment for MEC as well. In this sense, similarly to what edge computing (and MEC) can do for 5GAA use-cases on C-V2X and connected cars, also in this case edge computing (and again, MEC standard) can support 5G-ACIA use-cases on the enablement of connected and automated solutions for the industry.

2.5.3   Wireless Broadband Alliance

Founded in 2003, the mission of the Wireless Broadband Alliance (WBA) is to champion the development of the converged wireless broadband ecosystem through seamless, secure and interoperable unlicensed wireless broadband services for delivering outstanding user experience.

Recently, WBA's 5G Project identified the evolution of Mobile Edge Computing toward Multi-Access Edge Computing (MEC) to better reflect noncellular operators' requirements, including Wi-Fi (https://wballiance.com/multi-access-edge-computing/). WBA, as a leading alliance aiming at defining a set of services for Wi-Fi, has started to look at ETSI MEC ISG to ensure that the MEC APIs are suitable for supporting Wi-Fi use-cases.

More recently, in October 2020 the WBA published a whitepaper titled "Wi-Fi Opportunities for Connected Vehicles—Demand for Data and Multi-Access Edge Computing" [WBAMEC]. In this work, the data demand of future connected vehicles was analyzed, emphasizing the need for multi-access edge computing architectures for connected vehicles; moreover, the paper identified the gaps and opportunities for Wi-Fi in such architectures, and finally outlined the potential role of WBA in addressing them.

2.6   Other Fora and Projects

Apart from standard bodies and industry groups, many initiatives are currently active in the field of edge computing, starting from specific projects belonging to huge fora and open source communities.

As an example of open source communities, we can mention **Akraino Edge Stack**, "*a Linux Foundation project creating an open source software stack that supports high-availability cloud services optimized for edge computing systems and applications*" [Akraino-web].

This project was created in August 2018 by relevant companies in the telecommunication and IT domains (Arm, AT&T, Dell EMC, Ericsson, Huawei, Intel Corporation, inwinSTACK, Juniper Networks, Nokia, Qualcomm, Radisys, Red Hat and Wind River), and now it's on its execution phase.

The work of Akraino Edge Stack started from the definition of use-cases relevant for multiple stakeholders (like operators, technology providers and OTT players) and the consequent creation of blueprints that consisted of validated hardware and software configurations, suitable for these defined use-case.

As an example, according to some use-cases, this open source software stack is aiming at improving the edge cloud infrastructure, e.g. by providing processing power closer to endpoint customer devices to meet application latency requirements of less than ~ 20 ms. The objective is also to offer new levels of flexibility to scale edge cloud services quickly, to maximize the number of applications or subscribers supported on each edge server, and by looking as well at the reliability of systems, that must be up at all times.

Some key focus areas of improvements in the edge infrastructure are:

- Enable near real-time processing
- Reduce latency
- Improve availability

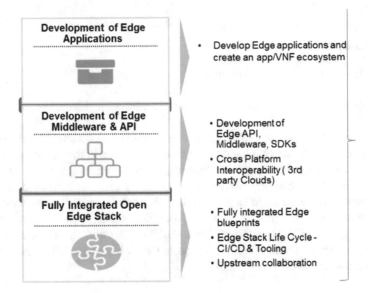

Fig. 2.14 Scope of Akraino Edge Stack. *Source* https://wiki.akraino.org; copyright © Akraino

- Lower operational overhead
- Provide scalability
- Address security needs
- Improve fault management (Fig. 2.14)

As we can see in the above figure, Akraino Edge stack community is focused on Edge APIs, Middleware, Software Development Kits (SDKs) and on allowing for cross-platform interoperability with third-party cloud providers. The edge stack will also enable the development of edge applications and create an application/Virtual Network Function (VNF) ecosystem.

Terminology:

The generic concept of **blueprint**, as defined by Wikipedia, is "a reproduction of a technical drawing, an architectural plan, or an engineering design, using a contact print process on light-sensitive sheets"

Regarding the applicability of this general concept to *edge cloud architecture*, Akraino defines a blueprint as "*a declarative configuration of the entire stack i.e., edge platform that can support edge workloads and edge APIs. In order to address specific use cases, a reference architecture is developed by the community. A declarative configuration is used to define all the components used within that reference architecture such as Hardware, Software, tools to manage the entire stack, and point of delivery i.e., the method used to deploy in a site. In Summary, the declarative configuration is a prescriptive configuration, which will be developed and tested by the community*".

(continued)

(continued)

Links:
https://en.wikipedia.org/wiki/Blueprint
Akraino Integration Projects (Blueprints):
https://wiki.akraino.org/display/AK/Blueprint

The figure below provides a pictorial definition of blueprints in Akraino (Fig. 2.15).

Apart from Akraino, many other software projects and communities are nowadays more or less explicitly related to edge computing, or to the implementation of key components of the MEC architecture (especially when integrating it with NFV infrastructure); some relevant examples are given by projects focused MEC management and orchestration. In fact, a number of different orchestration deployment options have emerged in this context from the industry and standardization: OpenNFV, CloudNFV, OpenBaton, OpenMANO, Cloudify, etc. For that purpose, more details can be found in Chap. 3 (and in Chap. 6), where MEC management and orchestration aspects are described.

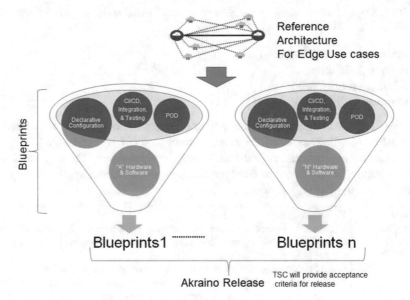

Fig. 2.15 Akraino blueprints (or Integration Projects) (source: https://wiki.akraino.org; copyright © Akraino)

2.7   Research Projects

The landscape of research projects and theoretical activities on edge computing is quite a waste, as the research articles and publications in this field span over a wide time period [Satya96–Barb13], starting many years before the introduction of MEC standard in 2015.

On the other hand, it's important to give an overview of EU-funded research projects, especially those running under the umbrella of the Horizon2020 framework (https://ec.europa.eu/programmes/horizon2020/) defined by the European Commission (EC). In fact, due to the nature of these activities, most of R&D centers of academia and industry players in Europe (and not only) have concentrated their research efforts for the advancement of communication technologies toward 5G systems, in alignment with standardization bodies (ITU, 3GPP, ETSI) and relevant activities in the field (ONF, Open Daylight, OPNFV, etc.).

The figure below is showing the 5G roadmap, according to the 5G PPP (5G Infrastructure Public Private Partnership) [5GPPP], where EU-funded projects are categorized in three different phases, corresponding to the related calls published by the EC:

1. Exploratory phase and specification (2015–16)
2. Detailed research and optimization (2017–18)
3. Experimentation and trials (2019–20) (Fig. 2.16)

Few relevant examples of 5G PPP projects related to MEC and edge computing aspects are provided by projects 5G-Coral, 5G-Miedge, 5G-Transformer. These projects are shortly described in the following tables.

5G-CORAL	
A 5G Convergent Virtualized Radio Access Network Living at the Edge	
Start-end dates	From 2017-09-01 to 2019-08-31, ongoing project
Call For proposal	ICT-08–2017 - 5G PPP Convergent Technologies
Links:	https://cordis.europa.eu/project/rcn/211421_en.html http://5g-coral.eu/
Topic	The 5G-CORAL project leverages on the pervasiveness of edge and fog computing in the Radio Access Network (RAN) to create a unique opportunity for access convergence. This is envisioned by the means of an integrated and virtualized networking and computing solution where virtualized functions, context-aware services, and user and third-party applications are blended together to offer enhanced connectivity and better quality of experience. The proposed solution contemplates two major building blocks, namely (i) the Edge and Fog computing System (EFS) subsuming all the edge and fog computing substrate offered as a shared hosting environment for virtualized functions, services, and applications; and (ii) the Orchestration and Control System (OCS) responsible for managing and controlling the EFS, including its interworking with other (non-EFS)

(continued)

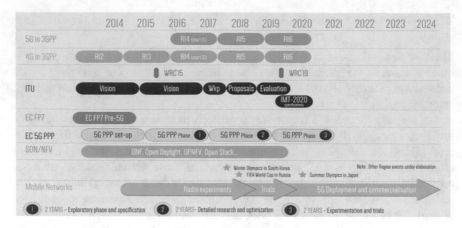

Fig. 2.16 5G roadmap and 5G PPP phases. *Source* https://5g-ppp.eu/

(continued)

5G-CORAL	
A 5G Convergent Virtualized Radio Access Network Living at the Edge	
	domains (e.g. transport and core networks, distant clouds, etc.). Through the 5G-CORAL solution, several Key Performance Indicators (KPIs) can be achieved, notably an ultra-low end-to-end latency in the order of milliseconds. Moreover, new business prospects arise with new stakeholders in the value chain, notably small players owning computing and networking assets in the local service area, such as in shopping malls, airports, trains and cars. These environments will be used to validate the system in three complementary end-to-end large-scale testbeds in Taiwan, supporting innovative applications such as augmented reality, car safety, and IoT gateway

This project 5G-CORAL, resulting from an EU-Taiwan collaboration, is considering edge and fog computing as key aspects, as well as the usage of orchestration function for the overall virtualized infrastructure.

5G-Miedge	
Millimeter-wave Edge Cloud as an Enabler for 5G Ecosystem	
Start-end dates	From 2016-07-01 to 2019-06-30, ongoing project
Call For proposal	EUJ-01–2016—5G—Next Generation Communication Networks
Links:	https://cordis.europa.eu/project/rcn/205497_en.html https://5g-miedge.eu/

(continued)

(continued)

5G-Miedge
Millimeter-wave Edge Cloud as an Enabler for 5G Ecosystem

Topic	Current research efforts on 5G Radio Access Networks (RANs) strongly focus on mmWave access to realize enhanced mobile broadband (eMBB) services. However, there are many unsolved issues to deploy mmWave 5G RAN. The most critical issue is backhauling since it is infeasible to provide 10 Gigabit Ethernet backhaul everywhere. Meanwhile, Mobile Edge Computing (MEC) have obtained much attention in 5G networks as a key technology to enable mission-critical (low latency) applications by allocating storage and computation resources at the edge of the network and circumvent the backhaul networks' limited capacity. However, in the case of mobile networks, it is not easy to reallocate computational resources on demand, while meeting the strict latency constraints foreseen in 5G networks. In this project, mmWave access and MEC are firstly combined as a perfect couple of technologies compensating each other's weaknesses. The new system is called mmWave edge cloud (MiEdge) henceforth. To facilitate MiEdge, the second main goal of the project is to develop a new cellular network control plane in which context information (location, traffic, classes of applications, etc.) of users are collected and processed to forecast traffic and users' requests, in order to enable a proactive resource allocation. This novel control plane is called liquid RAN C-plane since the services and connections can follow users like a liquid. Lastly, the users or application providers orchestrate MiEdge to create a user/application centric 5G network that supports both eMBB and mission-critical applications. In summary, MiEdge project will develop a feasible 5G ecosystem by combining MiEdge, liquid RAN C-plane, and user/application centric orchestration. The project will contribute to the standardization of mmWave access and liquid RAN C-plane in 3GPP and IEEE. At the end, the project will demonstrate a joint 5G test-bed in the city of Berlin and 2020 Tokyo Olympic Stadium

In this project, resulting from an EU-Japan collaboration, MEC is a key focus, and a relevant part of the research activities is related to the joint optimization of radio and computing resources. Optimal association of mobile users to base stations and clouds (based on many parameters, like channel state, backhaul state and application parameters) provides a mechanism for optimal instantiation and migration of containers (light version of virtual machines).

5G-TRANSFORMER
5G Mobile Transport Platform for Verticals

Start-end dates	2017-06-01 to 2019-11-30, ongoing project
Call For proposal	ICT-07–2017 - 5G PPP Research and Validation of critical technologies and systems
Links:	https://cordis.europa.eu/project/rcn/211067_en.html http://5g-transformer.eu

(continued)

(continued)

5G-TRANSFORMER	
5G Mobile Transport Platform for Verticals	
Topic	The vision of the 5G-TRANSFORMER project is that Mobile Transport Networks shall transform from today's rigid interconnection solutions into an SDN/NFV-based 5G Mobile Transport and Computing Platform (MTP) able of simultaneously supporting an extremely diverse range of networking and computing requirements to meet in particular the specific needs of vertical industries. A new networking paradigm known as Network Slicing has emerged for 5G as the most promising approach to address this challenge by enabling per slice management of virtualized resources. 5G-TRANSFORMER aims to bring the "Network Slicing" paradigm into mobile transport networks by provisioning and managing MTP slices tailored to the needs of vertical industries. Specifically: automotive, healthcare and media. The technical approach is twofold: (1) Enable Vertical Industries to meet their service requirements within customized MTP slices; and (2) Aggregate and Federate transport networking and computing fabric, from the edge up to the core and cloud, to create and manage MTP slices throughout a federated virtualized infrastructure. The proposed solution defines three novel building blocks that will be developed and demonstrated integrating the aforementioned three vertical industries: (1) Vertical Slicer as the logical entry point (i.e. one-stop shop) for verticals to support the creation of their respective transport slices in a short time scale (in the order of minutes) (2) Service Orchestrator to orchestrate the federation of transport networking and computing resources from multiple domains and manage their allocation to slices (3) Mobile Transport and Computing Platform as the underlying unified transport stratum for integrated fronthaul and backhaul networks The 5G-TRANSFORMER project addresses "ICT 07–2017: 5G PPP Research and Validation of critical technologies and systems" with a special focus on the TA11: Converged 5G FlexHaul Network objectives of the pre-structuring model

The 5G-Transformer project can be relevant for its focus on network slicing since network slices have an impact on the instantiation of MEC applications.

 Exercises/examples—Chapter 2

Exercise 2.1 Goal of the exercise: Understand the task offloading use-case, benefits and constraints, and the related opportunity to exploit MEC.

Overview of Task Offloading

Consider the following figure, with a resource-constrained mobile device (UE) asking for fully or partially offloading a computation-intensive task to a MEC host (e.g. co-located with the BS). The general motivation for computation offloading is mainly energy saving (i.e. to improve the device battery lifetime), or because end device is actually unable to process a computation-heavy application. In addition, the advantages of using MEC (instead of a remote server, e.g. located somewhere on Internet) are clearly given by the proximity to the user, with consequent benefits both in terms of delay and energy.

Several approaches for task offloading can be found in the scientific literature [Mach17], and a comprehensive overview of computation offloading considering small cell edge computing is provided in [IEEEBarb14]. As an example, in [Orsini15] a cloud-aware framework is proposed for application offloading based on some optimization parameters (network availability, radio signal quality and accessible surrogate computing resources). The framework is conceived in order to break down applications into different components and create an offloading online strategy based on these parameters. Moreover, many software frameworks enabling computation offloading are already available in the space: CloneCloud [Chun11], performing automatic code partitioning at the thread level; Cuckoo, using component-level partitioning [Kemp10]; MAUI, a relevant system implementation for task offloading [Cuervo10].

Task Offloading: System Modeling

In principle, computation offloading is a convenient use-case, due to energy-saving benefits and the exploitation of computation capabilities at the edge of the network. Nevertheless, several aspects and constraints should be taken into account for the evaluation of the actual convenience of task offloading, e.g.:

- The task to be offloaded needs inputs, that should be transferred to the server doing the actual computation; the time (and/or energy) needed for this data transfer ($T_{radio,IN}$) depends on the amount of input data to be transferred, and also on the actual radio conditions of the communication link between the device and the MEC server;

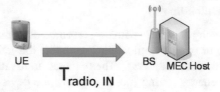

- The computation time at MEC side (T_{elab}) should be evaluated against the actual time needed by the original device requesting the offloading; this should include also the time needed to find in the system the actual MEC server capable to acquire and satisfy the request coming from the device;

- At the end of the computation, the MEC server should send back the output to the device; the time (and/or energy) needed for this data transfer ($T_{radio,OUT}$) depends on the amount of output data to be transferred, and also on the actual radio conditions of the communication link between the device and the MEC server;

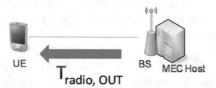

In summary, all these aspects and parameters identify a fundamental tradeoff between the time (/energy) spent by the device to run the task, and (on the other side) the overall time (/energy) spent by the device to transmit inputs to the MEC system, and by the MEC server to compute the task and sent back outputs. Of course, this convenience should be evaluated, also depending on the specific service constraints associated to the specific task (e.g. overall time when the constraints are in terms of end-to-end latency, or energy when improving battery lifetime is a key requirement).

Task Offloading: Exemplary Calculations

Consider a computation task (e.g. antivirus) with the following characteristics:

- CPU workload: 0.5 Mega Cycle
- End-to-end delay constraints: T_{MAX} = 55 ms
- UE computation capabilities: 1 Mega Cycle / s

Consider also a task offloading opportunity to a MEC server co-located with a BS (serving the UE) and with the following characteristics:

- Average SNR of the radio link: SNR_{dB} = 15 dB
- Bandwidth channel: BW = 5 MHz
- MEC server computation capabilities: 10 Mega Cycles / s
- Input Data (needed to run the task): d_{in} = 100 kbit (or 0.1 Mbit)
- Output Data (coming from the task): d_{out} = 50 kbit (or 0.05 Mbit)

We may be interested in understanding: (1) whether (in these conditions) the offloading is actually an opportunity to meet the end-to-end delay constraints; (2) what is the minimum SNR that makes convenient offloading the task to the MEC server (instead of running the task at UE side).

The first question can be easily answered by means of few simple calculations:

- The time T_{UE} needed by the UE to perform the task locally is 0.5 s (since the UE capability is of 1 Mega Cycle/s and the CPU workload is of 0.5 Mega Cycles)
- This time should be compared with $T_{offload} = T_{radio,IN} + T_{elab} + T_{radio,OUT}$
- T_{elab} = 0.05 s (since the MEC server computation capability is ten times better than the UE) = 50 ms
- The capacity of the radio link can be calculated through the Shannon formula:

$$C = BW \cdot log_2(1 + SNR)$$

- So, first of all, we have to convert the SNR from dB to linear scale:

$SNR_{dB} = 10 \cdot log_{10}(SNR)$ that means $SNR = 10^{(SNR_{dB}/10)}$
- SNR = 31.62 → C = 25.14 Mbps ≅ 25 Mbps

- $T_{radio,IN}$ is given by d_{in} / C = 4 ms
- Similarly, $T_{radio,OUT}$ = d_{out} / C = 2 ms
- Finally, $T_{offload}$ = 4 ms + 50 ms + 2 ms = 56 ms
- As a consequence, even if performing the task locally at the UE costs 0.5 s, the usage of MEC server for task offloading is requiring 56 ms, which is still too much compared to the end-to-end delay constraints (55 ms).

Now, in order to determine the target SNR that makes it convenient offloading the task to the MEC server (question n.2), if sufficient to:

- calculate the required capacity C from the end-to-end delay formula:

$$T_{MAX} = \frac{d_{in}}{C} + T_{elab} + \frac{d_{out}}{C}$$

$$C = \frac{d_{in} + d_{out}}{T_{MAX} - T_{elab}}$$

- invert the Shannon formula, to calculate the required SNR

$$SNR = 2^{(C/BW)} - 1$$

- as a consequence, we have C = 150 kbit / 5 ms = 30 Mbps
- and finally, SNR = 63, or better, $SNR_{dB} \cong 18$ dB

The following figure provides also the trend of the data transfer time (considering input and output data), varying with the SNR_{dB} of the radio link.

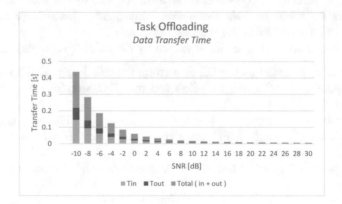

In summary, we have calculated the delay performances of a task offloading case when using MEC. As a side note, given the end-to-end delay constraints, in most of the cases, it is impossible to offload the task to a remote server. In fact, in practical cases, the delay due to the transport and backhauling network is measurable in tens or hundreds of milliseconds [SabeVTC18].

Another variant of the present exercise (left to the willingness of the reader) could consist in the modeling of energy consumption for task offloading.

Quiz—Part 1
(Chapters 1–2)

(1) **The lowest level of cloud computing service model is**

    a. PaaS
    b. IaaS
    c. SaaS
    d. Depends on the OS

(2) **According to Mathis formula, the QoE performance of a TCP network is limited by:**

    a. Latency and bandwidth
    b. Latency and throughput
    c. Packet loss and round-trip-time
    d. Packet Loss and bandwidth
    e. None of the above

(3) **Let's consider the tables below, showing latency and bandwidth distributions. We can say that:**

    a. home C and A are probably connected to the same ISP
    b. the link between home A & B is the best in terms of CDF of Latency and Bandwidth
    c. the link between home A & B is the worst in terms of CDF of Latency and Bandwidth
    d. home A and B are connected to ISP1 and home C with ISP2
    e. considering the median value, the link A&B is equivalent to B&C
    f. None of the above

	5%	10%	50%	90%	95%
Home A & B	18.5	19.2	26.4	77.8	133.6
Home B & C	36.4	37.2	44.9	87.2	98.0
Home C & A	38.8	39.3	44.9	75.1	92.6

(a) Latency Distribution (milliseconds)

	5%	10%	50%	90%	95%
Home A & B	0.5	0.6	1.9	2.3	2.3
Home B & C	0.5	0.7	0.8	0.9	0.9
Home C & A	0.5	0.5	0.8	0.9	0.9

(b) Upload Bandwidth Distribution (Mbps)

(4) **In the NIST's Cloud Computing Reference Architecture (CCRA) a Cloud Broker is:**

    a. an entity that manages the use, performance and delivery of cloud services, and negotiates relationships between Cloud Providers and Cloud Consumers

    b. an intermediary that provides connectivity and transport of cloud services from Cloud Providers to Cloud Consumers

    c. a party that can conduct independent assessment of cloud services, information system operations, performance and security of the cloud implementation

    d. a person or organization that maintains a business relationship with, and uses service from, Cloud Providers

(5) **MEC means "multi-access" edge computing because:**

    a. MEC users must use multiple access technologies

    b. ETSI is defining an access agnostic technology, thus applicable in principle to 4G/5G, Wi-Fi or fixed access

    c. ETSI is defining an access agnostic technology, thus applicable to 5G networks connected contemporarily to Wi-Fi and fixed access

    d. ETSI MEC system can be connected to multiple Wi-Fi access points

(6) **MEC is based on virtualized infrastructure. Please select the correct sentence below:**

    a. MEC is only intended to be deployed in Virtual Machines (VMs)

    b. MEC 027 is clarifying that support that needs to be provided by MEC when containers are used for MEC platforms

    c. MEC 027 is identifying the additional support that needs to be provided by MEC when MEC applications run as containers

    d. as ETSI NFV is also working on containers, the MEC work is aligned with this group, where applicable. Thus, MEC is not studying containers support.

(7) **MEC in NFV is:**

    a. a MEC architecture variant, applicable when MEC is not standalone but deployed in a network based on the ETSI NFV framework

    b. a MEC architecture flavor, applicable when MEC standalone is deployed in a NFV environment

    c. a MEC architecture flavor, applicable when MEC is not standalone but deployed in a network based on the ETSI NFV framework

    d. a MEC architecture variant, applicable when MEC standalone is deployed in a NFV environment

(8) **5GAA (5G Automotive Association) considers MEC as a key technology for C-V2X use-cases. in particular:**

    a. an example of highly relevant 5GAA use-case is given by "Software Updates" because they are frequently and continuously updating the whole in-car software package;

    b. a relevant example is provided by VRU (Vulnerable Road user) discovery use-case, where the VRU detection can be updated by means of frequent Software Updates in the car;

    c. "Cooperative Lane Change (CLC) of Automated Vehicles" can be done by switching frequently MEC platforms in different lanes, which are serving the same MEC application.

    d. an example of MEC relevance is given by the use-case on "High Definition (Local) Maps", due to the real-time and local nature of the data transferred;

(9) **Akraino blueprints are:**

    a. Research projects leveraging open source implementations from LF Edge

    b. A definition from Akraino Edge Stack community of "declarative configurations of the entire stack i.e. edge platform that can support edge workloads and edge APIs".

    c. A commonly recognized definition of "reproductions of a technical drawing, an architectural plan, or an engineering design, using a contact print process on light-sensitive sheets"

    d. A way, defined by LF Edge and properly covered by copyright, to print the word "Akraino" with a special blue painting and well-specified style

(10) **5G-ACIA**

    a. is an SDO, similar to 5GAA but focused on connected vehicles and automated driving

    b. is not an SDO, similar to 5GAA but focused on connected industries and automation

    c. is not an SDO, similar to 5GAA but focused on connected vehicles and automated driving

    d. is an SDO, similar to 5GAA but focused on connected industries and automation.

References

[MEC-web]   ETSI MEC, https://www.etsi.org/technologies-clusters/technologies/multi-access-edge-computing

[ETSIWP11]   ETSI White Paper: "Mobile Edge Computing: A key technology towards 5G", September 2015; Link: https://www.etsi.org/images/files/ETSIWhitePapers/etsi_wp11_mec_a_key_technology_towards_5g.pdf

[5GPPP]   5GPPP, "5G Vision: The 5G Infrastructure Public Private Partnership: The Next
                Generation of Communication Networks and Services", 2015. Link: https://5g-
                ppp.eu/wp-content/uploads/2015/02/5G-Vision-Brochure-v1.pdf
[TS23.501]   3GPP TS 23.501 V15.2.0 (2018–06): System Architecture for the 5G System, June
                2018, Link: http://www.3gpp.org/ftp/Specs/archive/23_series/23.501/23501-f20.zip
[TS23.502]   3GPP TS 23.502 V15.2.0 (2018–06): Procedures for the 5G System; Stage 2
                (Release 15)", June 2018, Link: http://www.3gpp.org/ftp/Specs/archive/23_
                series/23.502/23502-f20.zip
[5GAA-web]   5GAA website, http://5gaa.org
[5GAAWP]   5GAA white paper: "Toward fully connected vehicles: Edge computing for
                advanced automotive Communications", Dec. 2017, http://5gaa.org/wp-content/
                uploads/2017/12/5GAA_T-170219-whitepaper-EdgeComputing_5GAA.pdf
[TR22.804]   3GPP TR 22.804 V1.2.0 (2018–05) 3rd Generation Partnership Project;
                Technical Specification Group Services and System Aspects; Study on
                Communication for Automation in Vertical Domains;
[5GACIAWP]   5G-ACIA White Paper: "5G for Connected Industries and Automation", second
                edition,   February   2019,   https://www.5g-acia.org/fileadmin/5G-ACIA/
                Publikationen/Whitepaper_5G_for_Connected_Industries_and_Automation/
                WP_5G_for_Connected_Industries_and_Automation_Download_19.03.19.pdf
[TS23.222]   3GPP TS 23.222 version 15.2.0 Release 15, "Common API Framework for
                3GPP Northbound APIs", published as ETSI TS 123 222 V15.2.0 (2018–07)
                https://www.etsi.org/deliver/etsi_ts/123200_123299/123222/15.02.00_60/ts_
                123222v150200p.pdf
[SP-180390]   3GPP SA5, Tdoc on new Study Item TR28.803 on "Study on management
                aspects of edge computing", June 2018, Link: https://portal.3gpp.org/ngppapp/
                CreateTdoc.aspx?mode=view&contributionUid=SP-180390
[MEC001p1]   ]ETSI GS MEC 001 V1.1.1 (2016–03) Mobile Edge Computing
                (MEC) Terminology http://www.etsi.org/deliver/etsi_gs/MEC/001_099/001/
                01.01.01_60/gs_MEC001v010101p.pdf
[MEC001p2]   ETSI GS MEC 001 V2.1.1 (2019–01) Multi-access Edge Computing
                (MEC) Terminology https://www.etsi.org/deliver/etsi_gs/mec/001_099/001/02.
                01.01_60/gs_mec001v020101p.pdf
[MEC002p1]   ]ETSI GS MEC 002 V1.1.1 (2016–03) Mobile Edge Computing (MEC);
                Technical   Requirements   https://www.etsi.org/deliver/etsi_gs/MEC/001_099/
                002/01.01.01_60/gs_MEC002v010101p.pdf
[MEC002p2]   ETSI GS MEC 002 V2.1.1 (2018–10) Multi-access Edge Computing (MEC);
                Technical   Requirements   https://www.etsi.org/deliver/etsi_gs/MEC/001_099/
                002/02.01.01_60/gs_MEC002v020101p.pdf
[MEC003p1   ]ETSI GS MEC 003 V1.1.1 (2016–03) Mobile Edge Computing (MEC);
                Framework and Reference Architecture http://www.etsi.org/deliver/etsi_gs/
                MEC/001_099/003/01.01.01_60/gs_MEC003v010101p.pdf
[MEC003p2]]   ETSI GS MEC 003 V2.1.1 (2019–01) Multi-access Edge Computing (MEC);
                Framework and Reference Architecture https://www.etsi.org/deliver/etsi_gs/
                MEC/001_099/003/02.01.01_60/gs_MEC003v020101p.pdf
[IEG004]   ETSI GS MEC-IEG 004 V1.1.1 (2015–11) Mobile-Edge Computing (MEC);
                Service Scenarios http://www.etsi.org/deliver/etsi_gs/MEC-IEG/001_099/004/
                01.01.01_60/gs_MEC-IEG004v010101p.pdf
[IEG005]   ETSI GS MEC-IEG 005 V1.1.1 (2015–08) Mobile-Edge Computing (MEC);
                Proof of Concept Framework http://www.etsi.org/deliver/etsi_gs/MEC-IEG/
                001_099/005/01.01.01_60/gs_MEC-IEG005v010101p.pdf
[IEG006]   ETSI GS MEC-IEG 006 V1.1.1 (2017–01) Mobile Edge Computing; Market
                Acceleration; MEC Metrics Best Practice and Guidelines http://www.etsi.org/deliver/
                etsi_gs/MEC-IEG/001_099/006/01.01.01_60/gs_MEC-IEG006v010101p.pdf

[MEC009]   ETSI GS MEC 009 "General principles for Mobile Edge Service APIs" http://
              www.etsi.org/deliver/etsi_gs/MEC/001_099/009/01.01.01_60/gs_
              MEC009v010101p.pdf
[MEC010–1]   ETSI GS MEC 010–1 V1.1.1 (2017–10) "Mobile Edge Management; Part 1:
              System, host and platform management" https://www.etsi.org/deliver/etsi_gs/
              MEC/001_099/01001/01.01.01_60/gs_MEC01001v010101p.pdf
[MEC010–2]   ETSI GS MEC 010–2 V1.1.1 (2017–07) "Mobile Edge Management; Part 2:
              Application lifecycle, rules and requirements management" http://www.etsi.org/
              deliver/etsi_gs/MEC/001_099/01002/01.01.01_60/gs_MEC01002v010101p.pdf
[MEC011]   ETSI GS MEC 011 V1.1.1 (2017–07) "Mobile Edge Platform Application
              Enablement" http://www.etsi.org/deliver/etsi_gs/MEC/001_099/011/01.01.01_
              60/gs_MEC011v010101p.pdf
[MEC012]   ETSI GS MEC 012 "Radio Network Information API" http://www.etsi.org/
              deliver/etsi_gs/MEC/001_099/012/01.01.01_60/gs_MEC012v010101p.pdf
[MEC013]   ETSI GS MEC 013 "Location API" http://www.etsi.org/deliver/etsi_gs/MEC/
              001_099/013/01.01.01_60/gs_MEC013v010101p.pdf
[MEC014]   ETSI GS MEC 014 "UE identity API" http://www.etsi.org/deliver/etsi_gs/
              MEC/001_099/013/01.01.01_60/gs_MEC013v010101p.pdf
[MEC015]   ETSI GS MEC 015 "Bandwidth management API" http://www.etsi.org/deliver/
              etsi_gs/MEC/001_099/015/01.01.01_60/gs_MEC015v010101p.pdf
[MEC016]   ETSI GS MEC 016 "UE application interface" http://www.etsi.org/deliver/etsi_
              gs/MEC/001_099/016/01.01.01_60/gs_MEC016v010101p.pdf
[NFV-web]   ETSI NFV website, https://www.etsi.org/technologies-clusters/technologies/nfv
[MEC017]   ETSI GR MEC 017 "MEC in NFV environment" https://www.etsi.org/deliver/
              etsi_gr/MEC/001_099/017/01.01.01_60/gr_MEC017v010101p.pdf
[MEC018]   ETSI GR MEC 018 "End to End Mobility Aspects" http://www.etsi.org/deliver/
              etsi_gr/MEC/001_099/018/01.01.01_60/gr_MEC018v010101p.pdf
[MEC021]   ETSI GS MEC 021 V2.1.1 (2020–01) "MEC Application Mobility Service
              API" Link: https://www.etsi.org/deliver/etsi_gs/MEC/001_099/021/02.01.01_
              60/gs_MEC021v020101p.pdf
[MEC022]   ETSI GR MEC 022 "Study on MEC Support for V2X Use Cases"; Link to the
              GR/MEC-0022V2X.Support   Work   Item: https://portal.etsi.org/webapp/
              workProgram/Report_Schedule.asp?WKI_ID=52949
[MEC023]   ETSI GR MEC 023 "Describing ETSI MEC RESTful APIs using the OpenAPI
              specification"; Link to the DGR/MEC-0023OpenApi Work Item: https://portal.
              etsi.org/webapp/workProgram/Report_Schedule.asp?WKI_ID=51416
[MEC-forge]   ETSI Forge website, https://forge.etsi.org/
[MEC-open]   MEC open area, https://docbox.etsi.org/ISG/MEC/Open/
[MEC-wiki]   MEC wiki page, https://mecwiki.etsi.org
[MEC024]   ETSI GR MEC 024 "MEC support for network slicing", Link to the DGR/
              MEC-0024NWslicing Work Item: https://portal.etsi.org/webapp/workProgram/
              Report_Schedule.asp?WKI_ID=53580
[MEC025]   ETSI GR MEC 025 "MEC Testing Framework", Link to the DGR/
              MEC-0025TestingFramework   Work   Item: https://portal.etsi.org/webapp/
              workProgram/Report_Schedule.asp?WKI_ID=53612
[MEC026]   ETSI GS MEC 026 "Support for regulatory requirements", Link to the DGS/
              MEC-0026LI Work Item: https://portal.etsi.org/webapp/workProgram/Report_
              Schedule.asp?WKI_ID=53965
[MEC027]   ETSI GR MEC 027 "MEC support for containers", Link to the DGR/
              MEC-0027ContainerStudy   Work   Item: https://portal.etsi.org/webapp/
              workProgram/Report_Schedule.asp?WKI_ID=53966
[MEC028]   ETSI GS MEC 028 V2.1.1 (2020–06) "MEC WLAN information API"; Link:
              https://www.etsi.org/deliver/etsi_gs/MEC/001_099/028/02.01.01_60/gs_
              MEC028v020101p.pdf

[MEC029]   ETSI GS MEC 029 V2.1.1 (2019–07) "MEC Fixed Access information API";
                 Link: https://www.etsi.org/deliver/etsi_gs/MEC/001_099/029/02.01.01_60/gs_
                 MEC029v020101p.pdf
[MEC030]   ETSI GS MEC 030 V2.1.1 (2020–04) "MEC V2X information Service API";
                 Link: https://www.etsi.org/deliver/etsi_gs/MEC/001_099/030/02.01.01_60/gs_
                 MEC030v020101p.pdf
[Akraino-web]   Akraino Edge Stack website, https://www.akraino.org/
[Satya96]   M. Satyanarayanan, "Fundamental Challenges in Mobile Computing", in Proc.
                 of 15th Annual ACM Symposium on Principles of Distributed Computing,
                 Philadelphia, May 1996.
[Satya09]   M. Satyanarayanan, P. Bahl, R. Caceres, and N. Davies, "The Case for
                 VM-Based Cloudlets in Mobile Computing", IEEE Pervasive Computing,
                 Vol.8, No.4, Oct.-Dec. 2009, pp. 2–11
[Barb13]   S. Barbarossa, S. Sardellitti, P. Di Lorenzo "Joint allocation of computation and
                 communication resources in multiuser mobile cloud computing" Proc. of IEEE
                 Work. Signal Process. Adv. Wir. Commun (SPAWC 2013) pp. 16–19 Jun. 2013.
[Chun11]   Barbarossa S, Sardellitti S, Di Lorenzo P (Nov. 2014) Communicating while
                 Computing: Distributed Mobile Cloud Computing over 5G Heterogeneous
                 Networks. IEEE Signal Process Mag 31(6):45–55
[Chun11]   B.G. Chun, S. Ihm, P. Maniatis, M. Naik, and A. Patti, "CloneCloud: Elastic
                 Execution between Mobile Device and Cloud", in Proc. of the 6th ACM
                 European Conference on Computer Systems, Salzburg, Apr. 2011.
[Kemp10]   R. Kemp, N. Palmer, T. Kielmann, and H. Bal, "Cuckoo: A Computation
                 Offloading Framework for Smartphones", Mobile Computing, Applications and
                 Services, Vol.76, LNICST, Springer, Oct. 2010. pp. 59–79.
[Cuervo10]   E. Cuervo, A. Balasubramanian, D. Cho, A. Wolman, S. Saroiu, R. Chandra,
                 and P. Bahl, "MAUI: Making Smartphones last Longer with Code Offload", in
                 Proc. of ACM MobiSys, San Francisco, Jun. 2010.
[Mach17]   Pavel Mach and Zdenek Becvar, "Mobile Edge Computing: A Survey on
                 Architecture and Computation Offloading", IEEE Communications Surveys &
                 Tutorials, Volume: 19, Issue: 3, 2017, pp. 1628 – 1656
[Orsini15]   G. Orsini, D. Bade, and W. Lamersdorf, "Computing at the Mobile Edge:
                 Designing Elastic Android Applications for Computation Offloading", in Proc.
                 of IEEE 8th IFIP Wireless and Mobile Networking Conference (WMNC),
                 Munich, Oct. 2015.
[SabeVTC18]   D. Sabella, N. Nikaein ; A. Huang ; J. Xhembulla ; G. Malnati ; S. Scarpina: "A
                 Hierarchical MEC Architecture: Experimenting the RAVEN Use-Case", 2018
                 IEEE 87th Vehicular Technology Conference (VTC Spring), June 2018, Link:
                 https://ieeexplore.ieee.org/abstract/document/8417826/
[ETSIWP28]   ETSI White Paper #28: "MEC in 5G networks", June 2018; Link:https://www.
                 etsi.org/images/files/ETSIWhitePapers/etsi_wp28_mec_in_5G_FINAL.pdf
[ETSIWP36]   ETSI White Paper #36: "Harmonizing standards for edge computing - A
                 synergized architecture leveraging ETSI ISG MEC and 3GPP specifications",
                 first   edition,   July   2020;   Link: https://www.etsi.org/images/files/
                 ETSIWhitePapers/ETSI_wp36_Harmonizing-standards-for-edge-computing.pdf
[SabRezFraz]   D. Sabella, Alez Reznik, Rui Frazao, "Multi-Access Edge Computing in
                 Action", Taylor & Francis Group, September 2019, Link: https://doi.org/10.
                 1201/9780429056499
[WBAMEC]   WBA whitepaper, "Wi-Fi Opportunities for Connected Vehicles – Demand for
                 Data and Multi-Access Edge Computing", October 2020, Link: https://
                 wballiance.com/wi-fi-opportunities-for-connected-vehicles/

Part II
Edge Computing Standards

Chapter 3
MEC Standards on Edge Platforms

The present chapter describes the work done in European Telecommunication Standardization Institute (ETSI), by the MEC Industry Standard Group (ISG) [MEC-web], with a special focus on the description of the edge platform related specifications (while MEC services related specifications will be treated in Chap. 4). In the following sections, we will first describe the MEC Architecture and continue with the MEC Management and the Application Enablement framework.

As already anticipated in Chap. 2, in the following we will refer (by default) to the latest version of published specifications, unless specified otherwise. This is a convenient and reasonable choice, for two reasons: (1) alignment with the updated terminology and (2) presence of bug fixes and additional features in the newest MEC publications.

3.1   The Central Role of MEC Host

Figure 3.1 is showing the generic scenario considered by MEC, where the MEC host is essentially adding cloud computing capabilities at the edge of the mobile network, in addition to remote clouds and private clouds located somewhere else. The UE (and the application running on it) is thus capable to communicate not only with a remote application server (like in all traditional client-server systems) but also with another (intermediate) end-point (i.e. the MEC application), which is leveraging from a low-latency environment offered by the MEC Host.

It is thus clear how the **MEC Host** plays a central role in the MEC system, as this is the main element introduced in the communication from a user application perspective (and in an access-agnostic manner). In the following sections, we will analyze the architectural elements introduced by ETSI ISG MEC (by referencing the updated terminology and v2 specifications published so far in phase 2), identifying the MEC Host and the other components of the MEC system.

© Springer Nature Switzerland AG 2021   59
D. Sabella, *Multi-access Edge Computing: Software Development
at the Network Edge*, Textbooks in Telecommunication Engineering,
https://doi.org/10.1007/978-3-030-79618-1_3

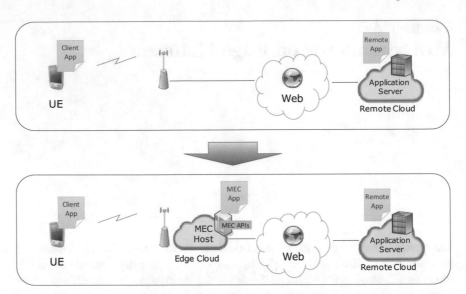

Fig. 3.1 Reference scenario for MEC (software developer perspective)

3.2   MEC Architecture

The main specification defining the Multi-access Edge Computing framework is MEC-003, where all general entities involved are grouped into system-level, host-level and network-level entities [MEC003v2].[1]

 Useful Reference
ETSI GS MEC 003 V2.1.1 (2019–01): Multi-access Edge Computing (MEC); Framework and Reference Architecture

Download (phase 2 spec):
https://www.etsi.org/deliver/etsi_gs/MEC/001_099/003/02.01.01_60/gs_MEC003v020101p.pdf
Further updates in the MEC open area (e.g. temporary preview of the stable drafts for subsequent versions):
https://docbox.etsi.org/ISG/MEC/Open/

---

[1]The version v2 referenced here is essentially based on v1, but with some amendment on the terminology and additions related to MEC in NFV deployments.

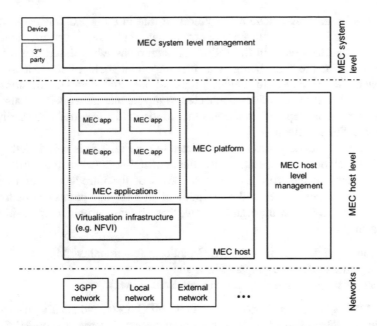

Fig. 3.2 Multi-access edge computing framework. *Source* [MEC003v2]—© European Telecommunications Standards Institute 2020. Further use, modification, copy and/or distribution are strictly prohibited

Figure 3.2 illustrates the MEC framework, which is consisting of the following parts:

- MEC host (sometimes indicated as mobile edge host, with the old phase 1 terminology), including a MEC platform and a virtualization infrastructure which provides compute, storage and network resources for the MEC applications;
- MEC applications, running on top of the virtualization infrastructure provided by the MEC host, and can interact with the MEC platform to consume and provide MEC services;
- MEC system-level management;
- MEC host-level management;
- External related entities, i.e. network-level entities, devices and third parties.

This is only a generic framework, that helps understanding the main functional blocks involved in the system, but at least it should be clear from this picture that MEC system is conceived in principle to be access-agnostic (as we also anticipated in Chap. 1), as this framework leverages the connectivity offered by the underlying network. So, even in Phase 1, MEC was focusing on mobile networks, all the main design work related to the architecture was abstracted from the particular access network (4G or 5G,...) and thus in principle applicable also to other kinds of networks (e.g. Wi-Fi, fixed networks, ...).

A careful reader may have already noticed that in the MEC framework only a generic device is present, but no mention of any application from user side. This aspect will be made clear in the following paragraphs when describing the MEC architecture. For the time being it is just important to highlight that the MEC Framework defines a generic "device", thus not only including UEs (as considered typically in 3GPP for terminals connected to the mobile networks) but also other kinds of terminals (laptops, but also other kinds of devices not necessarily connected to cellular networks).

The MEC reference architecture (depicted in the Fig. 3.3) is based on the above general framework, and specifies also the connection between the various functional elements of the MEC system, by means of the introduction of **reference points** between them (a detailed description of all reference points can be found in Annex B). As shown in the figure below, these reference points are divided into three main groups:

- reference points regarding the MEC platform functionality (Mp);
- management reference points (Mm); and
- reference points connecting to external entities (Mx).

The **MEC system** consists of one or more MEC hosts, and a MEC management necessary to set up and orchestrate the entire system, in order to run MEC applications within an operator network or a subset of an operator network. A MEC host is the actual server hosting the MEC platform and MEC applications. The device app is an application instance running on the device and communicating with the MEC system through Mx2; in addition to that, the above figure is showing also a remote app,[2] which is the traditional end-point in a client-server application relationship (even without the MEC in between). It should be noticed that the Device App is NOT what we would call Client App (i.e. the client side of an application, running in the user terminal). This differentiation is important because essentially the Client App is the instance installed by the end-customer, who typically is not a developer. A device App is instead likely to be managed by the developer (or application provider) and can run in the same terminal of the Client App or in a different terminal. This aspect is also clarified in a recent ETSI white paper [ETSIWP20], whereas the Fig. 3.4 captures the distinction between Device App and Client App and their relationship with the MEC system.

Now, looking again at Fig. 3.3, it is clear how the three blocks colored in light blue (device app, MEC app and remote app) are corresponding to software instances, respectively, in a device (managed by the application provider), in a MEC host and in a remote cloud (outside the MEC system).

The reader may also notice that in principle other software entities can be hosted and run co-located in the same server, especially if the overall software is based on the same virtualized infrastructure, e.g. virtual RAN, or other VNFs following the

---

[2]Nevertheless, it's worth noting that remote app is not explicitly present in the MEC specification [MEC003], and formally should be considered as outside the MEC system.

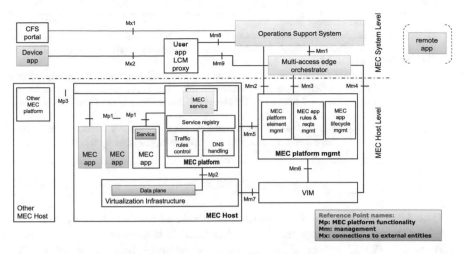

Fig. 3.3 Multi-access Edge Computing reference architecture [MEC003v2]

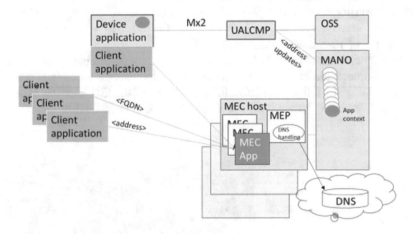

Fig. 3.4 Interaction between terminals and the MEC system [ETSIWP20]

ETSI NFV framework [NFV-web] [NFV002]. Moreover, MEC entities running on a virtualized infrastructure can be hosted by the same edge cloud together with other entities (including other VNFs). This is obviously compatible with the general concept of cloud (already explained in Chap. 1), but the specific MEC architecture resulting from the deployment is different from that one shown in the above figure and will be described in Chap. 6.

Terminology (v2):
The **MEC host** is an entity that contains a MEC platform and a virtualization infrastructure which provides compute, storage and network resources, for the purpose of running MEC applications
The **MEC platform** is the collection of essential functionality required to run MEC applications on a particular virtualization infrastructure and enable them to provide and consume MEC services. The MEC platform can also provide services
MEC applications are instantiated on the virtualization infrastructure of the MEC host based on configuration or requests validated by the MEC management

Download:
https://www.etsi.org/deliver/etsi_gs/MEC/001_099/001/02.01.01_60/gs_mec001v020101p.pdf

The MEC management shown on the right side of Fig. 3.5, encompasses both the MEC system-level management and the MEC host-level management (coherently with the MEC framework levels, separated by dotted lines in the figure). The first one includes the **MEC orchestrator** as its core component, which has an overview of the complete MEC system. The second level comprises the **MEC platform manager** and the Virtualization Infrastructure Manager (**VIM**) and handles the management of the MEC-specific functionality of a particular MEC host and the applications running on it.

These blocks are defined in [MEC003v2] only at a functional level, while the "stage 3" implementation details can be found in other specifications (e.g. related to mobility) and/or are left to detailed definition outside of the ISG (e.g. this is the case of the OSS, which is better specified in TM Forum, ITU and 3GPP).

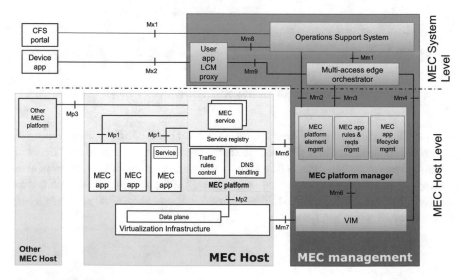

Fig. 3.5 MEC management, MEC hosts and external entities

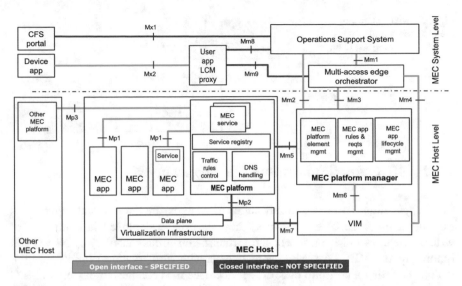

Fig. 3.6 Open and Closed interfaces in MEC

One other important aspect is related to the definition of the interfaces. In fact, MEC is specifying only a portion of the interfaces introduced in the architecture (indicated in green in Fig. 3.6 as open interfaces), while the others (indicated in red) are left to the specific vendor implementation. This approach is coherent with the general principle to introduce in a standard only the minimum amount of features to implement a system by enabling the interoperability between different players while leaving to implementers the possibility to compete with different products and technical solutions.

As an example, MEC architecture is specifying in detail Mp1 reference point, as this is the main interface between MEC applications (produced by App developers) and the MEC platform (that generally is provided by a third party). On the other hand, the interaction between the MEC platform and its manager (through Mm5 reference point) is generally realized by the same MEC platform vendor, thus the relative interface is considered proprietary and doesn't need to be specified in detail. Nevertheless, the above Fig. 3.6 is only an instantaneous photo picture of the current situation in ETSI ISG MEC, as it is possible that in the future phase 3 of standardization, some interfaces (here depicted in "red") will be specified (becoming "green"). This could be the case of some reference points connecting blocks owned by different stakeholders, and the relative specification in the standard is motivated by the need of interoperability between the blocks (and the stakeholders).

Fig. 3.7 MEC platform,
MEC services and
service-producing MEC apps

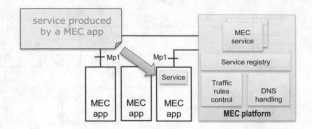

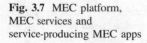

Fig. 3.7 MEC platform, MEC services and service-producing MEC apps

3.2.1   MEC Applications and UE Applications

MEC applications (here also simply called MEC apps) are the essential applications running at the edge of the network; they are usually deployed in containers or virtual machines (VM) on top of the virtualization infrastructure (e.g. NFVI) and running on the MEC host. MEC apps interact with MEC platform (through Mp1), which is offering a portfolio of MEC services (see next chapter). In its turn, a MEC app can provide services which are consumed by another MEC app (thus operating at a higher level of abstraction). In addition, MEC apps may consume also MEC services located in other MEC platforms, running on non-collocated MEC hosts (of course, this case may result in additional delay given by the transfer time of the messages between MEC app and a non-collocated MEC platform). Moreover, strictly speaking from a standardization perspective, support of the Mp1 reference point is not mandatory for a MEC app (which can run on the MEC host without connection to the MEC platform). In summary, all three kinds of MEC apps are depicted in Fig. 3.7.

As a consequence of the introduction of the MEC system at the edge of the network, a client application running on the UE (that we may call Client App) is able to communicate not only with a remote server but also with the MEC app running on the MEC host. The ETSI MEC standard defines another kind of application, called Device App or also UE app,[3] able to communicate with the MEC system. The User Application Life Cycle Management (LCM) Proxy (acronym: UALCMP) is the entity in the MEC system that permits this communication (through Mx2 reference point) so that the UE app can request onboarding, instantiation, termination of MEC applications and when supported, relocation of MEC applications in and out of the MEC system. This UALCMP entity is of course acting as a proxy for all requests coming from the UE app toward the MEC orchestrator or the OSS.

As a further interface for end-users, there is the Customer-Facing Service (CFS) Portal, which allows operators' third-party customers (e.g. commercial enterprises) to select and order a set of MEC applications that meet their particular

---

[3]Although the term Device app is coherent with the MEC-003 standard and the official ETSI MEC terminology in phase 2, the term UE app could help to better distinguish that application (running on the UE) from a remote app running somewhere else in the cloud.

needs, and to receive back service level information from the provisioned applications. This aspect is particularly important when talking about 5G systems, vertical segments and network slicing concepts (see Chap. 7).

3.2.2 MEC Platform and MEC Host

The MEC platform is the core function of MEC systems, as it is hosting MEC services, connecting to MEC applications (MEC apps) and offering an application enablement framework to developers. More in detail, the MEC platform is able to offer an environment where the MEC applications can discover, advertise, consume and offer MEC services. The MEC platform can also instruct the data plane according to the traffic rules (received from the MEC platform manager, applications or services), for a proper traffic routing to/from MEC apps.

As an important added value of the MEC technology, MEC applications can both *consume* services available in the MEC system and also *produce* services, which are made available by the MEC platform to other applications. This Application Enablement Framework is an essential functionality of the MEC platform, described more in detail later in Sect. 3.3.

The above Fig. 3.7 is thus showing two kinds of services offered by the MEC system, i.e. services produced by MEC apps, and standard MEC services (depicted in the MEC platform box) and offered to SW developers through standardized APIs (for further details, see Chap. 4). At this stage, it is just worth deriving an important definition for these service *producers* in MEC:

Terminology:
Service producing MEC application: a MEC application producing a service for other MEC applications (that in the book we can call as higher level MEC applications)

The MEC host is not only containing a MEC platform but also the virtualization infrastructure, which is essentially similar to the NFVI defined in ETSI NFV but with a substantial difference: the presence of data plane. This is an important element, as it has an impact on the end-to-end (E2E) performance of the MEC system. In fact, the **data plane** executes the traffic rules received by the MEC platform and routes the traffic among applications, services, DNS server/proxy, 3GPP network, local networks and external networks. This functional block will be also described in Chap. 7 when talking about MEC in 5G systems.

Another important functionality offered by MEC systems is the DNS service: in fact, in principle, the Client App (residing in the UE) could be totally unaware of the presence of MEC at the edge of the network, and simply communicate with a certain server (associated with a certain DNS name). The MEC platform is thus capable to configure a DNS proxy/server accordingly so that the user traffic is redirected to a MEC app running on the MEC host.

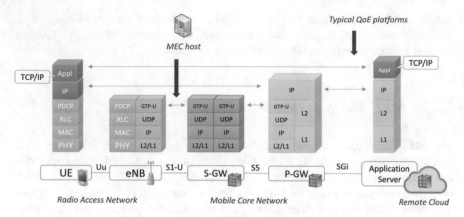

Fig. 3.8 U-plane protocol stack of the different entities in the mobile network

Traffic routing is an essential capability of the MEC host: this functionality is specifically implemented by the Data Plane, which is instructed and configured by the MEC platform in terms of traffic rules. In particular for some deployment options when the user traffic is encapsulated, the MEC host [MEC002v2] supports both de-capsulation of UL packets (and their routing to the authorized MEC apps), and encapsulation of DL packets (received from authorized MEC apps) before routing them to the network. In fact, in case of considering a MEC deployment in mobile networks, e.g. through S1 interface, these packets (IP traffic) are typically encapsulated with GTP header.

The above Fig. 3.8 shows the protocol stack of all entities of a mobile network involvement in the path of IP traffic from UE to an application server. As the reader can easily notice, IP layer is generally not visible deep into network (being encapsuled inside tunnels), and typical QoE platforms work at the level of the remote cloud (i.e. after the SGi/Gi interface) due to the traditional mobile protocol encapsulation, thus not in a close proximity to the end user. Also, in typical mobile networks, Cache and TCP optimizers are installed only after the Gi/SGi interface on the Core network (and not on radio edge, e.g. eNB). Instead, when some GTP tunnel decapsulation/encapsulation and traffic routing capabilities are implemented in a MEC host, a MEC app running on it can perform these QoE improvements with a shorter path and end-to-end delay. More details in GTP tunneling are analyzed in Exercise 3.1 at the end of the chapter.

3.2.3 MEC Orchestrator

The multi-access edge orchestrator (in this book often called also MEC orchestrator, for the sake of readability) is the "brain" of the MEC system-level management. In fact, it is responsible for maintaining an overall view of the MEC system based on deployed MEC hosts, available resources, available MEC services and topology; it is also responsible for selecting appropriate MEC host(s) for application instantiation based on constraints, such as latency, available resources, and available services; moreover, an orchestrator is managing applications, from their instantiation, relocation, termination, including the onboarding of application packages, validating application rules and requirements and preparation of the VIM (s) to handle the applications. Other functionalities are also required for the MEC system [MEC002v2] to host a MEC platform in cloud resources owned, operated and orchestrated by third-party edge owners. In these cases, the MEC orchestrator is responsible for selecting the most suitable MEC host to run the application and making the decision of the application instantiation.

It's important to point out that implementation details about the actual realization of the orchestrator, are not defined by the standard (this is by the way true also for other key elements of the MEC system, that are intentionally not fully specified at "stage 3", in order to permit proprietary implementations and leave to the different companies in the market the possibility to compete). In this perspective, a number of different orchestration deployment options have emerged so far from the industry and standardization (mainly based on open source frameworks).

The most significant orchestration frameworks are related to the list (here below) of the main open source frameworks based on Cloud and NFV technologies (a more detailed description of these frameworks is done in Chap. 6):

- OPNFV, https://www.opnfv.org/
- OpenMANO, https://github.com/nfvlabs/openmano
- OSM, https://osm.etsi.org/
- CloudNFV, http://www.cloudnfv.com/
- OpenBaton, https://openbaton.github.io
- ZOOM, https://www.tmforum.org/collaboration/zoom-project/
- Cloudify, https://github.com/cloudify-cosmo
- ONAP, https://www.onap.org/

For further deepening on orchestration, the reader is kindly invited to have a look at Exercise 6.1, helping a better understanding of these SW frameworks.

3.3 MEC Platform Application Enablement

ETSI ISG MEC defines MEC-011 a generic framework which is applicable to every environment aimed at opening up the network and exposing information toward authorized third-party applications. The scope of this document [MEC011] focuses on the Mp1 reference point (as depicted in Fig. 3.9), thus between MEC apps and MEC platform (Fig. 3.10).[4]

---

[4]The version 2 of this specification is produced in [MEC011v2].

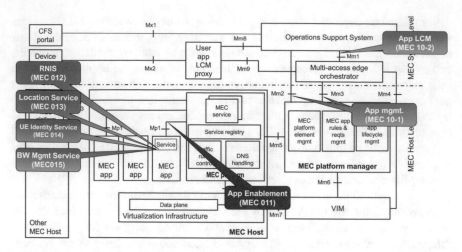

Fig. 3.9 Scope of the ETSI GS MEC-011 standard

This framework is essentially defining the main functionalities of the MEC platform, offering an environment where mobile edge applications may discover, advertise, consume and offer mobile edge services. Service-related functionality includes registration, discovery, event notifications; and other functionality includes application availability, traffic rules and DNS. In particular, traffic rules are pre-configured by the MEC management. The MEC application instance may request the MEC platform to activate, deactivate or update the traffic rules. Then the MEC platform instructs the data plane accordingly. Similarly, the MEC application instance may request the MEC platform to activate or deactivate a DNS rule(s). If the MEC platform succeeds in finding, based on the information contained in the request, the corresponding DNS rule(s) that have been pre-configured by the MEC management, the MEC platform will install or remove the DNS rule(s) into or from the DNS server/proxy.

The following subsections describe the main functions and procedures supported by the MEC platform and the resource structure defined in MEC-011. For more details, please have a look at the ETSI MEC specification (links below).

Useful References:

Version 1:

- ETSI GS MEC 011 V1.1.1 (2017-07): Mobile Edge Computing (MEC); Mobile Edge Platform Application Enablement; http://www.etsi.org/deliver/etsi_gs/ MEC/001_099/011/01.01.01_60/gs_MEC011v010101p.pdf

Version 2:

- ETSI GS MEC 011 V2.1.1 (2019-11): "Multi-access Edge Computing (MEC); Edge Platform Application Enablement"; https://www.etsi.org/deliver/etsi_gs/ MEC/001_099/011/02.01.01_60/gs_mec011v020101p.pdf

---

3.3.1   MEC App Assistance: MEC App Start-Up Procedure

The MEC application startup is an important procedure because it defines the communication between MEC App (after its instantiation) and the MEC platform.

The sequence of messages identifying the MEC application startup procedure is depicted in the above chart (see Fig. 3.10) and starts with a communication from the MEC application instance informing the MEC platform that it is running. The way the MEC App instance discovers the MEC platform is not defined by the standard, thus is left to the implementation (e.g. DNS). Also, authorization and authentication mechanisms are not specified, as they may depend on operator policies.

After a MEC application is instantiated, the MEC platform configures the MEC application and registers it internally. These configuration steps (performed via Mp1) include: "Traffic rule activation/deactivation/update", "DNS rule activation/deactivation" and "service availability update and new service registration", as described in the next subsections. The related information about this MEC application instance is stored in the MEC platform. More specifically, this application instance is registered with related information including:

- the required and optional services,
- the services to be offered by this application instance (and the associated transport dependency)
- the traffic rules associated with this application instance
- the DNS rules.

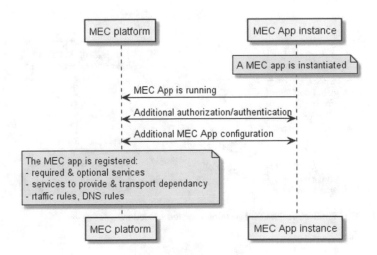

Fig. 3.10 MEC application startup procedure

The reader is invited to notice that this set of information is also identifying the data stored (for each registered MEC application instance) in the **service registry** located within the MEC platform (see also MEC architecture in Sect. 3.2).

3.3.2 MEC App Assistance: MEC App Graceful Termination/Stop

After the MEC platform receives a request to terminate or stop a MEC application instance, the MEC platform notifies the MEC application instance that it will be terminated or stopped soon if graceful termination/stop is required. The simple mechanism is also depicted in the chart below (Fig. 3.11).

The MEC platform may also indicate to the MEC application instance the length of a time interval during which the application may perform application-specific termination/stop. When the timer expires, the MEC platform continues the termination flow of the MEC application instance or stop application instance flow.

Dually to the startup procedure described in the previous section, during the termination/stop procedure the MEC platform is thus:

- deactivating the traffic rules
- deactivating the DNS rules
- removing the application instance from the list of instances to be notified about service availability
- removing the services provided by the application instance from the service registry
- sending service availability notification to the MEC applications that consume the services produced by the terminating MEC application instance.

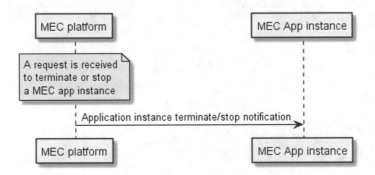

Fig. 3.11 MEC application graceful termination/stop procedure

3.3.3   Service Availability Update and New Service Registration

As we have already anticipated, the MEC platform is storing information about all MEC services available. This service registry is kept updated, and all related communication procedures for these updates are defined by MEC standards. These procedures include two cases: new services registrations (see Fig. 3.12) and service availability updates (Fig. 3.13). This second case is very important because related to the so-called *service-producing MEC applications* (already introduced in Sect. 3.2.2).

CASE 1 (*new service registration*).

In this case, a MEC application instance informs the MEC platform that the service (s) provided by this application instance becomes available for the first time. Then the MEC platform is notifying the other applications about the availability of this new service. Notice that this notification is sent only to the relevant authorized applications, i.e. those authorized application instances indicating the service(s) as "optional" or "required". It is also possible for applications to consume services available in other platforms, i.e. running on different MEC hosts (the figure below still holds). Nevertheless, the standard doesn't define how a certain platform A may

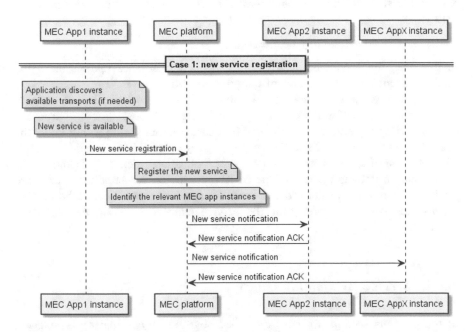

Fig. 3.12 MEC: new service registration

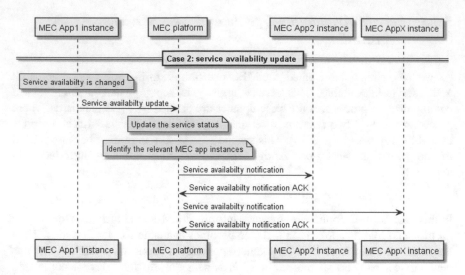

Fig. 3.13 MEC: service availability update

identify other platforms B or C in the MEC system, in order to inform them about
the changes in service availability. Anyway, all MEC platforms keep a trace in their
registry of a flag indicating whether a MEC service instance is running on the other
MEC host.

CASE 2 (*service availability updates*).

In this case, the service-producing MEC application instance updates the MEC
platform about the status change of the produced MEC services.

Also, in this case, the scheme is valid in case of consumption of services
available in other platforms, i.e. running on different MEC hosts. For this purpose, a
careful reader may notice that an inter-MEC host communication would imply the
overall knowledge of the MEC system topology (i.e. MEC hosts locations, logical
addresses, list of services hosted and relevant applications, etc....). This knowledge
is implicitly involving the role of the MEC orchestrator. Nevertheless, once more
(and for the already mentioned reasons), the standard doesn't define how MEC
orchestrator is kept informed of the service status updates in remote MEC
platforms.

3.3.4   Service Availability Query

In the previous section, we have described how the service registry is updated,
based on messages coming from MEC app instances. On the other hand, an

Fig. 3.14 MEC: service
availability query

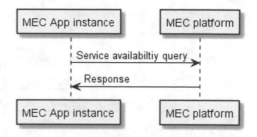

application acting as a *consumer* of MEC services is able to actively request the
availability of a service (or a list of MEC services) to the MEC platform.[5]

This is done through a "Service availability query" message, and related
"Response" coming from the MEC platform (with the message body containing the
information about the available service(s), including the information needed to
access the available service(s)). Figure 3.14 is depicting this simple procedure.

3.3.5 Service Availability Notification

Notification mechanisms are an important tool in MEC to automate the communication between MEC platform and MEC applications. In fact, the standard
foresees the possibility for MEC applications to subscribe/unsubscribe to event
notification, and get periodic information from the MEC platform, without asking
each time the information. Notification mechanisms and the related creation of
subscriptions for the delivery of notifications are specified in MEC-009 and will be
better explained in Chap. 4. For the purpose of the present procedure (related to
"service availability notification"), two simple messages are introduced by the
standard, respectively, to subscribe/unsubscribe to event notification messages
coming from the MEC platform (see Fig. 3.15).

Please notice that, after this subscription, the authorized relevant MEC applications will be notified both in case of a newly available service and any service
availability changes.

3.3.6 Traffic Rule Activation/Deactivation/Update

The management of traffic rules is another very important component of the MEC
platform. In fact, in order to correctly implement communication between Client

---

[5]In this perspective, it is worth to remember that this service availability information in stored in
the service registry of the MEC platform, and its status is anyway depending on the updates
coming from the service producing MEC application instances to the MEC platform.

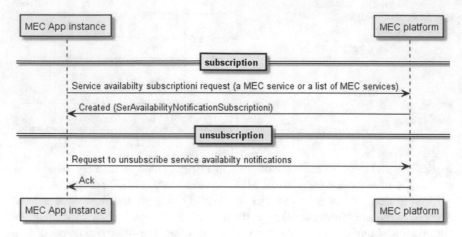

Fig. 3.15 MEC application requesting service availability event notifications subscription/ unsubscription

App (running on the UE) and MEC App (at the edge of the network), the MEC platform needs to provide the following functionalities [MEC002v2]:

- allow authorized MEC apps to **send/receive** user plane traffic to/from UEs
- **route** selected uplink and/or downlink user plane traffic (both from network to authorized MEC apps and vice versa)
- **inspect, modify, shape** selected uplink and/or downlink user plane traffic
- **route** selected uplink and/or downlink user plane traffic between authorized MEC apps.

Moreover, in case there is a specific traffic to be routed to more than one MEC application instances, the MEC platform selects one or more applications and assigns priorities to them (based on traffic rules, indeed defined per MEC app).

In addition to that, MEC foresees also the support forwarding and processing of the IP packets for IP multicast group management, even if the related normative work is not specified, and left to proprietary implementations.

The parameters related to traffic rules are thus allowing packet filtering. Relevant examples of these parameters are network address and/or IP protocol, Tunnel Endpoint ID (TEID), Subscriber Profile ID (SPID) and Quality Class Indicator (QCI) value(s). These parameters are stored in the MEC platform, and the MEC application can request to update them (in case of considering an existing traffic rule) or simply to activate/deactivate one or more traffic rules (in case they are already available and associated to that MEC application).

As an important consequence of this flow (depicted in the above chart, in Fig. 3.16), if the request is authorized, the MEC platform will update the data plane via Mp2 reference point, thus by actually instructing the data plane to execute these traffic rules, for a correct management of user plane packets to/from the MEC application.

Fig. 3.16 MEC: traffic rules activation/deactivation/update

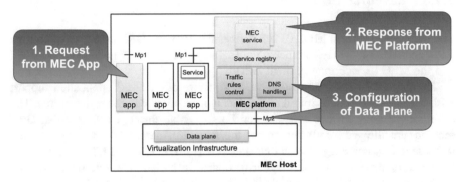

Fig. 3.17 MEC entities involved in traffic/DNS rules procedures

The overall sequence, with corresponding logical entities in the MEC architecture, is showed in Fig. 3.17 (note: the sequence is also valid when considering DNS rules, described in the next section).

3.3.7 DNS Rule Activation/Deactivation

As already said, the DNS is a key functionality provided by the MEC platform, including a name server and a proxy/cache function. The DNS functionality is also useful in case of need from an APP instance to dynamically discover a MEC platform. According to MEC specifications [MEC002] [MEC003, MEC011], the platform is able to receive DNS records from the MEC platform manager and to configure a DNS proxy/server accordingly. The DNS rules are thus activated/ deactivated by the MEC application, through a specific request message via Mp1 reference point, as depicted in Fig. 3.18. Thus, based on configuration or following an activation request from the MEC application, the MEC platform installs or removes the DNS rule(s) from the DNS server/proxy, and configures the mapping between an IP address and its FQDN (fully qualified domain name) into the DNS based on these rules.

Fig. 3.18 MEC: DNS rules
activation/deactivation/update

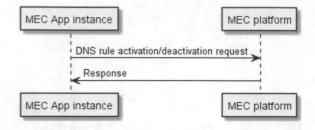

3.3.8 Resource Structures

The MEC specification MEC 011v2 defines the resource URI structures of the two
APIs applicable to Mp1 reference point (namely, the MEC application support API
and the MEC service management API).

The Fig. 3.19 shows the resource URI structure of the MEC application support
API, together with a description of the data types available in the resource tree. The
different elements of this structure can be accessible through HTTP verbs: GET
methods are allowed for all elements of this structure, in order to retrieve the related
information contained. Other HTTP methods (PUT/POST/DELETE) are allowed
for some of these elements, in order to update information, create or delete new
resources (more details are contained in MEC-011) specification.

The Fig. 3.20 shows the resource URI structure of the MEC service management
API, together with a description of the data types available in the resource tree. Also, in
this case, the different elements of this structure can be accessible through HTTP verbs.

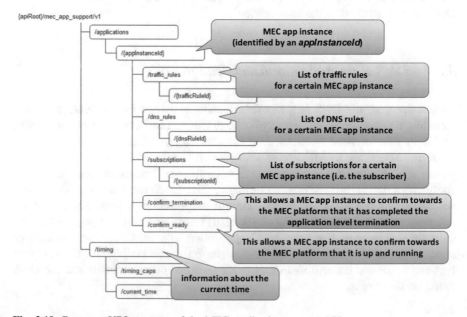

Fig. 3.19 Resource URI structure of the MEC application support API

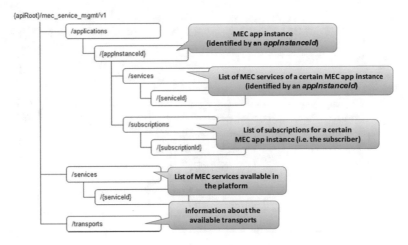

Fig. 3.20   Resource URI structure of the MEC service management API

3.4   ETSI MEC Standard: MEC Management

MEC management specification MEC 010 is divided into two documents:

- MEC 010–1: focusing on the management of the MEC system, MEC hosts and MEC platforms. It includes platform configuration and application monitoring. It defines also Mm2 and Mm3 reference points.
- MEC 010–2: focusing on the management of MEC applications, such as application lifecycle and application-related policies. It includes the description of application-related information, application lifecycle, and application-related events, mobility handling. It defines also Mm1 reference point.

The scope of the ETSI GS MEC-010–1 and MEC-010–2 standards is also depicted in Fig. 3.21.

 Useful References:
- ETSI GS MEC 010-1 V1.1.1 (2017-10): Mobile Edge Computing (MEC); Mobile Edge Management; Part 1: System, host and platform management
- ETSI GS MEC 010-2 V2.1.1 (2019-07): Multi-access Edge Computing (MEC); MEC Management; Part 2: Application lifecycle, rules and requirements management

Download:
http://www.etsi.org/deliver/etsi_gs/MEC/001_099/01001/01.01.01_60/gs_
MEC01001v010101p.pdf
https://www.etsi.org/deliver/etsi_gs/MEC/001_099/01002/02.01.01_60/gs_
MEC01002v020101p.pdf

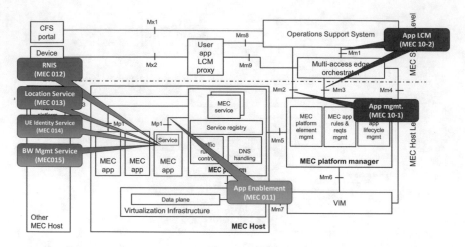

Fig. 3.21 Scope of the ETSI GS MEC-010–1 and MEC-010–2 standards

Through the **Mm2** reference the following main capabilities are available:

- the OSS is allowed to:

  - retrieve the information model of the MEC host, or parts thereof, from the MEPM (which is able to notify changes related to this information model).
  - configure the MEC host, DNS rules, traffic rules.
  - create and delete managed object instances representing MEC application instances in the MEPM.
  - activate and deactivate the DNS rules and traffic rules related to a certain MEC application instance.

- the MEPM is allowed to:

  - send MEC platform-related alarms to the OSS (which is able to retrieve and manage them).
  - notify changes of managed object instances representing mobile edge application instances to the OSS.
  - notify object creation and deletion events of managed object instances representing MEC application instances to the OSS.
  - expose the operational state of instantiated MEC applications to the OSS.

The MEC 010–1 standard defines also many other things, including how to represent a MEC platform, a MEC application or a MEC service, through proper definitions of Information Object Class, a set of related attributes identifying them. For instance, regarding the MEC app representation, some examples of attributes are appDId (the identifier of the MEC application descriptor), appName

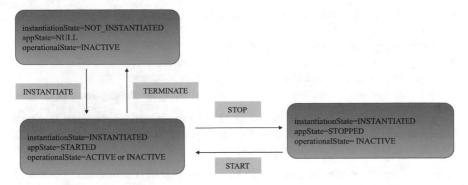

Fig. 3.22 Applications state diagram

(human-readable name of the MEC application), `appProvider` (provider of the MEC application). Applications should have also three states: `instantiationState`, `appState`, `operationalState`. According to the different transitions, the application state diagram is shown in Fig. 3.22.

Terminology:
application descriptor: descriptor provided by the application provider, and describes the application rules and requirements of a mobile edge application
application package: bundle of files provided by application provider, onboarded into mobile edge system and used by the mobile edge system for application instantiation, including an application descriptor, a VM image or a URL link to a VM image, a manifest file and other optional files

Web Link to the definition:
https://www.etsi.org/deliver/etsi_gs/MEC/001_099/001/01.01.01_60/gs_MEC001v010101p.pdf

The MEC 010–2 standard covers Mm1 reference point (used for onboarding application packages, triggering the instantiation and the termination of MEC applications in the MEC system) and Mm3 reference point (used for the management of the application lifecycle, application rules and requirements and keeping track of available mobile edge services). In particular, the following main capabilities are available:
(through the **Mm1** reference point, between OSS and MEO).

- onboarding, querying, deleting, enabling and disabling of an Application Package
- instantiating/terminating an Application, and related change the state of the application instance (i.e. starting/stopping the application instance) (through the **Mm3** reference point, between MEO and MEPM)
- support querying application package information
- providing notifications about the onboarding of application packages, and as a result of changes on application package states

- providing notifications fetching an application package, or selected files contained in a package
- instantiating/terminating an Application
- querying information about an application instance
- requesting to change the state of an application instance
- querying the status of an ongoing application lifecycle management operation

More in detail, the message flow related to the **onboard application package** (depicted in Fig. 3.23) is executed before an application is instantiated (Fig. 3.24). This flow consists of three main phases:

1. a first request, coming from the OSS to the MEO (MEC Orchestrator), including the MEC application package;
2. the MEO check this package, allocates a unique App package ID, and prepares the VIM with the application image;
3. finally, the MEO acknowledges to the OSS the onboarding of the app package. At this stage, the app package is available in the MEC system.

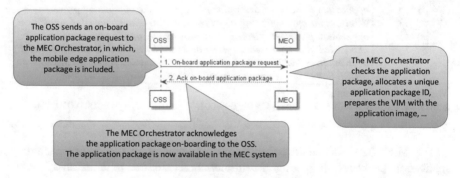

Fig. 3.23 Onboard of application package

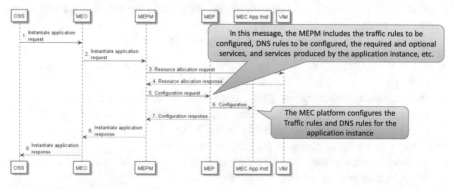

Fig. 3.24 Instantiation of a MEC application

The message flow of **application instantiation** is used to instantiate an application instance in the MEC system. As the reader can easily see in Fig. 3.23 and Fig. 3.24, this flow starts from a request of the OSS, generating a sequence of messages toward the MEO and the MEPM. After that, and the MEC platform is sending a resource allocation request to the VIM, and if the answer is positive, the MEC platform sends a configuration request (including traffic rules, DNS rules, the required and optional services, the services produced by the application instance, etc.). Finally, when these rules are configured, and the answer is sent back to the MEPM, and then again back to the MEO and to the OSS.

A similar message flow of **application instance termination** is used to terminate an application instance in the MEC system (see Figs. 3.25 and 3.26).

Finally, the figure below shows in a complete way the state model of application package in the MEO, essentially consisting of two main phases: onboarding phase and onboarded phases.

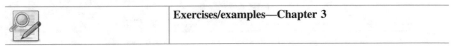

	Exercises/examples—Chapter 3

Exercise 3.1

Goal of the exercise: deepening on GTP tunneling for LTE mobile networks.

Let's consider an LTE mobile network, and an IP packet sent by a UE (user equipment), and delivered from an eNB to a P-GW through GTP tunnels and then to a PDN (the entire path is showed in Fig. 3.27).

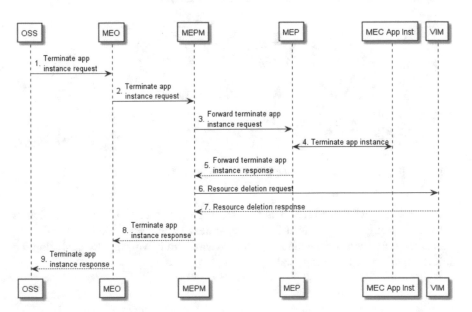

Fig. 3.25 Termination of a MEC application instance. *Source* [MEC-10–2]—© European Telecommunications Standards Institute 2019. Further use, modification, copy and/or distribution are strictly prohibited

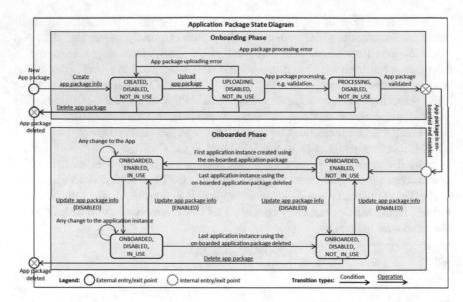

Fig. 3.26 Application package state model. *Source* [MEC-10–2]—© European Telecommunications Standards Institute 2019. Further use, modification, copy and/or distribution are strictly prohibited

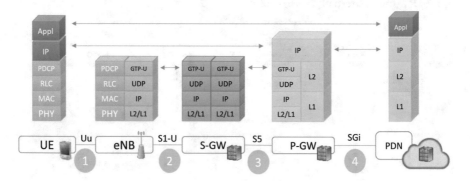

Fig. 3.27 IP packet traveling from the UE (1) to the PDN (4)

The meaning of these tunnels is that all IP packets that a UE sends are always delivered through an eNB to a P-GW regardless of their specified destination IP addresses. Lets try to understand how this mechanism is working, by analyzing the different phases of the traffic path.

1. From UE to eNB

An IP packet is sent by the UE with a certain payload (e.g. an HTTP request over TCP/IP) and an IP header (containing the destination IP address corresponding to a target PDN, e.g. www.google.com, and a source IP corresponding to the IP address of the UE).

IP header	IP payload

2. From eNB to S-GW

In this phase, a GTP tunnel header is added by the eNB on top of the received packet. This header is composed by a GTP header (relative to the link between eNB and S-GW), a UDP header and an outer IP header (composed by a Source IP = eNB and Destination IP = S-GW).

Outer IP header	UDP header	GTP header	IP header	IP payload

GTP tunnel header added by eNB

It should be noted that these IP addresses contained in the outer IP header are different from the IP addresses contained in the initial packet sent by the UE. In fact, we have 2 types of IP addresses:

– (outer IP): IP address of 3GPP nodes
– (Inner IP) + (Inner Protocol/Data): IP address and payload of UE

So, these 3GPP nodes are identified by IP addresses, and in case a network deployment consists of multiple nodes, this mechanism permits to reach the S-GW by performing IP routing based on the destination IP address of the packet (i.e. the IP address of the S-GW, the destination IP address shown in the outer IP header).

It should be noted that here the GTP header (related to the link between eNB and S-GW) is identified by means of a certain Tunnel Endpoint ID (e.g. TEID = X).

3. From S-GW to P-GW

In this phase, a GTP tunnel header is added by the S-GW on top of the received packet. This header is composed by a GTP header (relative to the link between S-GW and P-GW), a UDP header and an outer IP header (composed by a Source IP = eNB and Destination IP = S-GW).

GTP tunnel header modified by S-GW

It should be noted that here the GTP header (related to the link between S-GW and P-GW) is different from the previous one, and it is distinguished by means of a different Tunnel Endpoint ID (e.g. TEID = Y).

4. From P-GW to PDN
In this phase, the packet is delivered to the P-GW, which is removing the three headers (Outer IP header/UDP header/GTP header) and delivering the original IP packet sent by the UE to the Internet.

It should be also noted that in principle multiple traffic flows are present in a network, both in UL and DL directions. For that reason, in order to distinguish among different GTP tunnels (belonging to different UEs and to different traffic directions, i.e. UL and UL), a proper TEID (Tunnel Endpoint ID) is added in each GTP tunnel field of the GTP tunnel header. Fig. 3.28 shows how these numbers permit to identify uniquely each packet, without ambiguity.

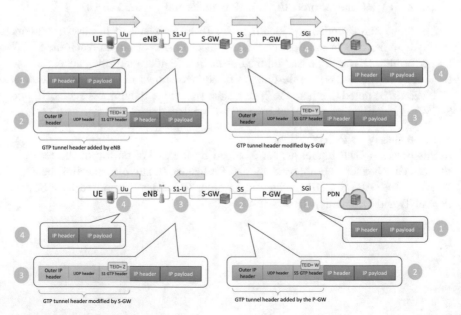

Fig. 3.28 Usage of TEID for the GTP tunneling in LTE networks

Moreover, the reader may notice that each TEID of each segment of the packet is different: in fact, a GTP tunnel (for a UL traffic from the UE to the internet) established by the eNB is marked with TEID = X, while the corresponding GTP tunnel in S5 (with header modified by the SGW) contains a different TEID = Y.

References

[MEC-web]   ETSI MEC, https://www.etsi.org/technologies-clusters/technologies/multi-access-edge-computing
[MEC002]   ETSI GS MEC 002 V1.1.1 (2016–03) Mobile Edge Computing (MEC); Technical Requirements http://www.etsi.org/deliver/etsi_gs/MEC/001_099/002/01.01.01_60/gs_MEC002v010101p.pdf
[MEC003p2]   ETSI GS MEC 003 V2.1.1 (2019–01) Multi-access Edge Computing (MEC); Framework and Reference Architecture https://www.etsi.org/deliver/etsi_gs/MEC/001_099/003/02.01.01_60/gs_MEC003v020101p.pdf
[NFV-web]   ETSI NFV website, https://www.etsi.org/technologies-clusters/technologies/nfv
[NFV002]   ETSI GS NFV 002 V1.1.1 (2013–10) Network Functions Virtualization (NFV); Architectural Framework https://www.etsi.org/deliver/etsi_gs/nfv/001_099/002/01.01.01_60/gs_nfv002v010101p.pdf
[ETSIWP20]   ETSI White Paper No. 20, "Developing Software for Multi-Access Edge Computing", 2nd edition, Feb 2019, Link: https://www.etsi.org/images/files/ETSIWhitePapers/etsi_wp20ed2_MEC_SoftwareDevelopment.pdf

Chapter 4
MEC Standards on Edge Services

In the previous chapter, we have described the MEC platform, and the procedures related to the communication with MEC applications. It was also explained (as anticipated already in Chap. 1) that a MEC application can both consume services available in the MEC system and also produce services, which are made available by the MEC platform to other applications. This concept is also depicted in the figure below, where a MEC app (the third one on the right) is exposing a service through Mp1 reference point (this Service, depicted as a blue box in the figure below, is thus registered by the MEC platform and advertised to applications authorized to use that Service). The ETSI ISG MEC is also standardizing MEC service APIs for the implementation of certain essential functionalities, which are the basis of the implementation of a MEC platform (although not necessarily mandatory for full compliance of MEC products[1]) Fig. 4.1.

The present chapter describes the work done by the MEC group [ETSI_MEC], to define the edge services (specified in MEC services APIs, which are accessible to app developers in a vendor-independent manner, in order to ensure app portability and interoperability at the application level).

Thus, after a first classification of MEC services in Sect. 4.1, in the subsequent sections, we will describe the "General Principles for MEC Service APIs" [MEC009], and continue with an overview of the MEC APIs standardized by ETSI ISG MEC (*note*: some of these specifications were introduced in Phase 1, and then updated/enhanced during the Phase 2 of standardization[2]; other APIs were instead introduced only in Phase 2; for all cases, in this chapter, we refer to the latest API versions available).

---

[1]For further details on ETSI MEC testing and compliance, see also Chap. 10.

[2]For more context on MEC Phases, see also Chap. 1 for the overall standardization plan.

© Springer Nature Switzerland AG 2021   89
D. Sabella, *Multi-access Edge Computing: Software Development at the Network Edge*, Textbooks in Telecommunication Engineering, https://doi.org/10.1007/978-3-030-79618-1_4

Fig. 4.1 MEC applications
and MEC services

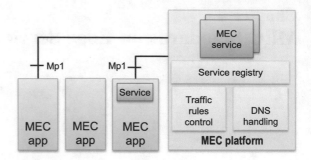

4.1   Classification of MEC Services

The ETSI GS MEC002 specification [MEC002] captures a lot of detailed use-cases, with related technical requirements for the implementation in the MEC system. This document summarizes also three categories for MEC use-cases since requirements of use-cases belonging to the same category are generally similar. These three categories are:

- **Consumer-oriented services.** These are innovative services that generally benefit directly the end user (using the device). Under this category we may find: gaming, remote desktop applications, augmented and assisted reality, cognitive assistance, stadium/retail real-time services, application computation offloading, etc.
- **Operator and third-party services**: Services that take advantage of computing and storage facilities close to the edge of the operator's network. These services don't usually benefit the end user, but can be operated in conjunction with third-party service companies: active device location tracking, big data, video analytics service chaining, connected vehicles, security, safety, enterprise services, etc.
- **Network performance and QoE improvements**: services generally aimed at improving network performance via application-specific or generic improvements. The user experience is generally improved, but these are not services provided directly to the end user. Examples for this category are content/DNS caching, performance optimization, video optimization acceleration, etc.

4.2   General Principles for MEC Service APIs

The ETSI MEC GS 009 specification [MEC-009] defines design principles for RESTful MEC service APIs, provides guidelines and templates for the documentation of these, and defines patterns. This is a generic specification, acting as central reference for all other MEC service APIs introduced by ETSI ISG MEC. In particular, it defines:

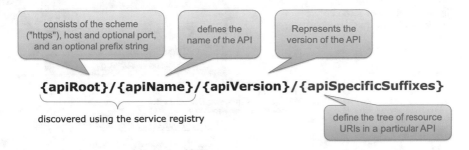

Fig. 4.2 RESTful MEC APIs definition—typical resource URI structure

- the typical sequence diagrams to be used for the APIs procedures (some of these relevant procedures are described in the following subsections);
- resource representations in MEC APIs and related data structures, and all typical resource URI structures of MEC APIs (an example is depicted in Fig. 4.2);
- REST methods for resource creation (e.g. through HTTP POST), resources reading (e.g. through HTTP GET) and for queries on a resource (with related parameters, e.g. for restricting a set of objects to a subset, based on filtering criteria, controlling the content of the result and reducing the content of the result, such as suppressing optional attributes).

An example of GET method to retrieve filtered information from a resource is given by the following command:

```
GET .../foo_list?vendor=MEC&ue_ids=ab1,cd2
```

where the answer from the server would return a `foo_list` representation that includes only those entries where the vendor is "`MEC`" and the UE IDs are "`ab1`" or "`cd2`". A more complete exercise on HTTP methods and RESTful APIs can be found at the end of the present chapter.

4.2.1   Methods to Update a Resource

Many operations can be done on a resource. A relevant example is provided by the methods defined to update a resource, e.g. through HTTP messages.

When a resource is updated using the PUT HTTP method, this operation has "replace" semantics, meaning that the new changed resource structure is passed as an attribute of the PUT method, replacing the old resource structure (Fig. 4.3).

The reader should notice, that the approach illustrated above can suffer from race conditions. In fact, in the presence of multiple requests from different clients, it may occur the case of another client modifying the resource before the first client could

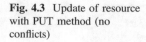

Fig. 4.3 Update of resource
with PUT method (no
conflicts)

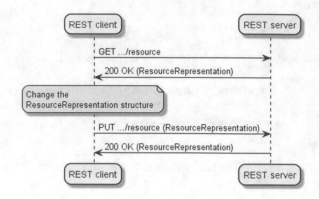

send the PUT request (which is known as the *lost update phenomenon* in concurrent systems). In this case, that modification gets lost, but HTTP (see IETF RFC 7232) supports means to detect such a situation and to give the client the opportunity to deal with it. In fact, each version of a resource gets assigned an "entity-tag" (ETag), which is inserted in the PUT message from the client (in a suitable "If-Match" HTTP header). This ETag is modified by the server whenever it changes the resource, by delivering also the new "ETag" HTTP header of future HTTP response messages (e.g. GET responses). That way, each subsequent PUT command (sent by the client) can be executed by the server only if the current ETag is matching with that one in the last GET response generated (otherwise a "412 Precondition Failed" response is sent by the server). This mechanism is illustrated in Fig. 4.4.

Another mean to update a resource is the usage of "Update by PATCH" typically more efficient than "Update by PUT" when partially updating a large resource. The main difference is that the PATCH command modifies the resource on top of the existing resource state, while resource update using PUT overwrites the resource. In the Fig. 4.5, showing the simple message sequence, the client request is not carrying a representation of the resource (as in the PUT), but a document that tells the server how to modify the resource.

Also in this case, in order to avoid conflicts between concurrent requests from different clients, the ETag is inserted in the PATCH message from the client (in a suitable "If-Match" HTTP header). As an exercise, the reader may try to derive the related message sequence chart used for PATCH in case of concurrent updates (e.g. by following the example of the above Fig. 4.3).

4.2.2 Asynchronous Operations

Not all the operations invoked via a RESTful interface (e.g. through POST/PUT/ PATCH/DELETE methods) can be completed instantaneously by the server. Some of them, in fact, may trigger processing tasks that may take a long time (e.g. from

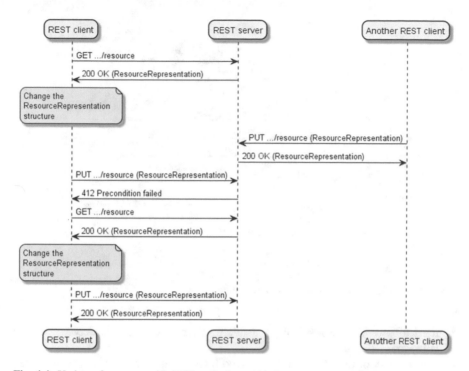

Fig. 4.4 Update of resource with PUT method, considering concurrent updates

Fig. 4.5 Update of resource with PATCH method

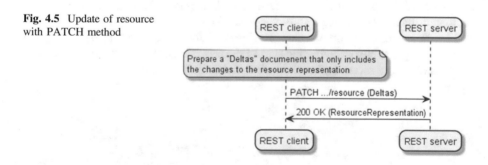

minutes over hours to even days). In this case, it is not appropriate for the REST client to keep the HTTP connection open while waiting for the result of the response. So, in order to cope with this problem, it is possible to work in asynchronous mode. The idea is the following:

- the REST server immediately returns the provisional response "202 Accepted" to indicate that the client request was understood and that the related processing has started;
- the REST client then can check the status of the operation by polling (as showed in Fig. 4.6); additionally or alternatively, the subscribe-notify mechanism can be

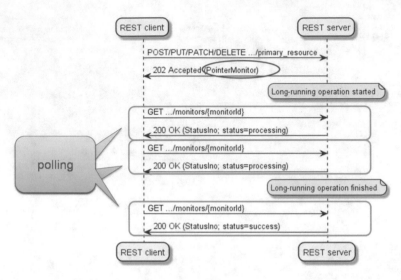

Fig. 4.6 Asynchronous operation flow, with polling

used to provide the result once available (this mechanism is depicted in the subsequent Fig. 4.7);

• the progress of the operation is reflected by a proper monitor resource, invoked by the client during the polling messages.

It should be noted that, in the case of polling (Fig. 4.7), every new request is initiated by the client, requesting for an update about the current ongoing task, while the subscribe/notify mechanisms (Fig. 4.8) foresees, first of all, a subscription requested by the client (before the start of the task itself), and a notification message produced by the server (when the task is finished).

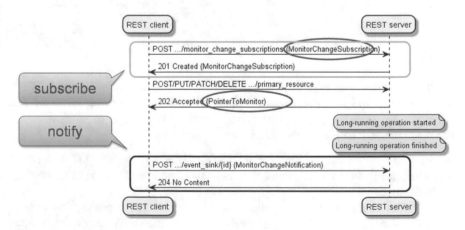

Fig. 4.7 Asynchronous operation flow, with subscribe/notify

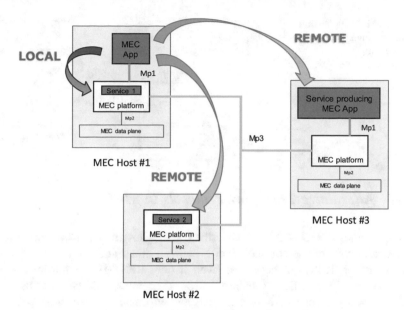

Fig. 4.8 Local and remote MEC service consumption

The second mechanism (subscribe/notify) has the obvious advantage to avoid the generation of multiple requests from the client (as instead done in the polling) and to receive directly an update from the server, at the completion of the task.

4.3 Remote MEC Service Consumption

As we already said, MEC applications can be consumers and/or producers of services. Let's consider a 5G system with MEC hosts deployed across a geographical area. For the sake of simplicity, we consider only one MEC system composed of different MEC hosts. The most general case corresponds to a MEC app running on a MEC host, which needs to consume MEC services instantiated within the (same) MEC system. The queried services are assumed available in the MEC system, however, they may run at different localities.

For more details on remote service consumption, the reader is invited to look at the paper "Flexible MEC service consumption through edge host zoning in 5G networks" [ZN-CL19], where authors investigate the general case of a MEC application consuming a MEC service running on a different MEC host, by focusing on control plane packet traffic. The paper also defines latency-aware proximity zones around MEC servers hosting MEC application instances, in order to enable flexible MEC platform service consumption at different localities, based on a state-of-the-art statistical model of the total processing time.

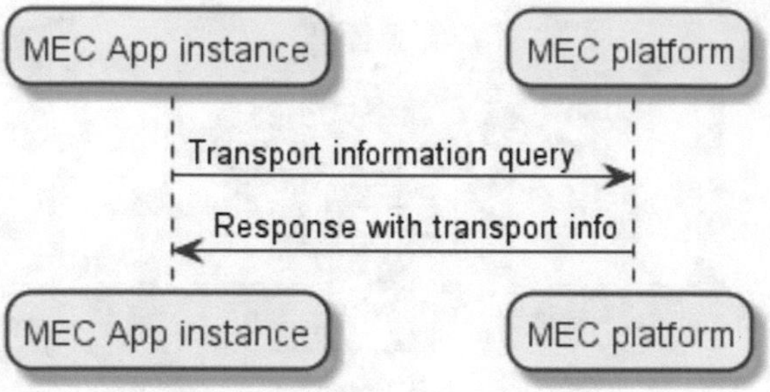

Fig. 4.9 MEC application requesting transport information

4.3.1 Alternative Transport Protocols in MEC

In the previous Sect. 4.2, we described some key mechanisms (and related sequence flows) for the communication between client and server via Mp1 interface, based on RESTful messages. Even if only REST-HTTP transport is fully specified by ETSI MEC standard, also other transports can be used (ref. [MEC011]), by means of a proper transport information query, that provides a standardized means to MEC applications to discover the available transports.

An example of simple message flow is depicted in Fig. 4.9, where a MEC application instance requests transport information, by sending a request to query the information about transports provided by the MEC platform. The related response (sent back by the platform) contains in the message body the transports information.

As said, alternative transport protocols can also be used, e.g. for certain services that require higher throughput and lower latency than a REST-based mechanism can provide. Examples of alternative transports are topic-based message buses (e.g. MQTT or Apache™ Kafka) and Remote Procedure Call frameworks (e.g. GRPC™). In addition, Zenoh is a transport protocol recently introduced by Eclipse Edge Native Working Group, as described in the following section.[3]

4.3.2 Zenoh: An Alternative MEC Transport Protocol

HTTP-REST is a widespread transport protocol many developers are familiar with. However, the client-server paradigm adopted by this protocol may not be well suited for provisioning advanced MEC services requiring data to be transparently shared, stored or even computed on-demand. In light of this, the Eclipse Edge Native Working Group proposed Zenoh, a transport protocol explicitly designed to tackle these challenges. Specifically, Zenoh blends traditional pub/sub mechanisms with geo-distributed storages, queries and remote computations to provide a unified

[3]The Zenoh section is courtesy of Luca Cominardi, Angelo Corsaro, Olivier Hecart, Julien Enoch.

data management capable of scaling out, up and down. This results in a simplified and streamlined development of distributed applications at any scale.

Zenoh adopts a Named Data Networking (NDN) paradigm to support access transparency in a distributed environment of both data at rest (e.g. databases) and data in motion (e.g. pub/sub). This is achieved by a routing infrastructure designed to run at an Internet scale that allows Zenoh to deliver the right data in the right place based on the set of available applications. With this approach, applications only need to focus on the data they are interested in rather than their location. To that end, applications publishing data need to name data (i.e. assign a *key*) and associate a value to it (i.e. the *data payload*). Similarly, applications interested in certain data subscribe to a given key expression.

In Zenoh, the keys take the structure of a Uniform Resource Identifier (URI), which represents a hierarchical organization of data. By carefully designing the key space (i.e. the key structure associated with the data), it is possible to aggregate and transparently access the data via specific queries and key selectors. When querying for data, the routing infrastructure takes care of doing the necessary pattern matching between keys, selectors, publishers and subscribers, in order to retrieve the data from the closest point in the network. By doing so, the applications do not need to be concerned with the location of data, allowing developers to focus on the application logic itself.

Figure 4.10 shows an example of an edge-based smart home scenario where a MEC service for remote temperature sensing is provided. A local storage is deployed at each home for privacy-sensitive data in addition to regional storages (e.g. city) for public data. The keys are structured as follows: `/region-id/house-id/room-id/temperature`. The special character "/" is used as a separator, while "*" can be used as a wildcard for a single field. The special character "**" can be then used as a wildcard for multiple contiguous fields. Getting the values for the key `/region01/*/*/temperature` allows to retrieve the temperature of every room of every house in region 01. Similarly, getting the values for the key `/**/temperature` will return the temperature of every home in every region.

In order to enable the above behavior, Zenoh defines four key abstractions, namely Zenoh.net abstractions, as shown in Fig. 4.11:

1. **resource**: a named data, i.e. a key-value. An example of resource is given by the `(key, value): = (/house/room/temp, 25 °C)`;
2. **publisher**: a spring of values for a key expression. E.g. a sensor publishing the temperature value associated to the key `/house/room/temp`;
3. **subscriber**: a sink of values for a key expression. E.g. a monitoring application interested in the temperature of any rooms in a given house and thus subscribing to the key expression `/house/*/temp`;
4. **queryable**: a well of values for a key expression. E.g. a sensor that when queried at `/house/room/temp` will return the current temperature value.

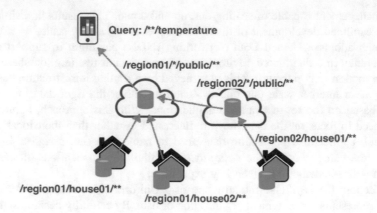

Fig. 4.10 Exemplary Zenoh application scenario

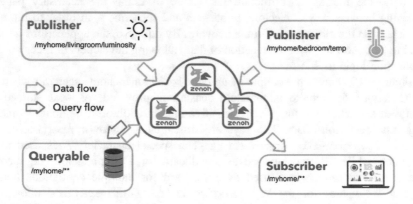

Fig. 4.11 Zenoh key abstractions: Zenoh.net

These fundamental abstractions can be natively combined to deal with data in motion, data at rest and remote computations. Publishers and Subscribers are the natural abstractions for dealing with data in movement. Likewise, the combination of a subscriber and a queryable models a distributed storage. Queryable can also be used to model distributed computations.

Along with the abstractions introduced above, Zenoh provides the following primitives:

- **scout:** it performs a discovery on the network looking for other Zenoh entities (e.g. routers or peers). Once an entity is discovered, it is possible to establish a Zenoh session;
- **open/close:** it opens/closes a Zenoh session (e.g. with a router or a peer);
- **declare/undeclare:** it declares/undeclares a resource, a publisher, a subscriber or a queryable. For subscribers, the declared primitive registers a user-provided

call-back that will be triggered when data is available. For queryable, the declare primitive register a user-provided call-back triggered whenever a query needs to be answered;

- **write:** writes data for a key expression. It can be seen as the publish operation;
- **pull:** pulls data for a pull subscriber, this allows the infrastructure to get the data as close as possible to the subscriber, in a proactive manner, so that when issuing the pull the data is usually one hop away;
- **query:** issues a distributed query and returns a stream of results. The query target, coverage and consolidation depend on configured policies.

Additionally, Eclipse Zenoh makes available a higher level API that combines core abstraction and primitives to provide a key/value geo-distributed storage and distributed computation abstraction. For individual storages, Zenoh's storage abstraction leverages a plug-in mechanism to integrate virtually any kind of database technology, such as Relational Databases (e.g. SQL), NoSQL Storages and time series databases. In spite of the back-end technology, Zenoh provides uniform access to data by transcoding data as necessary to deal with different internal representations.

As a further deep dive on alternative transport protocols such as Zenoh, the reader is invited to have a look at Exercise 4.2 which provides a hands-on tutorial on how to use the above primitives to partially implement the example illustrated in Fig. 4.8.

For other protocols, e.g. RPC (Remote Procedure Call), the reader can refer to gRPC, initially created by Google, and now a Cloud Native Computing Foundation incubating project (https://grpc.io/), as used by many organizations to power use-cases from microservices to the "last mile" of computing (mobile, web and Internet of Things).

4.4   Overview of MEC APIs

ETSI ISG MEC defines a set of MEC services, that are exposed authorized MEC Applications by means of standardized interfaces (indeed, the MEC APIs) following the general guidelines for API design [MEC009]. As a general note, MEC services can be implemented by MEC Apps or be part of the MEC platform (as already explained in Sect. 3.2.1). In both cases, the same MEC 009 general rules for API design apply. The reference point involved in the interaction between MEC Applications is Mp1 (see Fig. 4.12).

The table below summarizes the set of MEC service APIs published by ETSI ISG MEC. Currently, the group is working on other additional APIs, and on the extension/update of the existing ones. Nevertheless, it is important to highlight that the ETSI MEC standard foresees the possibility for third parties to develop (by following the MEC009 design principle) and expose their APIs in the MEC Platform. Consequently, many implementations of new APIs are possible, and for

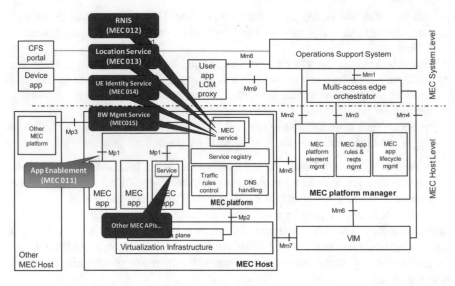

Fig. 4.12 Scope of the ETSI MEC standards related to MEC service APIs

this reason, the list of published APIs is intentionally not exhaustive (ETSI ISG MEC is standardizing exemplary functionalities that are particularly needed/required by the industry).

ETSI document	Overview
MEC 012—RNI API	Defines the Radio Network Information MEC service
MEC 013 —Location API	Defines the Location MEC service
MEC 014—UE identity API	Focuses on the UE Identity functionality
MEC 015 (v1) Bandwidth management API	Defines the Bandwidth Management API (BWM API)
MEC 015 (v2) Traffic Management APIs	Extends the previous v1 specification, by adding a Multi-access Traffic Steering API (MTS API)
MEC 028 WLAN Information API	Specifies the MEC service on Wireless LAN Information
MEC 029 Fixed Access Information API	Specifies the MEC service on Fixed Access Information for Fibre (PON,XG-PON,NG-PON), Cable (DOCSIS 3.1), xDSL, and Point-to-Point Fibre Ethernet access to MEC
MEC 030 MEC V2X Information service API	Introduces the Vehicular-to-Everything (V2X) MEC service, in order to facilitate V2X interoperability in a multi-vendor, multi-network and multi-access environment

Furthermore, the compliance of a MEC product doesn't necessarily require the implementation of all standardized MEC APIs listed in this table. The presence of a certain set of APIs in a MEC platform depends instead on business agreements between MEC technology providers and their customers.

This aspect will be clearer in Chap. 10 when talking about compliance and testing. Here is important to clarify the general mechanisms in common to all MEC APIs. In particular, according to the latest MEC 009 v2 and MEC 011 v2 specifications, the following applies to all RESTful MEC service APIs:

- **HTTP** provides a mechanism to negotiate the content format of a representation. Each ETSI MEC API specification defines the content formats that are mandatory or optional by the server to support that API; the client may use any of these. Examples of content types are **JSON** and **XML**.
- All MEC RESTful APIs use **resource representations** as an important concept in REST. A resource representation (which is included in the payload body of an HTTP request or response) is a serialization of the resource state in a particular content format. Of course, it depends on the particular **HTTP method** used whether a representation is required or not allowed in a request. If no representation is provided in a response, this shall be signaled by the "204 No Content" response code.

  Here, below a correspondence between HTTP methods and Payload body presence (according to IETF RFC 7231) is given:

  - GET → payload unspecified; not recommended
  - PUT → payload required
  - POST → payload required
  - PATCH → payload required
  - DELETE → payload unspecified; not recommended

- Usage of patterns (based on HTTP methods) to perform common operations and manage data types in the RESTful MEC service APIs. These patterns are recommended (if and where applicable) also in case of considering RESTful APIs exposed by MEC services designed by third parties.

Examples of patterns are provided in the table below, by matching operations with the related HTTP methods specified in ETSI MEC:

Pattern	HTTP method	Comments
Resource creation	POST	Used when the client requests the origin server to create a new resource under a parent resource
Resource creation	PUT	Used when the client requests the origin server to create a new resource (where the parent is implicit, i.e. can be derived from the resource URI)
Resource reading	GET	For most resources, the GET method should be supported, to read a resource. An exception is task resources (these cannot be read)
Queries on a resource	GET	query parameters are used for: (1) restricting a set of objects to a subset, based on filtering criteria; (2) controlling the content of the result; (3) reducing

(continued)

(continued)

Pattern	HTTP method	Comments
		the content of the result (such as suppressing optional attributes)
Resource update	PUT	This operation has "replace" semantics (i.e. previous resource state is discarded by the REST server when executing the PUT request)
Resource update	PATCH	Provides a "deltas document", not the whole resource. More efficient than "Update by PUT" in case of partially updating a large resource
Deleting a resource	DELETE	When a deleted resource is accessed subsequently by any HTTP method, typically the server responds with "404 Resource Not Found", or, if the server maintains knowledge about the URIs of formerly existing resources, "410 Gone"
Task resources	POST	The payload body (doesn't carry a resource representation but) provides input parameters to the process which is triggered by the request
Subscribe/ Notify	POST	Used by the client to create a subscription resource inside the subscription container resource
	DELETE	Used by the client to terminate a subscription resource inside the subscription container resource
	POST	Used by the server to deliver the notification to the client
Asynchronous operations	POST/PUT/ PATCH/ DELETE	applied to the primary resource (i.e. the resource that is about to be created/modified/deleted by the long-running operation)
Asynchronous operations	GET	Applicable to read the monitor resource (i.e. the resource that provides information about the long-running operation)
Hyperlinks (HATEOAS)	any	Links shall be embedded in that element (XML) or object (JSON) as child elements (XML) or contained objects (JSON)
Error responses	All methods	Response body in addition to the error code, provided by the server to deliver additional application specific error information

- Support **HTTP over TLS** (also known as **HTTPS**), and in particular, TLS version 1.2. In fact, HTTP without TLS is not recommended to use on production systems (if **HTTP/2** is supported, its use shall be negotiated as specified in IETF RFC 7540).
- Authorization of access to a RESTful MEC service API is done by using **OAuth 2.0**. OAuth 2.0 is the industry-standard protocol for authorization. It facilitates secure service provisioning, as well as secure interoperability among diverse

stakeholders. All OAuth 2.0 protocol flows result in the creation of an **access token**, which is then used by a user (i.e. the MEC API consumer) to request access to a protected resource (contained in the MEC service API itself).[4]

More details on OAuth2.0 can be found in Sect. 6.16 of MEC 009 specification. Anyway, since authorization of access is an important aspect for MEC Security, it's worth specifying that access rights are bound to access tokens, and typically configured at the granularity of methods applied to resources. This means, for any resource in the API, the use of every individual method can be allowed or disallowed. Additional policies can be bound to access tokens too, such as the frequency of API calls. Also, a token has a lifetime after which it is invalid.

In the next sections, a short description of all published MEC APIs is provided, while few selected services (e.g. RNI API, Location API, MTS API, V2X API) will be described more in detail in Chap. 5, where API usage will be coupled also with concrete use-cases of interest for MEC (e.g. from automotive market segment), and in some case also with practical examples and APIs implementations.

4.4.1 Radio Network Information API

This RNI API is introduced by MEC012 specification [MEC012] and defines Radio Network Information Service (RNIS) that provides information on current radio conditions to authorized MEC applications and MEC platforms.

The set of RNIS related information includes:

- up-to-date radio network information regarding radio network conditions;
- measurement information related to the user plane based on 3GPP specs;
- information about UEs connected to the radio node(s) associated with the MEC host, their UE context and the related radio access bearers;
- changes on information related to UEs connected to the radio node(s) associated with the MEC host, their UE context and the related radio access bearers.

The *granularity* of the radio network information may be adjusted based on parameters such as information per cell, per User Equipment, per QCI class or it may be requested over a period of time.

As an important note, this service can be used by MEC Apps and MEC platform to optimize existing services (e.g. video throughput guidance) or to enable new services (e.g. by improving their radio-aware adaptiveness). Chapter 5 will provide a useful exercise in that perspective, with the goal to understand how radio network information available from MEC server can support TCP traffic performance

---

[4]The definition of access tokens is transparent to the OAuth 2.0 protocol, which doesn't specify any particular token format, how tokens are generated, or how they are used. Instead, the OAuth 2.0 specification leaves all these as design choices for integrators. An example of token format is JWT (JSON Web Token).

improvements. In addition, RNI API may be also used by the MEC platform to optimize the mobility procedures required to support service continuity. The exercise at the end of the present chapter provides an example of how to exploit radio information (e.g. handover failure events) at the application level.

4.4.2   Location API

This API is introduced by MEC013 specification [MEC013] and defines a Location MEC service, which is producing location-related information to the MEC platform or authorized applications, in order to support several use-cases, such as active device location tracking, location-based service recommendation, etc.

The set of location-related information includes:

- the location information of specific UEs (or a certain set of UEs, or all UEs) currently served by the radio node(s) associated with the MEC host;
- a list of UEs in a particular location area;
- the specific UEs which move in or out of a particular location area;
- information about the location of all radio nodes currently associated with the MEC host.

As an important note, this service supports both geolocation (e.g. geographical coordinates) and logical location (e.g. Cell ID).

An important procedure (called UE Information Lookup) permits applications to acquire information of a list of UEs in a particular location: with this procedure, the MEC App sends a request application looks up UE information in a particular area by sending a request to the resource representing the UE Information, which includes location area information. The message body returned by the Location Service includes the list of UEs in the location area (if the UE information lookup is accepted).

The actual data model and data format which is functional for the Location API is reusing the definitions of the "RESTful Network API for Zonal Presence" published by the Open Mobile Alliance [OMA-ZnPr].

4.4.3   UE Identity API

This API is introduced by MEC014 specification [MEC014] and defines the UE Identity functionality, with the main purpose to allow UE-specific traffic rules in the MEC system.

When the UE identity API is supported, the MEC platform provides the functionality for a MEC application to register a tag (representing a UE) or a list of tags. The UE identity Tags are indeed the main information elements used in the context of this API. Essentially, each tag is mapped into a specific UE in the mobile network operator's system (note: this mapping is not standardized by ETSI ISG MEC). The MEC App instance can register in the MEC platform (with the message body containing the UeIdentityTagInfo data structure) so that the related traffic rules can then be activated. Dually, the MEC App can retrieve information about specific UeIdentityTagInfo (representing one or more UEs).

4.4.4   Bandwidth Management API

The BWM API (Bandwidth Management API) was already introduced in Phase 1 by MEC015 specification, which was then updated in Phase 2 [MEC015], and renamed as "Traffic Management APIs", to include also an additional MTS API (described in the next section). The main purpose of BWM API is to define the Bandwidth Management MEC service. It describes the related application policy information including authorization and access control, information flows, required information and service aggregation patterns.

The set of BWM API is for allocating/adjusting bandwidth resources and includes the following information:

- bandwidth size (the size of requested fixed BW allocation in [bps])
- bandwidth priority (indication of the allocation priority when dealing with several applications or sessions in parallel
- direction of the requested BW allocation (DL, UL or symmetrical).

4.4.5   Multi-access Traffic Steering API

The (Multi-access Traffic Steering) MTS service is introduced by MEC015 specification [MEC015] for seamlessly steering/splitting/duplicating application data traffic across multiple access network connections. The MTS allows:

(1) MEC applications to get informed of various MTS capabilities and multi-access network connection info.
(2) MEC applications to provide requirements, e.g. delay, throughput, loss, for influencing traffic management operations.

This API supports traffic steering/splitting/duplicating among a range of access network technologies, from Fixed access, to Wi-Fi (including Wi-Fi 6) and to 3GPP systems, i.e. 2G/3G, 4G/LTE and 5G NR.

The set of MTS related information includes:

- QoS requirement description of the MTS session, eventually including minimum Throughput, tolerable delay and jitter, tolerable packet loss.

- The specific MTS mode of operation
- the network access type

As an important note, MTS service can be application-specific, and also session-specific (where a session is identified by an IP or IP range, a source and destination port and a protocol).

4.4.6   WLAN Information API

This API is introduced by MEC028 specification [MEC028] and defines WLAN Access Information Service (WAIS) that provides WLAN access-related information to service consumers within MEC System. These are of course authorized MEC applications, and the service is discoverable via service registry over the Mp21 reference point specified by [MEC003].

The set of WLAN API related information includes:

- The list of Access Points (AP), with related attributes (MAC address, SSID and IP address), and WLAN capabilities (HT, VHT, HE, DMG, EDMG)
- AP radio parameters such as RSSI and BSS load attributes (information on the current STA population and traffic levels in the BSS)
- WAN backhaul link metrics such as transmission characteristics (DL and UL speed, DL and UL load, etc.)
- The Location on the Access Point (geospatial location and civic location)
- Information about neighbor Access Points

The *granularity* of the WLAN Access information may be adjusted based on parameters such as information per station (STA), per Access Point (AP) or per Multiple Access Points (Multi-AP).

As an important note, this service can be used by MEC Apps and MEC platform to optimize existing services or to enable a new type of services (e.g. based on up to date information from WLAN access possibly combined with the information such as RNI API or Fixed Access API from the other access technologies).

4.4.7   Fixed Access API

This API is introduced by MEC029 specification [MEC029] and defines Fixed Access Information Service (FAIS) that provides Fixed access related information to service consumers within MEC System. More in detail, this API is related to a MEC service on Fixed Access Information for Fiber (PON, XG-PON, NG-PON), Cable (DOCSIS 3.1), xDSL and Point-to-Point Fiber Ethernet access to MEC.

The set of FAIS related information includes:

- The kind of available Fixed access and related information (e.g. info on last-mile technology, physical interface used for the end customer site, bandwidth from/to the network towards the customer site, the maximum baseline latency between customer site and service edge node)
- The device(s) information (i.e. the info of the device that is connected to a fixed access network, e.g. customer GW info, device ID and its status, etc.).
- The cable line information (i.e. cable modem ID and its connectivity state, number of times the CM reset or initialized this interface, latency measurements, statistics, etc.)
- The optical network information (physical location, specific optical technology used, ONU operational state, downstream/upstream line rate)
- Eventual notifications (e.g. about the ONU alarms to the service consumer, or about the device information, cable modem connectivity state, ANI alarms, etc.)

4.4.8 V2X Api

This API is introduced by MEC030 specification [MEC030] and defines the Vehicle-to-Everything (V2X) MEC service, that provides information to authorized MEC consumers in order to facilitate V2X interoperability in a multi vendor, multi-network and multi-access environment.

The set of V2X API related information includes:

- provisioning information for V2X communication over Uu unicast/MBMS for a particular location
- provisioning information for V2X communication over PC5 for a particular location (by gathering PC5 V2X relevant information from the 3GPP network)
- Exposure of relevant (i.e. journey-specific) information about the QoS prediction to authorized UEs (e.g. info on potential routes of a vehicular UE, including location, estimated time at that location, with related RSRP and RSRQ predictions)
- publication/subscription/notification of V2X messages [ITS102894] which come from different vehicle OEMs or operators

As an important note, this service enables Inter-MEC system V2X application communication (e.g. exposure to MEC apps also potentially belonging to different MEC systems). This is an important aspect for automotive use-cases, often characterized by the presence of multiple MNOs, multiple MEC vendors and multiple car OEMs. In particular, since each MNO is expected to have its own MEC system, an Inter-MEC system communication is particularly needed, to ensure end-to-end proper working of V2X applications in such scenarios.

| **Exercises/examples—Chapter 4**

Exercise 4.1 Goal of the exercise: Practice with RESTful APIs and HTTP methods, for MEC service consumption. Design a MEC application that uses the information provided by the Radio Network Information (RNI) MEC Service (RNIS) defined in the ETSI GS MEC 012 specification [REF] to compute the percentage of failed handovers that occurred for a specific User Equipment (UE) in an area covered by the MEC service (e.g. a shopping center).

Note: the actual coding of the designed MEC application (making use of the OpenAPI representation of the RNI service) will be covered by Exercise 9.1.

Proposed solution: Failed handovers (HOfail) are identified in MEC 012 as rejected or canceled handovers so that their percentage can be expressed as follows:

$$\text{HOfail} \% = 100 \cdot \frac{\text{HOrej} + \text{HOcanc}}{\text{HOrej} + \text{HOcanc} + \text{HOsucc}}$$

Our MEC application will use this equation to compute the percentage of failed handovers, by retrieving the required handover values from the RNI MEC service. Just as an example, let us assume the target UE can be identified through its IPv4 address.

Information about handovers can be collected through notifications about the cell change procedures related to our target UE. As first step, we will verify the presence of the target UE in our MEC environment querying the Radio Access Bearer (RAB) Information using a filter on the UE's IPv4 address. Once verified that the target UE is under the coverage of a cell of our environment, we will subscribe to the RNIS MEC service to receive the cell change notifications associated to this UE and we will use their content to compute the percentage of failed handovers. Let's go into the details.

Step 1. Query of RAB information for the target UE.

The specification of the RAB Info Query method is provided in clause 7.3.3.1 of the ETSI GS MEC 012. In particular, the RAB information can be retrieved with an HTTP GET request to the *{apiRoot}/rni/v2/queries/rab_info* resource URI, with the possibility to filter on the query parameter *ue_ipv4_address* that identifies the IPv4 address of our target UE (ref. Table 7.3.3.1-1). If the UE is correctly found, the RNIS will reply with a 200 OK HTTP response and the RAB information encoded in the response body. The structure of the RabInfo information model is specified in clause 6.2.3 of the ETSI GS MEC 012. Here we can find the cell where the UE is currently connected (the E-UTRAN Cell Global Identifier) and all the details of the radio access bearer for the given UE.

Step 2. Subscription to cell change notifications for the target UE.

As mentioned before, information about handovers related to a UE can be received following a publish-subscribe pattern and, in particular, subscribing to the cell change notifications offered by the RNIS. The information models for the *CellChangeSubscription* and *CellChangeNotification* are specified in clauses 6.3.2 and 6.4.2 of the ETSI GS MEC 012, respectively. In particular, for our purposes, the *CellChangeSubscription* must include the URI of the callback where our MEC application will receive the notifications and a number of filter parameters to identify the types of notification we want to receive. In our case, the filter will specify the UE's IPv4 address (*associateId* field) and the list of handover status we want to be notified of (*hoStatus* field), i.e. completed, rejected and canceled ones.

The specification of the RNIS subscriptions method is provided in clause 7.6 of the ETSI GS MEC 012. In particular, the creation of a new subscription for cell change notifications can be requested with an HTTP POST message to the *{apiRoot}/rni/v2/subscriptions* resource URI, providing the *CellChangeSubscription* details in the message body (ref. table 7.6.3.4 - 1). If the subscription is correctly created, the RNIS will reply with a 201 Created HTTP response, sending back the subscription information encoded in the response body.

Step 3: Reception and processing of cell change notifications.

If the subscription described in the previous step is successfully created, the RNIS will start to send cell change notifications to our MEC application whenever a handover for the target UE is successfully completed, rejected or canceled. In order to receive and process these notifications, our application must implement a REST server that accepts HTTP POST messages on the resource URI provided in the callback field of the subscription created in step 2. These POST messages will encode the *CellChangeNotification* information in the message body (see clause 6.4.2 of the ETSI GS MEC 012). Processing the *hoStatus* field of the notification it is possible to determine the result of the handover and increment the related counters, which can be used to compute the percentage of failed handovers according to the equation defined at the beginning of the exercise.

Exercise 4.2 Write a MEC service and application with Zenoh.

Goal of the exercise[5]: implement two variants of a MEC application which uses Zenoh as transport protocol to (1) publish/consume and (2) query data as illustrated in Fig. 4.8.

Prerequisites: the reader is expected to have a basic knowledge of pub/sub programming. This exercise uses basic constructs of the Rust programming language.

Useful links: Zenoh documentation is available at: http://zenoh.io/. Rust documentation is available at: https://doc.rust-lang.org/stable/book/. The source code of

---

[5]Courtesy of Luca Cominardi, Angelo Corsaro, Olivier Hecart, Julien Enoch.

this exercise is available at: https://github.com/atolab/zenoh-educational/tree/master/mec-book.

Preparatory steps: install Rust on your machine: https://www.rust-lang.org/tools/install. Once installed, set the default Rust toolchain to *nightly* as follows:

```
$ ~: rustup default nightly
```

First version: pub/sub approach.

The first version considers a monitoring application that subscribes to a given key expression to receive any new temperature data within a region. In this case, the monitoring application is the subscriber, while the sensors are the publishers.

Create two Rust projects using *cargo*: one for the subscriber and one for the publisher.

```
$ ~: cargo new zenoh-subscriber && cargo new zenoh-publisher
```

Edit the *Cargo.toml* file of both projects to include the necessary dependencies as follows:

```
[dependencies]
async-std = { version = "=1.6.2", features = ["attributes"] }
zenoh = "=0.5.0-beta.1"
```

```rust
use async_std::prelude::*;
use zenoh::net::*;
#[async_std::main]
async fn main() {
    // Open a Zenoh session using the default configuration
    let session = open(Config::default(), None).await.unwrap();
    // Declare a subscriber for a given key selector
    let selector = "/region01/**";
    let mut subscriber = session
        .declare_subscriber(&selector.into(),  &SubInfo::default()).await.unwrap();
    // Process the incoming publications
    while let Some(sample) = subscriber.stream().next().await {
        let bytes = sample.payload.to_vec();
        let value = String::from_utf8_lossy(&bytes);
        println!("<< Received ('{}': '{}')", sample.res_name, value);
    }
}
```

The following block of code implements the monitoring application as **Zenoh subscriber**. It has to be included in the src/main.rs file of the zenoh-subscriber project.

The following block of code implements the sensor application as **Zenoh publisher**. It has to be included in the *src/main.rs* file of the *zenoh-publisher* project.

```
use zenoh::net::*;
#[async_std::main]
async fn main() {
    // Open a Zenoh session using the default configuration
    let session = open(Config::default(), None).await.unwrap();
    // Periodically publish the house temperature
    loop {
        let (res, val) = ("/region01/house01/temp", "25");
        let bytes = val.as_bytes().into();
        session.write(&res.into(), bytes).await.unwrap();
        println!(">> Published ('{}': '{}')", res, val);
        async_std::task::sleep(std::time::Duration::from_secs(5)).await;
    }
}
```

Run the **Zenoh publisher** (i.e. monitoring app) and **Zenoh subscriber** (i.e. sensor app) with *cargo run* command. A publication should be periodically printed on screen.

```
$ /zenoh-publisher. cargo run          $ ~/zenoh-subscriber: cargo run
>> Published ('/region01/house01/temp': '25')   << Received ('/region01/house01/temp': '25')
```

Second variant: queryable/querier approach.

The seconds variant considers a monitoring application that periodically queries a given key expression to receive the on-demand generated temperature data within a region. The monitoring application is the querier while the sensors are the queryables.

Create two Rust projects using *cargo*: one for the queryable and one for the querier and.

edit the *Cargo.toml* of both projects in the same way as described for the first variant.

```
$ ~: cargo new zenoh-queryable && cargo new zenoh-querier
```

The following block of code implements the sensor application as a **Zenoh queryable**. It has to be included in the *src/main.rs* file of the *zenoh-queryable* project.

```
use async_std::prelude::*;
use zenoh::net::*;
#[async_std::main]
async fn main() {
    // Open a Zenoh session using the default configuration
    let session = open(Config::default(), None).await.unwrap();
    // Declare a queryable resource to react on
    let (res, val) = ("/region01/house01/temp", "25");
    let mut queryable = session
        .declare_queryable(&res.into(), queryable::EVAL).await.unwrap();
    // Process the incoming queries
    while let Some(query) = queryable.stream().next().await {
        println!(">> Replying to '{}': ('{}': '{}')", query.res_name, res,
val);
        // Generate and send back the response
        let sample = Sample {
            res_name: res.to_string(),
            payload: val.as_bytes().into(),
            data_info: None,
        };
        query.reply(sample).await;
    }
}
```

The following block of code implements the monitoring application as a **Zenoh querier**. It has to be included in the *src/main.rs* file of the *zenoh-queryable* project.

```
use async_std::prelude::*;
use zenoh::net::*;
#[async_std::main]
async fn main() {
    let session = open(Config::default(), None).await.unwrap();

    loop {
        let selector = "/region01/**";
        println!(">> Query '{}'", selector);
        let predicate = "";
        let target = QueryTarget::default();
        let consol = QueryConsolidation::default();
        let mut replies = session
            .query(&selector.into(),  predicate,  target,   consol).await.un-
wrap();
        while let Some(reply) = replies.next().await {
            let bytes = reply.data.payload.to_vec();
            let value = String::from_utf8_lossy(&bytes);
            println!("<< Received ('{}': '{}')", reply.data.res_name, value);
        }
        async_std::task::sleep(std::time::Duration::from_secs(5)).await;
    }
}
```

Run the **Zenoh querier** (i.e. monitoring app) and **Zenoh queryable** (i.e. sensor app) with *cargo run* command. A query should be periodically printed on screen.

```
$ ~/zenoh-querier: cargo run          $ ~/zenoh-queryable: cargo run
>> Query '/region01/**'               >> Replying to '/region01/**':
<< Received ('/region01/house01/temp': '25')    ('/region01/house01/temp': '25')
```

References

[MEC-web]   ETSI   MEC.   https://www.etsi.org/technologies-clusters/technologies/multi-
               access-edge-computing
[MEC002]   ETSI GS MEC 002 V2.1.1 (2018–10) Multi-access Edge Computing (MEC);
               Phase 2: Use cases and Requirements. https://www.etsi.org/deliver/etsi_gs/
               MEC/001_099/002/02.01.01_60/gs_MEC002v020101p.pdf
[MEC003]   ETSI GS MEC 003 V2.1.1 (2019–01) Multi-access edge computing (MEC);
               Framework and reference architecture. https://www.etsi.org/deliver/etsi_gs/
               MEC/001_099/003/02.01.01_60/gs_MEC003v020101p.pdf
[MEC009]   ETSI GS MEC 009 V2.1.1 (2019–01) Multi-access edge computing (MEC);
               General principles for MEC service APIs. https://www.etsi.org/deliver/etsi_gs/
               MEC/001_099/009/02.01.01_60/gs_MEC009v020101p.pdf
[MEC011]   ETSI GS MEC 011 V2.1.1 (2019–11) - Multi-access edge computing (MEC);
               Edge platform application enablement. https://www.etsi.org/deliver/etsi_gs/
               MEC/001_099/011/02.01.01_60/gs_MEC011v020101p.pdf
[MEC012]   ETSI GS MEC 012 V2.1.1 (2019–12) - Multi-access edge computing (MEC);
               radio network information API. https://www.etsi.org/deliver/etsi_gs/MEC/001_
               099/012/02.01.01_60/gs_MEC012v020101p.pdf
[MEC013]   ETSI GS MEC 013 V2.1.1 (2019–09) - Multi-access edge computing (MEC);
               Location API. https://www.etsi.org/deliver/etsi_gs/MEC/001_099/013/02.01.
               01_60/gs_MEC013v020101p.pdf
[OMA-ZnPr]   OMA-TS-REST-NetAPI-ZonalPresence-V1–0–20160308-C: "RESTful
               Network API for Zonal Presence"
[MEC14]   ETSI GS MEC 014 V1.1.1 (2018–02) Mobile edge computing (MEC); UE
               identity API. https://www.etsi.org/deliver/etsi_gs/MEC/001_099/014/01.01.01_
               60/gs_mec014v010101p.pdf
[MEC015]   ETSI GS MEC 015 V2.1.1 (2020–06) Multi-access edge computing (MEC);
               traffic management APIs. https://www.etsi.org/deliver/etsi_gs/MEC/001_099/
               015/02.01.01_60/gs_MEC015v020101p.pdf
[MEC028]   ETSI GS MEC 028 V2.1.1 (2020–06) Multi-access edge computing (MEC);
               WLAN information API. https://www.etsi.org/deliver/etsi_gs/MEC/001_099/
               028/02.01.01_60/gs_MEC028v020101p.pdf
[MEC029]   ETSI GS MEC 029 V2.1.1 (2019–07) - Multi-access edge computing (MEC);
               fixed access information API. https://www.etsi.org/deliver/etsi_gs/MEC/001_
               099/029/02.01.01_60/gs_MEC029v020101p.pdf
[MEC030]   ETSI GS MEC 030 V2.1.1 (2020–04) - Multi-access edge computing (MEC);
               V2X information service API. https://www.etsi.org/deliver/etsi_gs/MEC/001_
               099/030/02.01.01_60/gs_MEC030v020101p.pdf
[ITS102894]   ETSI TS 102 894–2 V1.3.1 (2018–08) - Intelligent transport systems (ITS);
               users and applications requirements; Part 2: Applications and facilities layer

            common data dictionary. https://www.etsi.org/deliver/etsi_ts/102800_102899/
            10289402/01.03.01_60/ts_10289402v010301p.pdf

[ZN-CL19]   Filippou M, Sabella D, Riccobene V (CLEEN2019) "Flexible MEC service
            consumption through edge host zoning in 5G networks". https://arxiv.org/abs/
            1903.01794

Chapter 5
MEC Service APIs in Action

This chapter will describe more in detail few selected MEC services. The specific APIs usage will be coupled also with concrete use cases of interest for MEC (e.g. from automotive market segment), and in some case also with practical examples and APIs implementations.

More in detail, this chapter covers the following APIs:

- Radio Network Information (RNI) API.
- Location Service (LS) API.
- Multi-access Traffic Steering (MTS) API.
- V2X information service (VIS) API.

For each of them, a specific exercise will be proposed at the end of the chapter, highlighting the usage of these APIs from an application development perspective.

5.1 Radio Network Information (RNI) API

One of the main advantages of the edge computing technology is of course not only proximity to end users, but also the availability at the edge of additional information that can be exposed at application level.

The information related to radio network quality (specified by the ETSI GS MEC 012 [MEC012]) is of paramount importance, because it can enable, e.g. further upper-layer adaption (e.g. TCP throughput guidance mechanisms [GSMA-MTG, IETF-MTG]), or radio-aware applications (e.g. for video adaptation [RAVEN18]) or again more sophisticated cross-layer algorithms based on the combination with information coming from other APIs (this is the case, e.g. of the MTS API and V2X API). Figure 5.1 depicts a typical 5G scenario, where multiple cells are associated to a MEC server, and radio network information is exposed through MEC RNI API.

© Springer Nature Switzerland AG 2021   115
D. Sabella, *Multi-access Edge Computing: Software Development at the Network Edge*, Textbooks in Telecommunication Engineering,
https://doi.org/10.1007/978-3-030-79618-1_5

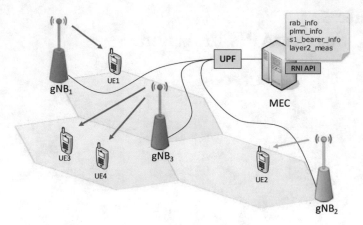

Fig. 5.1 MEC RNI API exposure in a typical multi-cell 5G deployment

As anticipated in Chap. 4, the set of RNIS related information includes:

- up-to-date radio network information regarding radio network conditions;
- measurement information related to the user plane based on 3GPP specs;
- information about UEs connected to the radio node(s) associated with the MEC host, their UE context and the related radio access bearers;
- changes on information related to UEs connected to the radio node(s) associated with the MEC host, their UE context and the related radio access bearers.

5.1.1   Resource URI Structure of RNI API

Figure 5.2 below illustrates the resource tree of the RNI API.

As we can see from the resource tree, Service Consumers can retrieve current status of Radio Access Bearer (RAB) information, PLMN information, S1 Bearer information, Layer 2 measurements (Ref. [TS-36.314]) or again perform operations on active subscriptions.

5.1.2   Services Offered to RNI API Consumers

The ETSI GS MEC 012 specifies the interactions between the Service Consumers (i.e. MEC applications or MEC platform) and the Radio Network Information Service (RNIS) through the RNI API.

Table 5.1 summarizes the possible interactions between a Service Consumer (e.g. a MEC application or a MEC platform) and the RNI API providing details about the information provided in each interaction.

Fig. 5.2 Resource URI
structure of the RNI API.
Source [MEC012]—©
European
Telecommunications
Standards Institute 2019.
Further use, modification,
copy and/or distribution are
strictly prohibited

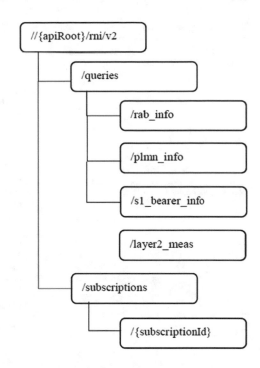

Table 5.1 List of information and URIs provided to RNI API consumers

Procedures	Description	URI Base URI: {apiRoot}/rni/v2
Request RAB (Radio Access Bearer) information	The Service Consumer requests receive a cell level RAB information from the cells that are associated with the requested MEC application instance	
	{apiRoot}/rni/v2/queries/rab_info The following additional URI query parameters can be used to filter the request: • *app_ins_id*, (comma separated) list of application instance identifiers; • *cell_id*, list of E-UTRAN Cell Identities; • *ue_ipv4_address*, list of UE IPv4 addresses • *ue_ipv6_ address*, list of UE IPv6 addresses • *nated_ip_address*, list of NATed IP addresses • *gtp_teid*, list of GTP TEID addresses • *erab_id*, E-RAB identifier • *qci*, QoS Class Identifier (Ref. [TS-23.401]) • *erab_mbr_dl*, Max downlink E-RAB Bit Rate • *erab_mbr_ul*, Max uplink E-RAB Bit Rate • *erab_gbr_dl*, Guaranteed DL E-RAB Bit Rate • *erab_gbr_ul*, Guaranteed UL E-RAB Bit Rate	
	Example: **{apiRoot}/rni/v2/queries/plmn_info?** **app_ins_id=appListString**	

(continued)

Table 5.1 (continued)

Procedures	Description	URI Base URI: {apiRoot}/rni/v2
Request PLM information	The Service Consumer requests receive cell level PLMN information related to specific MEC application instance(s)	
	{apiRoot}/rni/v2/queries/plmn_info The following additional URI query parameter can be used to filter the request: • *app_ins_id*, (comma separated) list of application instance identifiers	
	Example: **{apiRoot}/rni/v2/queries/plmn_info? app_ins_id=appListString**	
Request S1 bearer information	With this information, the Service Consumer can, e.g. optimize the relocation of MEC applications, or uses the acquired information for managing the traffic rules for the related application instances	
	{apiRoot}/rni/v2/queries/s1_bearer_info The following additional URI query parameters can be used to filter the request: • *temp_ue_id*, (comma separated) list of temporary identifiers allocated for the specific UEs • *ue_ipv4_address*, list of UE IPv4 addresses • *ue_ipv6_ address*, list of UE IPv6 addresses • *nated_ip_address*, list of NATed IP addresses • *gtp_teid*, list of GTP TEID addresses • *cell_id*, list of E-UTRAN Cell Identities; • *erab_id,* E-RAB identifier	
	Example: **{apiRoot}/rni/v2/queries/ s1_bearer_info?cell_id=cellIdList**	
Request layer 2 measurements	The Service Consumer requests receive the Layer 2 measurements information from one or more eNBs associated with the requested MEC application instance	
	{apiRoot}/rni/v2/queries/layer2_meas The following additional URI query parameters can be used to filter the request: • *app_ins_id, cell_id, ue_ipv4_address*, • *ue_ipv6_ address, nated_ip_address, gtp_teid* • *dl_gbr_prb_usage_cell, ul_gbr_prb_usage_cell*, PRB usage (in %) for • downlink/uplink GBR traffic (Ref. [TS-36.314]) • *dl_nongbr_prb_usage_cell, ul_nongbr_prb_usage_cell* • *dl_total_prb_usage_cell, ul_total_prb_usage_cell* • *received_dedicated_preambles_cel* • *received_randomly_selected_preambles_low_ range_cells* • *received_randomly_selected_preambles_ high_range_cells* • *number_of_active_ue_dl_gbr_cell, number_of_active_ue_ul_gbr_cell* • *number_of_active_ue_dl_nongbr_cell,* • *number_of_active_ue_ul_nongbr_cell* • *dl_gbr_pdr_cell, ul_gbr_pdr_cell* • *dl_nongbr_pdr_cell, ul_nongbr_pdr_cell*	

(continued)

Table 5.1   (continued)

Procedures	Description	URI Base URI: {apiRoot}/rni/v2
	• *dl_gbr_delay_ue, ul_gbr_delay_ue* • *dl_nongbr_delay_ue, ul_nongbr_delay_ue* • *dl_gbr_pdr_ue, ul_gbr_pdr_ue*, packet discard rate in percentage of the • downlink/uplink GBR traffic of a UE • *dl_nongbr_pdr_ue, ul_nongbr_pdr_ue* • *dl_gbr_throughpout_ue, ul_gbr_throughpout_ue* • *dl_nongbr_throughpout_ue, ul_nongbr_throughpout_ue* • *dl_gbr_data_volume_ue, dl_gbr_data_volume_ue* • *dl_nongbr_data_volume_ue, ul_nongbr_data_volume_ue*	
	Example: **{apiRoot}/rni/v2/queries/ layer2_meas?cell_id=cellIdList**	
Operations on subscriptions	The Service Consumer can perform operations on a subscription to certain specific RNI event that is available at RNIS. Possible operations are request subscriptions information, create (subscribe), update, delete (unsubscribe) and also receive notification on expiry of the subscription	
	{apiRoot}/rni/v2/subscriptions The following additional URI query parameters can be used to filter the request: • *subscription_type*, query parameter to filter on a specific subscription type Permitted values are: *cell_change, rab_est, rab_mod, rab_rel, meas_rep_ue, nr_meas_rep_ue, timing_advance_ue, ca_reconf, s1_bearer*	

5.2   Location API

One of the main advantages of the edge computing network architecture stands in the possibility to become aware of the user context and, consequently, to deliver services in a more effective way. The position of User Equipment is one fundamental element to build this context-awareness. For this reason, the ETSI MEC group specified the MEC Location Service and the related API in ETSI GS MEC 013 [MEC013]. This service can provide information related to the UE's location, to any authorized application running on a certain edge node. The concept is based on the Zonal Presence standardized by the Small Cell Forum [SCF084, SCF152], and the information is accessible by RESTful APIs that were originally defined by the Open Mobile Alliance (OMA).

Fig. 5.3 Exemplary UE
activities, e.g. Zone Enter (1),
Zone Exit (2) and Zone
Transfer (3) (according to
[OMA-TS-ZP])

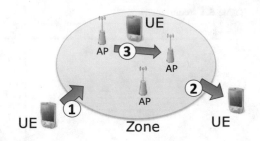

According to the definition used by OMA [OMA-TS-ZP], a zone is a set of access points (e.g. in general, any cell type accessible to the mobile device, e.g. cellular network base stations including small cells, and Wi-Fi access point).

Zones may be formed, for example, based on location, ownership, type of area they are deployed in, type of traffic they carry, etc. in any combination. They do not have to be geographically adjacent.

The Location Service can provide information on user activities within a "zone" (such as zone enter, zone exit and zone transfer, as depicted in Fig. 5.3) and notifications of specific user tracking or events of interest.

Additionally, the Location Service defines operations where applications can gain information about zones, access points and mobile devices.

In summary, the Location Service does not only provide, through the Location API, the position of a certain UE, but also to deliver a series of "geo-services", such as a notification on UE activities related to a certain area (zone) or the information about the UEs present in a certain area (zone). Importantly, the position reported by the Location Service can be anonymized to respect the privacy of the users.

The concept of "zone" defined in OMA is here re-used by the Location API specification in a MEC context, so that a zone lends itself to be used to group all the radio nodes that are associated to a MEC host, or a subset of them, according to the desired deployment.

5.2.1   Resource URI Structure of Location API

Figure 5.4 illustrates the resource tree of the Location API.

5.2.2   Services Offered to Location API Consumers

The ETSI GS MEC 013 specifies the interactions between the Service Consumers (i.e. MEC applications or MEC platform) and the Location Service (LS) through the Location Service API.

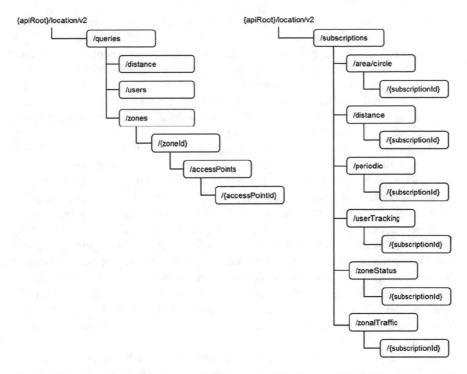

Fig. 5.4 Resource URI structure of the Location API. *Source* [MEC013]—© European Telecommunications Standards Institute 2019. Further use, modification, copy and/or distribution are strictly prohibited

A Service Consumer can perform two types of requests to the LS API:

- *A single time query*: the LS provides just once the information requested;
- *A subscription request*: the LS sends periodically the information requested to the Service Consumer, until it receives a cancellation of the subscription or until a time limit is reached.

Both queries and the subscriptions are based on RESTful APIs using a similar mechanism to the OMA APIs as specified in [OMA-ZnPr, OMA-TrLo]. The *Lookup* queries (i.e. single time interaction) are performed by the Service Consumer using an HTTP GET method requesting the desired information. The LS provides the information in the message body of the response to the request. The *subscription* to a specific information notification is instead performed by the Service Consumer through an HTTP POST method in which the Service Consumer provides the details about the desired subscription in the body of the POST method. The LS provides in the response the identifier of the subscription created, while it sends the information using the body of POST methods. A Service Consumer can cancel a subscription using an HTTP DELETE method, specifying in it the identifier of the subscription to be cancelled.

The content of the messages is in JSON format. The content type header shall then be set to "application/json" in the header of the HTTP methods. The JSON format is used because the elements of the data model are represented using JSON objects. The LS API re-uses the data model defined in [OMA-ZnPr, OMA-TrLo] with some small variations: the terminology of some data types can differ and some data elements have been extended to include the definitions of some additional data types. ETSI GS MEC 013 [MEC013] describes in detail the correct meaning of the re-used data types and the extended data types.

The Uniform Resource Identifier (URI) to be used in the GET and POST methods is made by an URI root and an URI specific to the requested resource. The URI root has the following format **{apiRoot}/location/{apiVersion}/**. The **{apiRoot}** contains the scheme (i.e. "http" or "https"), the host (and optionally the port) and an optional prefix string. The Service Consumer discovers the specific **{apiRoot}** through the service registry. The **{apiVersion}** is currently set to "v2" as specified in ETSI GS MEC 013 V2.1.1. The URI related to the requested resource strictly depends on which resource is requested. Table 5.1 provides the details about the resources (i.e. the information) that can be requested and the related URI.

The LS can indeed provide different types of information related to UE's location. The information can be merely the position of a single UE or it can be a more advanced information such as the distance between a UE and a specific geographical location.

Table 5.2 summarizes the possible interactions between a Service Consumer and the LS API providing details about the information provided in each interaction.

5.3   Multi-access Traffic Steering (MTS) API

As the mobile industry evolves toward 5G and beyond, it is becoming clear that no single access technology will be able to meet a great variety of requirements for human and machine communications. For example, the emerging Industry 4.0 application (e.g. robotics, remote control, etc.) require extremely low packet loss ratio, e.g. $< 10^{(-12)}$, and low latency (<1 ms). On the other hand, driving more data through a scarce and finite radio spectrum becomes a real challenge, and spectrum efficiency is approaching a plateau and will not deliver the needed increase in bandwidth improvement itself.

Let's use the above Table 5.3 as an example to explain the benefits of "Multi-Access". Consider N available (wireless) connections between a client and edge, each of which has some data rate (R) and packet loss rate (P). Table 5.3 shows the linear improvement in data rate, and the exponential improvement in reliability, by seamlessly combining multiple independent connections to deliver user traffic. However, it is not possible to achieve the maximum data rate (400 Mbps) and the highest reliability $\left(10^{(-12)}\right)$ at the same time. Therefore, it is critical to manage traffic across all available access networks dynamically based on

Table 5.2 List of information and URIs provided to Location API consumers

Procedures	Description	URI Base URI: {apiRoot}/location/v2
UE Location Lookup	The Service Consumer requests the current location information of a specific UE or a group of UEs	
		{apiRoot}/location/v2/users The additional URI query parameter *address* is used to request the location of a specific UE or group o UEs Example: **{apiRoot}/location/v2/users?address=acr %3A{ueAddress}**
UE Information Lookup	The LS provides the information about one or more UEs in a particular location that is specified by the Service Consumer in the query	
		{apiRoot}/location/v2/users The following additional URI query parameters can be used to filter the request: • *zoneId*, the string of the identifier of zone; • *accessPointId*, the string of the identifier of access point; Example:**{apiRoot}/location/v2/users?zoneId= {zoneIdString}**
UE Location Subscribe	The same information of UE Location Lookup are provided but the LS sends it periodically until the subscription is cancelled or the provided time limit is reached	
		{apiRoot}/location/v2/subscriptions/userTracking The message body of the request shall contain the information required for the subscription. This information is described by the *UserTrackingSubscription* data type. In the specific, the address of the UE shall be inserted with other optional fields specifying the events to be tracked
UE Information Subscribe	This procedure allows to periodically receive the information related to UE(s) in a particular location	
		{apiRoot}/location/v2/subscriptions/zonalTraffic The identifier of the zone to monitor shall be provided in the body of the request. In the specific, an instance of the *zonalTrafficSubscription* data type shall be inserted into request
Subscribe Cancellation	This procedure is used to cancel the subscription to a specific procedure	
		{apiRoot}/location/v2/subscriptions/ {subscriptionProcedure}/{subscriptionId} {subscriptionProcedure} is specific to the subscription procedure, e.g. *userTracking* for UE Location Subscribe, *zonalTraffic* is for UE Information Subscribe. The {subscriptionId} is provided to the Service Consumer when the subscription is performed

(continued)

Table 5.2 (continued)

Procedures	Description	URI Base URI: {apiRoot}/location/v2
Radio Node Location Lookup	The Service Consumer can gather the location information of the radio nodes that are currently associated with the MEC host	
	{apiRoot}/location/v2/zones/{zoneId}/accessPoints The {zoneId} corresponds to the identifier of the zone which the access point is associated. The filtering URI query parameter *interestRealm* can be used	
UE Tracking Subscribe	The UE information about a given UE is provided periodically or at some specific trigger events such as the status change of the UE when it performs a hand-over between cells	
	{apiRoot}/location/v2/subscriptions/periodic The body of the message shall contain the following information: the address of the UE to track, the requested accuracy and the frequency of the notification. The duration of the Subscription may also be specified. The data type to be used to provide this information is the *PeriodicNotificationSubscription* data type	
UE Distance Lookup	This procedure is used to retrieve the distance between a UE and a geographical location or another UE	
	{apiRoot}/location/v2/queries/distance The *address* parameter shall be used to identify the UE(s) and the parameters *latitude* and *longitude* shall be used to indicate the geographical locations Examples: **{apiRoot}/location/v2/queries/distance?address=acr%3A{ueAddress1}&address=acr%3A{ue2Address}** **{apiRoot}/location/v2/queries/distance?address=acr%3A{ueAddress}&latitude={latitudeValue}&longitude={longitudeValue}**	
UE Distance Subscribe	In this case, the distance information of the UE to a specific UE or location is periodically provided	
	{apiRoot}/location/v2/subscriptions/distance The address of the UE(s) of interest is provided in the body exploiting the *DistanceNotificationSubscription* data type	
UE Area Subscribe	The Service Consumer can use this procedure to ask for periodical notification of UE movement notifications for a defined geographic area	
	{apiRoot}/location/v2/subscriptions/area/circle The Service Consumer provides the information in the body of the request using the *CircleNotificationSubscription* data type in which the address of the UE(s) and a circular geographical area (i.e. the coordinates of the center and the radius) are provided	

Table 5.3 Example of QoS Improvement through "Multi-Access" (R = 100 Mbps, P = $10^{(-3)}$)

	N = 1	N = 2	N = 3	N = 4
Data rate	100 Mbps	200 Mbps	300 Mbps	400 Mbps
Reliability	10^{-3}	10^{-6}	10^{-9}	10^{-12}

applications requirements, e.g. delay, latency, etc. and maximize the total number of "happy" or QoS-satisfied users in time-varying/dynamic wireless environment.

Multi-access Traffic Steering (MTS) is part of Traffic Management (TM) services [MEC015] in the ETSI/MEC reference architecture, as illustrated in Fig. 5.5. Applications may request specific Bandwidth Management (BWM) or/and Multi-access Traffic Steering (MTS) requirements for the whole application instance or different requirements per session.

The TM services can aggregate all the requests and act in a manner that will help optimize the BW usage and improve Quality of user Experience for applications. Specifically, The MTS service enables seamlessly steering/splitting/duplicating application data traffic across multiple access network connection, and allows:

- MEC applications to get informed of various MTS capabilities and multi-access network connection info.
- MEC applications to provide requirements, e.g. delay, throughput, loss, for influencing traffic management operations.

There are many ways to implement MTS services on MEC platform. For example, Multiple Access Management Services (MAMS) [RFC8743] provides mechanisms for flexible selection of network paths in a multi-connection (access) communication environment. It leverages network intelligence and policies to dynamically adapt traffic distribution across selected paths and user plane treatment

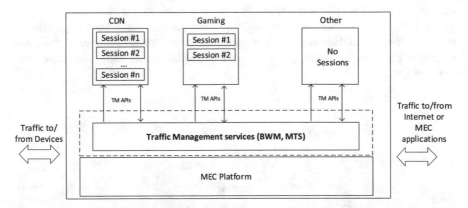

Fig. 5.5 Traffic Management services description. *Source* [MEC-015]—© European Telecommunications Standards Institute 2020. Further use, modification, copy and/or distribution are strictly prohibited

to changing network/link conditions. The network path selection and configuration messages are carried as user plane data between the functional elements in the network and the end user device, and thus without any impact to the control plane signaling schemes of individual access network. MAMS user plane protocol stack consists of two sublayers:

- Convergence: this layer performs multi-access specific tasks, e.g. access (path) selection, multi-link (path) aggregation, splitting/reordering, lossless switching, keep-alive, probing, etc.
- Adaptation: this layer performs functions to handle tunneling, network layer security and NAT (network address translation).

Two types of connections are defined in MAMS:

- Anchor Connection: refers to the E2E network path between the client and the Application Server.
- Delivery Connection: refers to the network path between the client and the MAMS network data proxy for delivering data traffic. Multiple delivery connections may be configured to distribute user traffic, which are combined and managed at the convergence sublayer.

Usually, the anchor connection is also a delivery connection. Figure 5.6 shows an example where LTE is the anchor & delivery connection and Wi-Fi is only for delivery. In this example, UDP tunneling is used in the adaptation sublayer for transporting LTE IP packets over the Wi-Fi connection.

Applications may have different requirements with respect to multi-access traffic steering. For example, a background data-centric application, e.g. email, drop-box, etc., may not care about delay or throughput, and therefore may prefer the connection with lower cost, e.g. Wi-Fi. In another example, a URLLC (ultra-reliable and low latency communications) application, e.g. robot control, industrial applications, etc., requires very high reliability and low latency, and therefore may prefer sending duplicated packets over multiple connections simultaneously. Also, a real-time high-definition video streaming application requires high throughput and therefore may prefer splitting traffic over multiple connections, aka bandwidth aggregation.

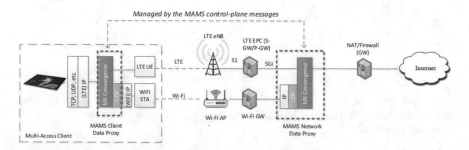

Fig. 5.6 MAMS-based Multi-Access Edge Network with LTE and Wi-Fi

Multi-access Traffic Steering (MTS) is then a service introduced to supports traffic steering/splitting/duplicating among a range of access network technologies. In that perspective, the MTS API [MEC015] supports a range of access network technologies, from Fixed access, to Wi-Fi (including Wi-Fi 6) and to 3GPP systems, i.e. 2G/3G, 4G/LTE and 5G NR. As anticipated, in Chap. 4, the set of MTS related information includes:

- QoS requirement description of the MTS session, eventually including minimum Throughput, tolerable delay and jitter, tolerable packet loss.
- The specific MTS mode of operation.
- the network access type.

Regarding the specific MTS modes of operation, several traffic management policies are available for this API (corresponding to the key attributes of "MtsCapabilityInfo" in the MTS API, which are used for the MEC platform to inform MEC applications of various MTS capabilities and multi-access network connection info). In particular, the following MTS operations are specified:

- **Low Cost**: using the unmetered access network whenever it is available.
- **Low Latency**: using the access network connection with lower latency.
- **High Throughput**: using the access network with higher throughput, or/and multiple access networks simultaneously.
- **Redundancy**: sending duplicated packets over multiple access network connections to improve reliability.
- **QoS**: performing MTS based on the specific QoS requirements of the app.

5.3.1 Resource URI Structure of MTS API

Figure 5.7 below illustrates the resource tree of the MTS API.

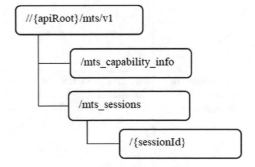

Fig. 5.7 Resource URI structure of the MTS API. *Source* [MEC015]—© European Telecommunications Standards Institute 2020. Further use, modification, copy and/or distribution are strictly prohibited

5.3.2   Services Offered to MTS API Consumers

The interactions between the Service Consumers (i.e. MEC applications or MEC platform) and the Multi-access Traffic Steering through the MTS API are specified in the ETSI GS MEC 015 deliverable.

Table 5.4 summarizes the possible interactions between a Service Consumer and the MTS API providing details about the information provided in each interaction.

In summary, the Multi-access Traffic Steering (MTS) service and API that was recently introduced in the ETSI MEC reference architecture address the need of supporting diverse applications by leveraging multiple access networks & connections that are deployed at the edge. Now, it is time for application developers to take advantages of such new capability in improving Quality of user Experience (QoE) for their MEC applications.

Table 5.4 List of information and URIs provided to MTS API consumers

Procedures	Description	URI Base URI: {apiRoot}/mts/v2
Get MTS service info	With this procedure the Service Consumer gets the available MTS service information from the MTS service	**{apiRoot}/mts/v1/** **mts_capability_info** This request doesn't use any URI query parameter. Example: {apiRoot}/mts/v1/ mts_capability_info?
Operations on MTS sessions	The Service Consumer retrieves information about a list of MTS sessions, or creates a new MTS session	**{apiRoot}/mts/v1/mts_sessions** This request may use "app_instance_id" or "app_name" or "session_id" or none of them as URI query parameter. Example: {apiRoot}/mts/v1/mts_sessions? app_instance_id=1234 {apiRoot}/mts/v1/mts_sessions? app_name=abcd {apiRoot}/mts/v1/mts_sessions? session_id=5678 {apiRoot}/mts/v1/mts_sessions?
Operations on an individual MTS session	With this procedure the Service Consumer retrieves information about a specific MTS session	**{apiRoot}/mts/v1/mts_sessions/** **{sessionId}** This request doesn't use any URI query parameter. Example: {apiRoot}/mts/v1/mts_sessions/ 5678?

5.4   V2X Information Service API

The Vehicle-to-Everything (V2X) Information Service API was introduced by ETSI ISG MEC specifically for scenarios involving automotive communications within a MEC-enabled deployed wireless communications network. In particular, the purpose of this V2X API is to help the ecosystem adopt MEC for automotive use cases via facilitating V2X interoperability in a multi-vendor, multi-network and multi-access environment. The considered reference automotive communication scenarios are the following:

- Vehicle Original Equipment Manufacturer (OEM) scenarios: in such scenarios, vehicle OEMs can manage some V2X services, hence, both single and multi-operator scenarios can be considered for these services. It should be noted that network operators in the same country and/or in different countries are able to provide such V2X services.
- Intelligent Transportation System (ITS) operator scenarios: in these scenarios, services for different vehicle OEMs may be additionally provided. An ITS operator may need to provide a nation-wide V2X service, based on MEC infrastructure deployed by different operators' networks, and offer this service to vehicles of different OEMs. In this case, V2X services are expected to be provided by different network operators in the same country and/or in different countries.

Vehicle OEM scenario, single MNO	ITS operator, single MNO	ITS operator, single OEM, single MNO
Vehicle OEM scenario, multiple MNO	ITS operator, multiple MNO	ITS operator, multiple OEM, multiple MNO

In a nutshell, the above-described use cases may involve a multitude of stakeholders, where the most general automotive scenario in real-world application implies the presence of multiple car OEMs and multiple MNOs. Therefore, it becomes imperative to fully specify the MEC related interoperability reference points involving all potential system entities. To accomplish that, the MEC V2X API has been standardized in ETSI GS MEC 030 [MEC030], including, as anticipated in Chap. 4, the following set of V2X related information:

- provisioning information for V2X communication over Uu unicast/Multimedia Broadcast Multicast Services (MBMS) for a particular location.
- provisioning information for V2X communication over PC5 for a particular location (by gathering of PC5 V2X relevant information from the 3GPP network).

- exposure of relevant (i.e. journey-specific) information about the QoS prediction to authorized UEs (e.g. info on potential routes of a vehicular UE, including location, estimated time at that location, with related Reference Signal Received Power (RSRP) and Reference Signal Received Quality (RSRQ) predictions).
- publication/subscription/notification of V2X messages [ITS102894] which come from different vehicle OEMs or operators.

As a special note on the V2X messages exposed by MEC V2X API, this information is referred to the V2X messages specified by ETSI TC ITS (ref. ETSI TS 102 894–2 [6], clause A.114) for the ITS application and facilities layers:

- DENM (Decentralized Environmental Notification Message).
- CAM (Cooperative Awareness Message).
- POI (Point of Interest) message.
- SPAT (Signal Phase And Timing) message.
- Map Data (MAP) message.
- In Vehicle Information (IVI) message.
- Electric vehicle recharging spot reservation (EV-RSR) message.
- Messages for Tyre Information System (TIS) and Tyre Pressure Gauge (TPG) interoperability.
- Traffic light Signal Request Message (SREM).
- Traffic light Signal Request Status Message (SSEM).
- Electrical Vehicle Charging Spot Notification (EVCSN) message.
- Services Announcement Extended Message.
- Radio Technical Commission for Maritime Services (RTCM) Message.

As depicted in Fig. 5.8, a vehicle is hosting a client application and is connected to a given MEC host and, therefore, to a respective MEC application. Also, other remote application servers can be instantiated elsewhere, e.g. private clouds owned by the operator or by the OEM. When multiple MEC hosts are deployed, the VIS facilitates information exposure among MEC applications running on different MEC hosts. It is noteworthy that the VIS may be produced either by a MEC platform or a service-producing MEC application.

One of the main features of the VIS is its capability of producing journey-specific predictive QoS notifications. Such a feature is essential, as accurate and timely predictions of the radio environment at locations planned to be visited by vehicles can either trigger, modify or postpone the application of certain V2X functionalities and/or the download of content delivery/ software packages. However, focusing on V2X system scenarios characterized by high mobility and dynamic topology (see Fig. 5.9), the challenge incurred is that the accuracy and the timeliness of information (e.g. radio network, location information, etc.) when

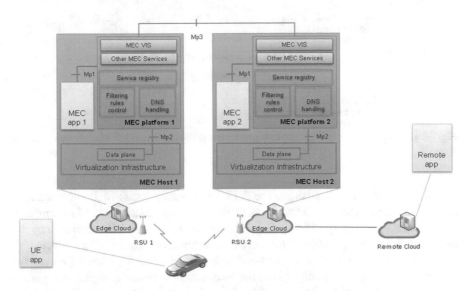

Fig. 5.8 Typical V2X system involving multiple MEC hosts and the use of VIS. *Source* [MEC030]—© European Telecommunications Standards Institute 2020. Further use, modification, copy and/or distribution are strictly prohibited

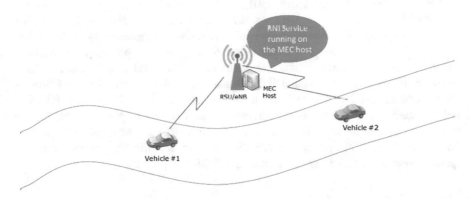

Fig. 5.9 Exemplary V2X system scenario involving use of the RNI service. *Source* [MEC030]— © European Telecommunications Standards Institute 2020. Further use, modification, copy and/or distribution are strictly prohibited

centrally collected by a MEC host are hampered by (i) the environmental situation, e.g. the occurrence of network congestion events when, for example, many vehicles attempt to provide radio measurements to the connected eNB/gNB, collocated with a MEC host, as well as by (ii) the deployment density of the cellular network, together with the capabilities of the deployed MEC infrastructure.

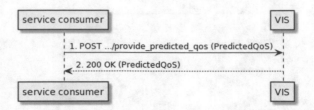

Fig. 5.10 Message flow of a V2X application requesting the predicted QoS of a UE with potential routes

To overcome such hardships, the VIS service may assist in implementing a framework for cooperative acquisition, partitioning and distribution of information for efficient, journey-specific QoS prediction. That is, the VIS service may be utilized to identify space/time correlations between radio quality data collected by different vehicles in a V2X system and a specific vehicle's planned journey for better prediction of the quality of the communication network along the designated route. Figure 5.10 shows a scenario where the service consumer (e.g. a V2X application) sends an HTTP POST request to VIS to receive the predicted QoS correspondent to potential routes of a vehicular UE. The response contains the required information.

Per ETSI GS MEC 030 [MEC030], the *PredictedQoS* data type represents the predicted QoS of a vehicular UE for a set of potential UE routes. Attributes of this data type are the time granularity of visiting a location, the granularity of the visited location (in meters) and the information relating to the potential routes of a vehicular UE. The latter attribute contains information relating to each specific potential route, i.e. the vehicular UE locations and the estimated arrival time at these locations. After the predictions are obtained via a Prediction Function (PF) instantiated at the network side, as part of the VIS response to the service consumer, the predicted RSRP and the predicted RSRQ are provided to the service consumer. The V2X application can then behave accordingly, given the signal quality conditions to be expected along the vehicle's journey.

5.4.1 Resource URI Structure of V2X API

Figure 5.11 below illustrates the resource tree of the V2X API.

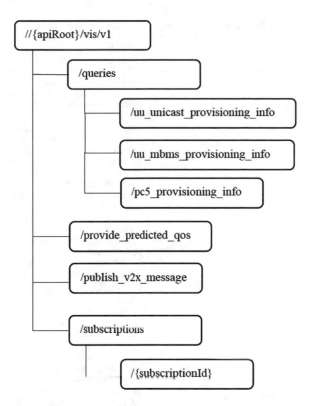

Fig. 5.11 Resource URI structure of the V2X API. *Source* [MEC030]—© European Telecommunications Standards Institute 2020. Further use, modification, copy and/or distribution are strictly prohibited

5.4.2   Services Offered to V2X API Consumers

The ETSI GS MEC 030 specifies the interactions between the Service Consumers (i.e. MEC applications or MEC platform) and the V2X information service through the V2X API.

Table 5.5 summarizes the possible interactions between a Service Consumer and the V2X API providing details about the information provided in each interaction.

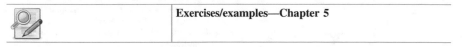

	Exercises/examples—Chapter 5

Exercise 5.1 *Goal of the exercise*: Understand how radio network information available from MEC server can support TCP traffic performance improvements (as a follow-up of the Exercise on Chap. 1).

Table 5.5 List of information and URIs provided to V2X API consumers

Procedure name	Description	URI Base URI: {apiRoot}/vis/v1
Request provisioning info for V2X over Uu unicast	The Service Consumer requests the provisioning information for V2X communication over Uu unicast for a particular location **{apiRoot}/vis/v1/queries/** **uu_unicast_provisioning_info** The additional URI query parameter *location_info* is represented by a comma separated list of locations, to identify a cell of a base station or a particular geographical area. There are thus two cases: 1. a list of cells (with E-UTRAN Cell Global Identifiers) …/uu_unicast_provisioning_info? location_info=ecgi,{String} 2. a list of geographical areas (with GPS coordinates) …/uu_unicast_provisioning_info? location_info=latitude,{String},longitude,{String}	
	Examples: 1. requesting provisioning information related to the two cells with E-UTRAN Cell Global Identifiers respectively 1357924680 and 1357924681 **{apiRoot}/vis/v1/queries/** **uu_unicast_provisioning_info?** **location_info=ecgi,** **1357924680,1357924681** 2. requesting provisioning information related to the two locations with GPS coordinates, respectively [000.000,000.000] and [001.000,001.000] **{apiRoot}/vis/v1/queries/** **uu_unicast_provisioning_info?** **location_info=latitude,000.000,001.000,** **longitude,000.000,001.000**	
Request provisioning info for V2X over Uu MBMS	The Service Consumer requests the provisioning information for V2X communication over Uu MBMS for a particular location	
	{apiRoot}/vis/v1/queries/ **mbms_unicast_provisioning_info** The additional URI query parameter *location_info* is represented by a comma separated list of locations, to identify a cell of a base station or a particular geographical area. There are thus two cases: 1. a list of cells (with E-UTRAN Cell Global Identifiers) …/mbms_unicast_provisioning_info? location_info=ecgi,{String} 2. a list of geographical areas (with GPS coordinates)	

(continued)

Table 5.5 (continued)

Procedure name	Description	URI Base URI: {apiRoot}/vis/v1
		…/mbms_unicast_provisioning_info? location_info=latitude,{String},longitude, {String}
		Examples: 1. requesting provisioning information related to the two cells with E-UTRAN Cell Global Identifiers, respectively 1357924680 and 1357924681 **{apiRoot}/vis/v1/queries/** **mbms_unicast_provisioning_info?** **location_info=ecgi,** **1357924680,1357924681** 2. requesting provisioning information related to the two locations with GPS coordinates, respectively [000.000,000.000] and [001.000,001.000] **{apiRoot}/vis/v1/queries/** **mbms_unicast_provisioning_info?** **location_info=latitude,000.000,001.000,** **longitude,000.000,001.000**
Request provisioning info for V2X over PC5	The Service Consumer requests the provisioning information for V2X communication over PC5 for a particular location	**{apiRoot}/vis/v1/queries/** **pc5_provisioning_info** The additional URI query parameter *location_info* is represented by a comma separated list of locations, to identify a cell of a base station or a particular geographical area. There are thus two cases: 1. a list of cells (with E-UTRAN Cell Global Identifiers) …/pc5_provisioning_info? location_info=ecgi,{String} 2. a list of geographical areas (with GPS coordinates) …/pc5_provisioning_info? location_info=latitude,{String},longitude,{String}
		Examples: 1. requesting provisioning information related to the two cells with E-UTRAN Cell Global Identifiers, respectively 1357924680 and 1357924681 **{apiRoot}/vis/v1/queries/** **pc5_provisioning_info?location_info=ecgi,** **1357924680,1357924681** 2. requesting provisioning information related to the two locations with GPS coordinates, respectively [000.000,000.000] and [001.000,001.000] **{apiRoot}/vis/v1/queries/** **pc5_provisioning_info?** **location_info=latitude,000.000,001.000,** **longitude,000.000,001.000**

(continued)

Table 5.5 (continued)

Procedure name	Description	URI Base URI: {apiRoot}/vis/v1
Request journey-specific predicted QoS information	The Service Consumer (e.g. a V2X application) requests to VIS to receive the predicted QoS correspondent to potential routes of a vehicular UE	
	{apiRoot}/vis/v1/queries/ provide_predicted_qos The request body contains the type PredictedQoS, which contains structured information relating to a specific route of a vehicular UE (route origin, route destination and potential intermediate waypoint locations). To each location provided is also associated an estimated arrival time at that location. The response body contains the type PredictedQoS, which contains information relating to QoS predictions, for each location of a specific route(at that particular estimated time of arrival). The predictions are expressed in terms of RSRP and RSRQ, as defined in [TS-36.214]	
Publish V2X messages	The V2X message publication is the procedure for a service consumer (e.g. a MEC application or a MEC platform) to publish a V2X message to VIS who will then notify the subscribed service consumers	
	{apiRoot}/vis/v1/publish_v2x_message The attributes of the V2xMsgPublication are *stdOrganization*, *msgType*, *msgEncodeFormat*, *msgContent* where the V2X message types of ETSI are used as specified by ETSI TC ITS [ITS102894]	
Operations on subscriptions	The Service Consumer can perform operations on a subscription to events that cause changes of certain specific V2X information that is available at VIS Possible operations are request subscriptions info, create (subscribe), update, delete (unsubscribe) and also receive notification on expiry of the subscription	
	{apiRoot}/vis/v1/subscriptions The following additional URI query parameter can be used to filter the request: •. *subscription_type*, query parameter to filter on a specific subscription type. Permitted values are: *prov_chg_uu_uni*, *prov_chg_uu_mbms*, *prov_chg_uu_pc5*, *v2x_msg*	

As explained in this chapter, the information related to radio network quality (specified in the MEC RNI API [MEC012]) is of paramount importance, because it can enable the creation of services with better performance at use level, e.g. TCP throughput guidance mechanisms that can improve upper layer adaption. Mobile Throughput Guidance was introduced in 2016 by IETF [IETF-MTG], and also further elaborated by GSMA, as a potential means to improve customer experience during mobile Internet sessions, by making explicit what range of bandwidth the mobile access link is likely to sustain. In fact, according to GSMA [GSMA-MTG], the video optimization gains achievable in live LTE networks consists typically in a significant reduction ($\sim 20\%$) in re-buffering time and an increase ($>5\%$) in video resolution.

In the first chapter we have already showed as the traffic behavior of TCP-based networks is described by the well-known Mathis formula [Mathis97], which is drawing a relation between the Quality of Experience (QoE) in IP networks with Network Latency (expressed by the Round Trip Time, *RTT*) and Packet Loss (P_{loss}):

$$Throughput \leq \frac{MSS}{RTT\sqrt{P_{loss}}} \tag{5.1}$$

This equation, since based on real-world measures in TCP-based networks, tells us also how to improve User Experience in practical cases. In fact, we can revert the formula and also calculate the maximum packet loss for a certain needed BW.

$$p \leq \left(\frac{MSS}{BW \cdot RTT}\right)^2 \tag{5.2}$$

This mean that, given a required Network Latency (expressed by the Round Trip Time, *RTT*) and bandwidth (BW), we can calculate (for a certain packet length) the needed maximum packet loss (P_{loss}).

Let's analyze now more in detail the TCP traffic behavior.

The Transmission Control Protocol (TCP) uses a sliding window mechanism to mitigate traffic problems between clients and servers trying to share segments of data that are too large or small, and therefore cannot be transmitted effectively. The receiver advertises how much data the sender should send before the sender must wait for a window update from the receiver. As a result, if this data can't be received correctly, there's a limit to how much data can be queued waiting for the receiving application. This TCP windowing mechanism (defined in 1974) allowed thus an efficient use of the memory of clients and servers, because only an affordable amount of data is generated at the client side to be transmitted to the server. Nevertheless, currently both bandwidth (BW) and devices memory (MSS) have improved of several magnitude, thus the windowing mechanisms risk not be efficient anymore, if not properly configured according to the effective capacity of the network to transfer the data.

Let's thus define Bandwidth-Delay Product (BDP), calculated as the BW multiplied by RTT, representing the optimal number of bits to send in order to fill the pipe. Consequently, a way to optimize TCP performance is to consider proper scaling of the TCP window size. This exercise consists in two basic steps:

1. Measure bandwidth and RTT to calculate BDP.
2. Compute BDP and use it as TCP window size for optimization.

Let's assume that a Client Application is running on a UE associated with a certain eNB, where a MEC Application instance is running (and peered with the Client App). Then, the two measures in Step 1 can be retrieved by the where the MEC application instance by querying the RNI API on Layer 2 measurements, as radio information coming from the eNB serving the UE (based on 3GPP specifications [TS-36.314]). For the purpose of the bandwidth and RTT measures, the following information can be used:

- *dl_gbr_delay_ue, ul_gbr_delay_ue,*
  dl_nongbr_delay_ue, ul_nongbr_delay_ue
- *dl_gbr_pdr_ue, ul_gbr_pdr_ue,*
  dl_nongbr_pdr_ue, ul_nongbr_pdr_ue
- *dl_gbr_throughpout_ue, ul_gbr_throughpout_ue,*
  dl_nongbr_throughpout_ue, ul_nongbr_throughpout_ue
- *dl_gbr_data_volume_ue, dl_gbr_data_volume_ue,*
  dl_nongbr_data_volume_ue, ul_nongbr_data_volume_ue

Let's assume that the UE is using a non-GBR traffic flow. Then, an estimate of the bandwidth can be given by *dl_nongbr_throughpout_ue*, which is a measure of the DL throughput experienced at the radio link (say, 100 Mbps).

On the other hand, the RTT can be calculated as the sum of (*dl_nongbr_delay_ue + ul_nongbr_delay_ue + rtt_radio_to_mec*), where the latter addendum is an estimate of the 2-ways delay between the eNB and the MEC App. Assuming, respectively, 15 ms, 15 ms and 10 ms for these three components, we have RTT = 40 ms.

Then, BDP can be calculated as the product BW x RTT, so that the actual TCP window size can be used to optimize the transmission:

$$(BW * RTT)/8bits = window \quad size$$

That way, by plugging in the example numbers, we have

$$(100 \ Mbps * 0.040 \ s) \ / \ 8 \ bits/byte = 0.5 \ MB$$

As a proof, when considering a standard of the TCP window size (65,535 bytes), we can only reach a smaller throughput (i.e. only about 13 Mbps):

$$(65535 \text{ bytes} * 8 \text{ bits/byte}) = \text{bandwidth} * 0.040 \text{ s}$$
$$\text{bandwidth} = (65535 \text{ bytes} * 8 \text{ bits/byte}) / 0.040 \text{ s}$$
$$\text{bandwidth} = 524280 \text{ bits} / 0.040 \text{ second} = 13107000 \text{ bits} / \text{s}$$

To state it another way, these values (obtained without TCP window size scaling) result in throughput that's a bit more than 13 Mbits per second, which is a small fraction of network's capability (100 Mpbs).

To resolve the performance limitations imposed by the original design of TCP window size, extensions to the TCP protocol were introduced that allow the window size to be scaled[1] to much larger values (about 1 GB).

A simple exemplary way to see if the feature is enabled on Linux-based systems is to use the following command:

```
sudo sysctl net.ipv4.tcp_window_scaling
```

where, if the feature is disabled, it can be enabled with the following command:

```
sudo sysctl -w net.ipv4.tcp_window_scaling=1
```

As a preliminary note, before changing the TCP window side, it is always recommended to save the current settings, in case needed to roll back any changes (of course, by making sure previously to have root privileges):

```
sudo sysctl -a | grep mem
```

then, as an example of possible settings for enabling TCP communication with, e.g. up to 8 MB window size, the reader can cut and paste the following into a Linux shell with root privileges:

---

[1]This feature is outlined in RFC 1323 as "TCP window scale option". For more information, see also https://tools.ietf.org/html/rfc1323#page-8.

```
sudo sysctl -w net.core.rmem_max=8388608
sudo sysctl -w net.core.wmem_max=8388608
sudo sysctl -w net.core.rmem_default=65536
sudo sysctl -w net.core.wmem_default=65536
sudo sysctl -w net.ipv4.tcp_rmem='4096 87380 8388608'
sudo sysctl -w net.ipv4.tcp_wmem='4096 65536 8388608'
sudo sysctl -w net.ipv4.tcp_mem='8388608 8388608 8388608'
sudo sysctl -w net.ipv4.route.flush=1
```

Where the above Linux tunables *net.core.rmem_max* and *net.core.wmem_max* affect the maximum TCP window size for applications that attempt to control the TCP window size directly, by limiting the applications' request to no more than those values. Similarly, the other two Linux tunables *net.ipv4.tcp_rmem* and *net. ipv4.tcp_wmem* affect the TCP window size for applications that let Linux auto-tuning do the work.

As simple way to test these changes is to run the following commands (by previously verifying to have made TCP window size changes to both local computer and remote computer, by setting the tunables on both machines):

```
dd if=/dev/urandom of=sample.txt bs=1M count=1024 iflag=fullblock
scp sample.txt your_username@remotehost.com:/some/remote/directory
```

Where the first command creates a 1 GB sample.txt file with random data, while the second command copies that file from the local machine to a remote machine (note: the scp command output on the console, which displays also the bandwidth in Kbps; thus, the reader should be able to see sizable difference in the results from before and after the TCP window size changes).

Exercise 5.2 *Goal of the exercise*: practice with MEC Location API by using a location service simulator.

For this exercise we use a location service simulator developed by LINKS Foundation, that is described in Annex C. The simulator is not a standard component, but rather a tool implemented to emulate the Location Service typically available from a mobile network.

In the example being discussed, it is a RESTful web service which provides a simple set of APIs to simulate the user movement over a certain path. The path is a GPX file, a GPS data format that contains coordinates, speed, heading and other valuable information as available by a GPS receiver (for more details see [GPX]). The simulator first loads the GPX file, selected by the user through a GUI (Graphical User Interface) to start a simulation; then, the Station ID is included; recursively, more UEs can be added to the simulation, assigning to each of them a distinct GPX file and a distinct id.

When the user sends the simulation request, the simulator reads the GPX file and starts transmitting data that are made available through the Location API web server.

Which APIs are implemented

Below the list of APIs implemented in the simulator:

- **POST**/simulate/locationApi/start. The request triggers the simulator to start transmitting data toward the location API. It must include a JSON object with all the simulation details, needed for the simulation set up, such as stationId and Path. For instance, a JSON object could be the following:

```
{
    "simulation": [{
        "stationId": "1",
        "path": "path1.gpx"
    }, {
        "stationId": "2",
        "path": "path2.gpx"
    }]
}
```

- **DELETE**/simulate/locationApi/stop. HTTP request needed for stopping a simulation earlier.
- **GET**/tracks. It allows exploring which tracks in the simulator are stored.
- **POST**/upload. HTTP request through which a user can upload its own GPX file.

How to work with the SW: step by step via CLI.

Location APIs can be used to retrieve information by interacting as a RESTful web service, therefore via GET, POST, PUT and DELETE HTTP requests. According to the ETSI standard, there are multiple responses based on the creation of a given HTTP request.

First, users should know which is the IP address, or the demanded domain, to contact and the port. In this case, the webserver will be reachable as localhost or from the local address at the port 8080. Therefore, to know if the webserver is up and reachable, just open a terminal (or a browser if preferred) and type [GPX]:

curl -v http://localhost:8080/

then, if 200 HTTP is received, the webserver is running. By typing the address on a web browser, a simple index.html file will show the same response.

Get a list of UEs

Now, it is possible to explore and retrieve information by following the ETSI standard requests.

For instance, to retrieve general information about all UEs for whom location details are available, the following command can be typed:

curl -v http://localhost:8080/location/v2/users

and a list of all UEs will be shown. It has to be noticed that there is no information about location in this response, as ETSI standard requires. Indeed, to know detailed information, a filter can be added considering the information obtained by this first HTTP response.

Get detailed information

To get detailed information of a single user, the GET request can be executed as the following:

curl -v http://localhost/location/v2/users?address=<UE_address>

and latitude, longitude and other information will be shown. Besides, the timestamp shows the last update; surely, performing again the same request, all the dynamic values will change.

The same result can be obtained for several UEs, including in the URL other filters, such as zone or access point information:

curl -v http://localhost/location/v2/users?zoneId=zone01

curl -v http://localhost/location/v2/users?accessPoint=01000001

receiving back a subgroup of the previous list of UEs associated with the given filter.

User tracking and subscription

Location API allows receiving periodical information about a single or multiple users passively by submitting a subscription to them. If there is the need to track users, a subscription can be performed creating a JSON request with all the requested information. For instance, a JSON request (as described in [MEC013] Sect. 7.3.8) could be:

```
// subscription.json
{
  "periodicNotificationSubscription": {
    "address": "acr:10.0.0.1",
    "callbackReference": {
      "callbackData": "Call back information",
      "notifyURL": "http://localhost:8030/"
    },
    "clientCorrelator": "my-hostname",
    "frequency": "10",
    "requestedAccuracy": "10"
  }
}
```

For sending this information with curl, the command is:

curl -v http://localhost:8080/location/v2/subscriptions/periodic-d@subscription.
json.

where the flag "-d @filename" allows to include the JSON file in the request. In this
way, the curl command will automatically perform a POST and the users will
receive back the resulting resource URL containing subscriptionId.

After that, whenever Location service will receive fresh new information about
UEs, it will check the subscription and then submit data to those subscribers.

The subscription can be deleted simply indicating the subscriptionId as
following:

curl -v http://localhost:8080/location/v2/subscription/periodic/{subscriptionId}.

where the subscriptionId is the one contained in the Location API JSON response at
the subscription step.

**How to build a simple application to visualize the movement of some vehicles
on a map**.

Now, a simple JavaScript application with the role of Location service consumer will
be shown. It will be able to get the list of UEs and then it continuously retrieves update
information of one of them. Finally, the UE movement will be displayed on a map.

Requirements:

- Python 3 and pip;
- HTTP Python module;
- Node, npm;
- Internet connection for external libraries;
- Web browser;
- Location API web server running locally.

Environment setup

The application has been tested on a Linux system, and package installation on this step refers to Ubuntu 16.04 and 18.04 LTS.

(1) Python 3, pip and HTTP module installation:
   a. sudo apt-get install Python3.
   b. sudo apt-get install Python3-pip.
   c. pip3 install http.

(2) Node and npm installation:
   a. sudo apt install nodejs.
   b. sudo apt install npm.

Application development

The application acts as an HTTP client and it is developed as a JavaScript application, which makes easier performing HTTP requests. From now on, the JavaScript file will be named as "MecApp.js".

First, a function could allow performing HTTP requests easier to the location API. The code below shows an example:

```javascript
var HttpClient = function () {
    this.get = function (aUrl, aCallback) {
        var anHttpRequest = new XMLHttpRequest();
        anHttpRequest.open("GET", aUrl, true);
        anHttpRequest.onreadystatechange = function () {
            if (anHttpRequest.readyState == 4 && anHttpRequest.status ==
200 && anHttpRequest.withCredentials != undefined)
                aCallback(anHttpRequest.responseText);
        }
        anHttpRequest.send(null);
    }}
```

```
var client = new HttpClient();
// send the first request to get a list of users
cli-
ent.get('https://127.0.0.1:8080/location/v2/users', function (response) {
    var obj = JSON.parse(response);
    //select arbitrary UEs for getting detailed information about them
    var clientAddress = obj['userList']['user'][0]['address'];
    var clientAddress2 = obj['userList']['user'][1]['address'];
    var clientAddress3 = obj['userList']['user'][2]['address'];
    // create an array of UEs
    var arr = [clientAddress];
    (function loop(i) {setTimeout(() => {
    arr.forEach(e => {
    cli-
ent.get('https://127.0.0.1:8080/location/v2/users?address=acr:' + e, functi
on (response) {
        var objUE = JSON.parse(response);
        var lat = objUE['latitude'];
        var log = objUE['longitude'];
        //display latitude and longitude in the browser console
        console.log('Longitude: ' + log );
        console.log('Latitude: ' + lat );
    });
    });
    if(--i) loop(i);
    }, 600)
    })(150);
});
```

Then, since the objective is to track and get continuously new information about a UE, the script could perform a request as in the following snippet of code, to get information about arbitrary UEs:

In this way is possible to send multiple requests, updating the request after an arbitrary waiting time, and receiving JSON responses. From these JSON responses, the user can retrieve all the needed information.

Display data on Map

First of all, a simple HTML file is needed, named "index.html", then both JavaScript and CSS Mapbox files have to be included in the <head> of the HTML file, like below:

```
    <script src="https://api.mapbox.com/mapbox-gl-js/v1.8.1/mapbox-
gl.js"></script>
    <link href="https://api.mapbox.com/mapbox-gl-js/v1.8.1/mapbox-
gl.css" rel="stylesheet" />
    <script src="MecApp.js"></script>
```

and then the following code in the <body> for including tag useful for Mapbox to anchor the map.

```
<body>
    <div id="mapview">
    <!-- MAPBOX INTEGRATION -->
    <div id='mapid'></div>
    </div>
</body>
```

Coming back to the "MecApp.js", the following code allows to create a Mapbox map:

```
mapboxgl.accessToken = "<your access token code>";
var map = new mapboxgl.Map({
  container: 'mapid',
  style: 'mapbox://styles/mapbox/streets-v11,
  zoom: 4,
  center: [12.5736, 41.2925], //Italy
});
```

Mapbox requires an access token[2] [MPBX-Tk] to use its service. Then, the function for creating the map requires a JSON object to set up several characteristics of a map, such as the HTML container associated with the HTML id; then it can include style, zoom, initial coordinates and many more [MPBX-Par].

The last step is to instantiate a marker for each UE and animate those markers while the Location API is sending information about UEs. It can be developed as follows:

---

[2]Requests have been run through a curl command (https://curl.haxx.se/) in a Linux system, but it is not a requirement. Any other tool can be used to test and perform the requests.

```
var marker = new mapboxgl.Marker();
var marker2 = new mapboxgl.Marker();
var marker3 = new mapboxgl.Marker();
var markers = [marker, marker2, marker3];


function animateMarker(map, marker, long, lat) {
        marker.setLngLat([long, lat]);
        marker.addTo(map);
        map.easeTo({
            center: [long, lat];
            easing: easing,
            pitch: 0,
            zoom: 16,
        });
}
```

Finally, this function can be added to previous GET function:

```
(function loop(i) {setTimeout(() => {
arr.forEach(e => {
cli-
ent.get('https://127.0.0.1:8080/location/v2/users?address=acr:' + e, functi
on (response) {
    var objUE = JSON.parse(response);
    var lat = objUE['latitude'];
    var log = objUE['longitude'];
    //trigger the UEs movements on Mapbox map
    animateMarker(map, marker, log, lat);
  });
  });
  if(--i) loop(i);
  }, 600)
  })(150);
});
```

Start button

To perform and execute the request, a simple button can be added to the index.html file and then called from the MecApp.js as below:

– index.html

```html
div id="demo">
    <button id="btn" type="button">Start requests</button>
</div>
```

– MecApp.js

```javascript
document.getElementById("btn").addEventListener('click', function () {
    // put here the GET request
}
```

Run the application

At this point, there are two developed files: index.html and MecApp.js. Now the application can be tested with the following step, and checking if it works properly:

1. Open a Terminal and locate in the same folder which contains the above-developed files.
2. Run the following command, it will start a web server on the preferred port (i.e. 8030)

$ Python3 -m http.server 8030.

3. Finally, open a browser and locate to http://localhost:8030.

A map and a button will be shown as in the Fig. 5.12. To start requesting data, just push the button and as many markers have been created in the MecApp.js will go through the map, continuously updating its position according to the received data [LS-video].

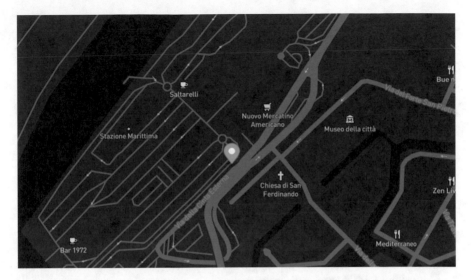

Fig. 5.12 Example of a MEC application showing UEs positions on a map

Exercise 5.3 *Goal of the exercise*: Understand how to configure MEC MTS Multi-Access Traffic Steering (MTS) service for meeting individual application QoS requirements, e.g. reliability, latency, data rate, etc.

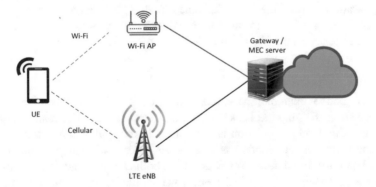

Consider the figure above, showing a typical indoor environment, e.g. enterprise, retail, hospital, etc. where both Wi-Fi and Cellular networks are available. In addition, a MEC server is deployed at the gateway to enable MTS service. Table 5.6 shows the connection characteristics between UE and MEC server via Wi-Fi and Cellular, respectively, where packet loss rate, latency and DL/data rate are referred to IP packet flows.

The UE has three applications running with QoS requirements as show in Table 5.7.

Table 5.6 Example of Wi-Fi and Cellular connection characteristics

	Packet loss rate	Latency (ms)	DL data rate (Mbps)	UL data rate (Mbps)
Wi-Fi	<0.01 (10^{-2})	<10	40	20
Cellular	<0.001 (10^{-3})	<5	20	10

Table 5.7 QoS requirements of applications

	Packet loss rate	Latency (ms)	DL data rate (Mbps)	UL data rate (Mbps)
App #A	<0.00001 (10^{-5})	<10	5	5
App #B	<0.001 (10^{-3})	<5	5	5
App #C	<0.01 (10^{-2})	<20	40	10

Then, the **questions** are:

1. *Can all three applications be supported with their QoS requirements?*
2. *How should MTS service be configured?*

Let's look at App #A first. Its reliability requirement of < 0.00001 is the most stringent among the three, and can't be met with Wi-Fi or LTE alone. Therefore, we should configure the app to use the MTS redundancy mode such that its packet will be duplicated over both connections. Let's use p1 and p2 to indicate the packet loss rate of Wi-Fi and LTE, respectively, and use p as the packet loss rate perceived by the application. With the MTS redundancy mode (where recombination is done at IP, e.g. if using MAMS to support MTS service), a packet is lost only if its duplicate is lost on both Wi-Fi and LTE, and we have

$$p = p1 \times p2$$

In the above example, p1 < 0.001 and p2 < 0.01, and therefore we have p < 0.00001 (satisfying the Packet Loss Rate requirement). The reader should also notice that this kind of operation is transparent to the application or transport layer, since the packet recombination is done at IP layer (and then p is the equivalent packet loss rate). In addition, Wi-Fi and LTE both can meet the data rate as well as latency requirements of App #A. Hence, we should configure App #A to use the MTS redundancy mode, and the available data rate on Wi-Fi and LTE are reduced as shown in Table 5.8.

Now, let's look at App #B. Its latency requirement of <5 ms is the most stringent among the three, and can only be met by the LTE link. Therefore, we should

Table 5.8 Available data rate of Wi-Fi and LTE after configuring App #A

	DL data rate (Mbps)	UL data rate (Mbps)
Wi-Fi	40 − 5 = 35	20 − 5 = 15
Cellular	20 − 5 = 15	10 − 5 = 5

Table 5.9 Available data rate of Wi-Fi and LTE after configuring App #B

	DL data rate (Mbps)	UL data rate (Mbps)
Wi-Fi	35	15
Cellular	15 − 5 = 10	5 − 5 = 0

Table 5.10 Summary of MTS configurations

	App #A	App #B	App #C
MTS mode	Redundancy	Low latency	High throughput

configure App #B to use the MTS low latency mode. That way, the available data rate on Wi-Fi and LTE are reduced as shown in Table 5.9.

Last, we will consider App #C. Its latency and reliability requirements can be met by both Wi-Fi and LTE. But it requires DL data rate of 40Mbps, which can't be met by Wi-Fi or LTE alone. As a result, the MTS high throughput mode should be used for App #C to combine both Wi-Fi and Cellular and achieve the required data rate (35 + 10 > 40, 15 + 0 > 10).

In summary, Table 5.10 shows how MTS should be configured for the three applications.

It is now left to the willingness of the reader the possibility to repeat the exercise, by using more realistic performance numbers from Wi-Fi and Cellular base stations, as per the following Table 5.11.

Exercise 5.4 *Goal of the exercise*: Understand the usefulness of the MEC V2X API in obtaining requests for predicted QoS across a planned vehicle route and dispatching the QoS predictions to the VIS service consumer aiming to influence decisions, such as, e.g. enabling/disabling autonomous driving features, etc.

Issuing the journey-specific predicted QoS request.

As a first step, the VIS service consumer (i.e. a V2X application, accompanied by an instantiated MEC application) communicates the journey information to the PF via the VIS service API—these points are the journey's origin point A, destination point B and K intermediate waypoints. An example of the structure describing the vehicle's route of interest is shown in Table 5.12

Table 5.11 Wi-Fi and Cellular connection characteristics (to be filled by the reader)

	Packet loss rate	Latency	DL data rate (Mbps)	UL data rate (Mbps)
Wi-Fi	−	−	−	−
Cellular	−	−	−	−

Table 5.12 Structured information (*routeInfo*) representing a specific vehicular route—the predicted RSRP/ RSRQ values will be provided as part of the response to the VIS service consumer

Potential route #n	Location	Time (UTC)	RSRP	RSRQ
Origin A	(x_A, y_A)	12:00	–	–
Intermediate waypoint 1	(x_1, y_1)	12:10	–	–
Intermediate waypoint 2	(x_2, y_2)	12:30	–	–
...	...		–	–
Intermediate waypoint K	(x_K, y_K)	12:45	–	–
Destination B	(x_B, y_B)	13:10	–	–

Obtaining journey-specific QoS predictions and structuring the response message to the VIS consumer.

Once the PF acquires the request for predicted QoS information, it will execute a prediction algorithm (outside the scope of this chapter). The output of the prediction algorithm is predicted RSRP and RSRQ values for all communicated journey waypoints and times of interest. These values will populate the (until then) empty attributes of the *routeInfo* structure and, consequently, the message body of the response message destined for the VIS service consumer (i.e. the V2X application interested in these predictions). After constructing the message body, the VIS will dispatch the query response to the VIS service consumer. The end consumer (i.e. the requesting V2X application) will then exploit the obtained predictions to take actions along the journey, e.g. starting/ postponing a Software/ Firmware (SW/FW) Over-the-Air (SOTA/ FOTA) update download, enable/ disable autonomous driving features, etc. The procedure is explained in Fig. 5.13.

Quiz—Part 2
(Chapters 3, 4 and 5)

(1) **According to the definition of Multi-access Edge Computing, the "core business" of ETSI MEC is**:
   a. Providing cloud computing and IT service environment to operators at the edge of the network.
   b. Providing cloud computing to application developers at the edge of the network.
   c. Providing to application developers cloud computing and IT service environment at the edge of the network.
   d. All the previous are correct.

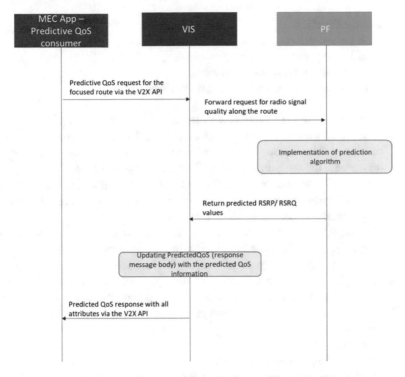

Fig. 5.13 Sequence chart describing the procedure of a MEC application (tailored to a V2X application) requesting and obtaining predictive QoS information along a vehicle route of interest

(2) **The MEC architecture, as described in the ETSI GS MEC-003 specification**:
   a. Is fully compatible with the ETSI NFV framework.
   b. Isn't based on a virtualization infrastructure, and cannot be implemented in a NFV environment.
   c. Is based on a virtualization infrastructure, and can be implemented in a NFV environment.
   d. Is built upon and consistent with NFV principles.
   e. None of the above.

(3) **The MEC reference point Mp3 is defined**:
   a. Between MEC app and MEC platform.
   b. Between MEO and ME platform manager.
   c. Between different MEC platforms.
   d. Between MEC platform and the Data Plane.
   e. None of the above.

(4) **MEC applications**:
    a. Can be consumers and/or producers of services
    b. Can be only consumers of services.
    c. Can be only producers of services.
    d. Cannot consume nor produce any services.

(5) **According to the ETSI MEC architecture**:
    a. The MEC host is an entity that contains a MEC platform and a virtual-ization infrastructure which provides compute, storage and network resources, for the purpose of running MEC applications.
    b. The MEC platform is the collection of essential functionality required to run MEC hosts on a particular virtualization infrastructure and enable them to provide and consume MEC services.
    c. The MEC application is an entity that contains a MEC platform and a virtualization infrastructure which provides compute, storage, and network resources, for the purpose of running MEC hosts.
    d. The MEC host is the collection of multiple MEC platforms, which are all running on the same virtualization infrastructure which provides compute, storage, and network resources, for the purpose of running MEC applications.

(6) **A MEC system is composed by**:
    a. A MEC host level with a single MEC host and multiple MEC platforms; a MEC system level, with a MEC orchestrator and a single MEC platform manager for the whole system.
    b. A MEC host level with one or more MEC hosts and MEC platforms; a MEC system level, with a MEC orchestrator and one or more MEC platform managers.
    c. A MEC system level with a single MEC host and multiple MEC platforms; a MEC host level, with a MEC orchestrator and one or more MEC platform managers.
    d. A MEC host level with one or more MEC hosts and MEC platforms; a MEC system level, with a MEC orchestrator and a single MEC platform manager for the whole system.

(7) **Mp1 reference point is**:
    a. Mandatory for all MEC applications (which can run on the MEC host without connection to the MEC platform).
    b. Not mandatory for a MEC App (which can run on the MEC host without connection to the MEC platform).
    c. Mandatory for all MEC applications (but without Mp1, the MEC app cannot run on the MEC host because there is no connection to the MEC platform).

d. Not mandatory for a MEC App (but without Mp1, the MEC app cannot run on the MEC host because there is no connection to the MEC platform).
e. Not even a reference point of ETSI MEC.

(8) **Mp2 reference point is**:
a. Connecting together all MEC applications with a single MEC platform.
b. Connecting together MEC applications running in different MEC hosts.
c. Connecting together the MEC platform and the Data Plane.
d. Connecting together the MEC application and the Data Plane.
e. Connecting together Data Planes running in different MEC hosts.
f. Not even a reference point of ETSI MEC.

(9) **Mp3 reference point is**:
a. Only defined for audio files and not in ETSI MEC.
b. Connecting together the MEC platform and the Data Plane.
c. Connecting together the MEC application and the Data Plane.
d. Connecting together Data Planes running in different MEC hosts.
e. Connecting together MEC platforms running in different MEC hosts.

(10) **Mx2 reference point is**:
f. Connecting together a MEC App and the MEC system via a proxy called UALCMP.
g. Connecting together a Device App and the MEC system via a proxy called UALCMP.
h. Connecting together a MEC App and the MEC system via a proxy called CFS.
i. Connecting together a Device App and the MEC system via a proxy called CFS.

References

[MEC012]   ETSI GS MEC 012 V2.1.1 (2019–12) - Multi-access Edge Computing (MEC); Radio Network Information API. https://www.etsi.org/deliver/etsi_gs/MEC/ 001_099/012/02.01.01_60/gs_MEC012v020101p.pdf
[RAVEN18]   Sabella D, Nikaein N, Huang A, Xhembulla J, Malnati G, Scarpina S (2018) A Hierarchical MEC architecture: experimenting the RAVEN use-case. In: 2018 IEEE 87th vehicular technology conference (VTC Spring), Porto, pp 1–5. https://doi.org/10.1109/VTCSpring.2018.8417826
[TS-36.314]   ETSI TS 136 314: Evolved Universal Terrestrial Radio Access (E-UTRA); layer 2 - measurements (3GPP TS 36.314)
[TS-23.401]   ETSI TS 123 401: LTE; General Packet Radio Service (GPRS) enhancements for Evolved Universal Terrestrial Radio Access Network (E-UTRAN) access (3GPP TS 23.401)

[MEC013]   ETSI GS MEC 013 V2.1.1 (2019–09) - Multi-access Edge Computing (MEC);
                Location API. https://www.etsi.org/deliver/etsi_gs/MEC/001_099/013/02.01.
                01_60/gs_MEC013v020101p.pdf

[SCF084]   SCF 084.07.01: Small cell zone services - RESTful bindings. https://scf.io/en/
                documents/084_-_Small_Cell_Zone_services_RESTful_Bindings.php.
                Accessed 7 Feb 2014

[SCF152]   SCF 152.07.01: Small cell services API. https://scf.io/en/documents/152_-_
                Small_cell_services_API.php. Accessed 7 Dec 2015

[OMA-ZnPr]   OMA-TS-REST-NetAPI-ZonalPresence-V1-0-20160308-C: RESTful
                Network API for Zonal Presence.

[OMA-TrLo]   OMA-TS-REST-NetAPI-TerminalLocation-V1-0-1-20151029-A: RESTful
                Network API for Terminal Location.

[MEC015]   ETSI GS MEC 015 V2.1.1 (2020-06) Multi-Access Edge Computing (MEC);
                Traffic Management APIs. https://www.etsi.org/deliver/etsi_gs/MEC/001_099/
                015/02.01.01_60/gs_MEC015v020101p.pdf

[RFC8743]   Multiple Access Management Services. https://www.rfc-editor.org/rfc/rfc8743.
                txt

[MEC030]   ETSI GS MEC 030 V2.1.1 (2020-04) - Multi-access Edge Computing (MEC);
                V2X Information Service API. https://www.etsi.org/deliver/etsi_gs/MEC/001_
                099/030/02.01.01_60/gs_MEC030v020101p.pdf

[ITS102894]   ETSI TS 102 894-2 V1.3.1 (2018-08) - Intelligent Transport Systems (ITS);
                Users and applications requirements; Part 2: Applications and facilities layer
                common data dictionary. https://www.etsi.org/deliver/etsi_ts/102800_102899/
                10289402/01.03.01_60/ts_10289402v010301p.pdf

[TS-36.214]   ETSI TS 136 214: LTE; Evolved Universal Terrestrial Radio Access
                (E-UTRA); Physical layer; Measurements (3GPP TS 36 214)

[GPX]   GPX format. https://www.topografix.com/gpx.asp

[MPBX-Tk]   Mapbox   token.   https://docs.mapbox.com/help/how-mapbox-works/access-
                tokens/

[17]   [MPBX-Par]Mapbox map parameters. https://docs.mapbox.com/mapbox-gl-js/
                api/map/

[LS-video]   Video that shows the main features of the Location Service in practice. https://
                www.youtube.com/watch?v=L93zlvVevzw

[GSMA-MTG]   GSMA white paper, Mobile Throughput Guidance, Version 1.0, 08 March
                2017.   https://www.gsma.com/newsroom/wp-content/uploads/WWG.17-v1.0.
                pdf

[IETF-MTG]   Sprecher N et al (2016) Requirements and reference architecture for Mobile
                Throughput Guidance, March 2016. https://tools.ietf.org/html/draft-sprecher-
                mobile-tg-exposure-req-arch-02#page-5

Part III
Edge Computing Deployments

Chapter 6
MEC in Virtualized Environments

The present chapter will describe some principles of virtualization, with particular reference to the work done by ETSI in the ISG NFV (Network Function Virtualization), as the basis for the MEC deployment in virtualized environments. Finally, a description of the recent concept of cloud native will conclude the chapter.

6.1 Principles of Virtualization

In the first chapter we have already described the various cloud computing models, and their association with the application stack (see figure below). In this section we will concentrate on the technologies used for the actual virtualization layer (i.e. moving from the physical hardware to the actual virtual hardware used at higher levels) (Fig. 6.1).

The consolidated trend of progressive virtualization of hardware resources is somehow a natural consequence of the well-known Moore's Law (the guideline developed by Gordon Moore, founder of Intel, predicting a doubling in the number of transistors density in a CPU every two years). In fact, virtualization, as the ability of a machine to run multiple virtualized servers or client operating systems simultaneously, is a consequence of this incredible availability of computational capacity, which is obviously a sort of waste of resources (on one hand), but it gives also an incredibly high number of advantages. First of all, the possibility to decouple the actual software from the underlying hardware in the hosting machine is providing a degree of flexibility in terms of possibility to develop and test applications and software independently from the specific HW version and characteristics. Moreover, costs associated to hardware upgrades can be decoupled from the SW using these resources. Furthermore, virtualization is providing huge advantages in terms of multi-tenancy, where multiple independent instances of one or multiple applications operate in a shared environment (note: the instances, or

© Springer Nature Switzerland AG 2021   159
D. Sabella, *Multi-access Edge Computing: Software Development
at the Network Edge*, Textbooks in Telecommunication Engineering,
https://doi.org/10.1007/978-3-030-79618-1_6

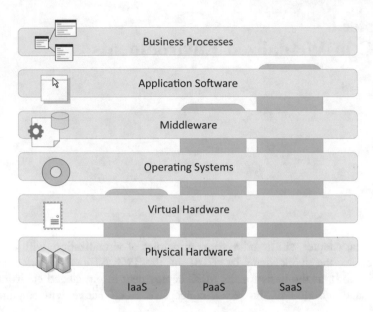

Fig. 6.1 Association between application stack and cloud service models

tenants, are logically isolated, but physically integrated). Finally, the adoption of virtualization enables a wide variety of ecosystems and encourages openness, as it opens the virtual appliance market to pure software entrants and small players, encouraging innovation and the introduction of new services.

But how this virtualization is realized? Essentially, a middleware between hardware resources and the OS is needed, and this is called *hypervisor*. This is a software, firmware or hardware that creates and runs VMs (virtual machines), and then virtualizes the underlying server, in order to permit applications to run on the OS independently on the hardware. Many VMs are available as today, e.g. Linux-based, Windows-based VMs, etc., each of them having its own binaries/ libraries and application(s) that it serves, and with different characteristics in terms of size (the VM may be many gigabytes large).

When talking about virtualization, the key enabling technology component is then the hypervisor. There are two types of hypervisors (as shown in Fig. 6.2):

- **Type 1**: hypervisors running directly on the system hardware, called also "native" or "bare metal" hypervisors. Examples of Type 1 hypervisors are VMware ESX/ESXi, Microsoft Hyper-V, KVM and Citrix Hypervisor.
- **Type 2**: hypervisors running on a host OS that provides virtualization services, such as I/O device support and memory management. They are called also "hosted" or "embedded" hypervisors. Examples of hosted hypervisors are VirtualBox, QEMU and VMware Workstation.

Fig. 6.2 Type 1 ("bare metal") and Type 2 ("hosted") hypervisors

As the reader can easily imagine, bare metal hypervisors are typically faster than Type 2 hypervisors, because they have direct access to the underlying hardware and don't need to go through the OS layer. On the other hand, hosted hypervisors are significantly easier to set up and get running (sometimes also equipped with a more user-friendly operating system) and can be more suitable for testing and development purposes, as they fully abstract guest operating systems from the host operating system. Conversely, Type 1 hypervisors tend to be more secure, because, without an operating system on the host, less attack surface is available for malicious intruders.

As we anticipated, a *virtual machine* (VM) is containing an operating system that shares the physical resources of one server. It is a guest on the host's hardware, which is why it is also called a guest machine, and it is completely isolated from the host operating system (see Fig. 6.2).

Terminology:
Virtual machine monitor (VMM): another name for the hypervisor
Host machine: the hardware on which the VM is installed
Guest machine: another name for the VM

On the other hand, recently the introduction of *containers* gained interest in the industry, as a lightweight way to run isolated software instances on a single server/ host OS. In fact, containers sit on top of a physical server and its host OS, e.g. Linux or Windows (see Fig. 6.2). Each container shares the host OS kernel in order to run all the individual apps within the container. The only elements required by each container are bins, libraries and other runtime components (sometimes, even some binaries and libraries are shared): all these characteristics make them exceptionally "light" (containers are only megabytes in size and take just seconds to start, versus minutes for a VM), and faster to launch, with respect to virtual machines (here we are talking about seconds vs minutes). On the other hand, containers don't offer the same security and stability of VMs. Since containers share the host's kernel, they cannot be as isolated as a virtual machine. Consequently,

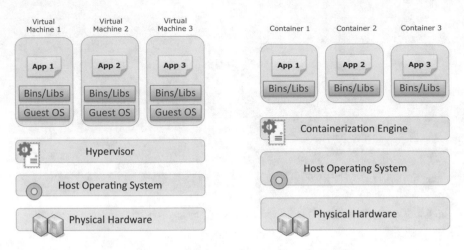

Fig. 6.3 Difference between virtual machines (VMs) and containers

containers are isolated at process level, and in principle a container can affect others by compromising the stability of the kernel (Fig. 6.3).

Conversely, Virtual machines may take up a lot of system resources of the host machine (not only in terms of disk space but also RAM and CPU cycles), since from a practical perspective a complete isolation of a single app on a virtual server means running a copy of an operating system as well as a virtual copy of all the hardware required for the system to run. Furthermore, the process of relocating an app running on a VM can also be more complicated as it is always attached to the operating system. In summary, the choice of using virtual machines or containers depends on the actual needs. VMs are better to manage a diverse set of OSs and multiple apps on a single server and provide better isolation and security. Containers, on the other side, and more efficient in terms of resources utilization, as lightweight solution for faster and more flexible way to develop applications. Nevertheless, we should also say that VMs and containers are not necessarily to be seen as rivals. Rather, both can be used, e.g. in mixed deployments, to balance the workload between the two.

At this point, the reader might be confused between containers and hypervisors (especially Type 1 hypervisors, i.e. those running without the interposition of OS). But the most important difference between hypervisors and containers is the way they boot up and consume resources. While hypervisors work directly on the hardware (bare metal hypervisors) or on top of the operating system (Type 2 hypervisors), containers, on the other hand, work on the host kernel itself. Taking as an example Docker, as a popular open source project used for automation for fast deployment of applications running on containers, a Docker engine is required to run the Docker containers, similarly as VMs run inside the hypervisor. The reader more interested in learning about Docker can have a look at the related website (https://docs.docker.com/get-started/overview/).

6.2   Network Functions Virtualization (NFV)

We have talked about virtualization as a general concept, thus involving all hardware and software resources. When it comes more specifically to network components, mobile operators are the main stakeholders driving the process of progressive virtualization of their network infrastructure. This process started in 2013 with the creation of the ETSI NFV (Network Functions Virtualization) ISG (Industry Specification Group).

NFV is about virtualizing network functions previously implemented as proprietary hardware appliances Fig. 6.4; hence, an important topic to address is how to realize virtualization from a network function provider's perspective. The above figure shows how ETSI ISG NFV intends to progressively transform Network Functions based on specialized hardware into virtual appliances, where Network Functions are software-based and multiple functionalities are running over same hardware. This approach of decoupling software implementations of network elements (called VNF—Virtualized Network Functions) from standard servers, storage and network components, permits operators to reduce operational costs (both equipment and power consumption), increase scalability and flexibility of their deployments, simplify management and increase openness by decoupling software from hardware upgrades, with reduced time-to-market to deploy new network services and also improved return on investment from new services.

As a result of a first phase of work (called pre-standardization), at the end of 2014 the ISG NFV delivered the Release 1 that defined high-level use cases and requirements and proposed a unified terminology for virtualization [NFV003]. Then, the growing NFV community produced the Release 2 normative specifications with the development of architecture, interfaces and information model

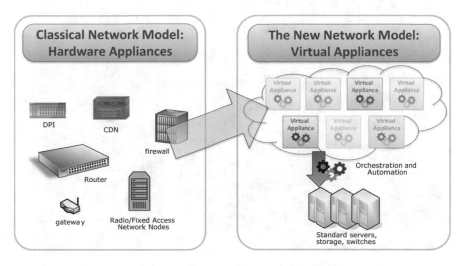

Fig. 6.4 The NFV approach: from hardware appliances to virtual appliances

aspects (aka stage 2 specifications). Release 3 added also the implementable protocol and data model solutions (aka stage 3) of interfaces, descriptors and other
artifacts.

Figure 6.5 shows four main components of the NFV framework: the MANO
(Management and Network Orchestration), the OSS/BSS, the Virtualized Network
Functions (VNFs) and the NFV Infrastructure (NFVI).

The MANO sub-system in the NFV system is similar to what we described for
the MEC system in the first part of the book, and it is composed of an NFV
Orchestrator, a VNF Manager (or better, a set of managers, each one capable of
managing a subset of the VNF types in the system) and a Virtualized Infrastructure
Manager (VIM). As an important note, the NFV Infrastructure can span across
several locations (NFVI-PoPs), i.e. places where data centers are operated. The
network providing connectivity between these locations is regarded to be part of the
NFVI.

Figure 6.6 shows the NFV Reference Architectural Framework, including all
functional blocks and reference points in the NFV framework. On important remark
is that Network Functions (NFs) can be virtualized or not, and in the latter case we
talk about PNFs (Physical Network Functions). In the figure only VNFs are shown.
From an NFV perspective, the functional behavior and the external operational
interfaces of a PNF and a VNF are out of scope. NFV sees a VNF as a collection of

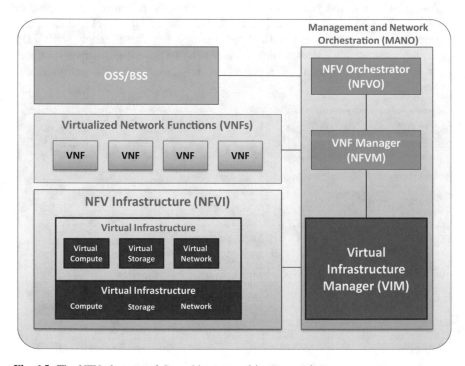

Fig. 6.5 The NFV elements of the architecture and interconnections

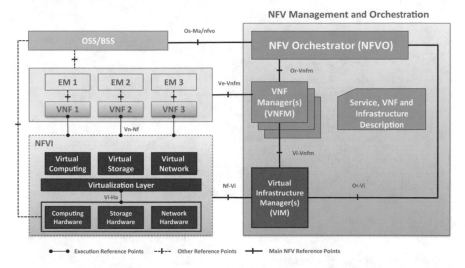

Fig. 6.6 NFV reference architectural framework

virtualized resources and network endpoints, and a PNF as a set of network end-points. Examples of NFs are 3GPP Core Network elements (e.g. MME, SGW/PGW), elements in a home network (e.g. Residential Gateway), conventional network functions (e.g. DHCP servers, firewalls) but also base stations (eNB, virtualized or not).

In order to build end-to-end network services (e.g. mobile voice/data, Internet access, a virtual private network) the NFV group introduced the concept of NF Forwarding Graph of interconnected Network Functions (NFs) and end points. For more details, the reader is invited to look at the ETSI GS NFV 002 specification [NFV002]. Moreover, the VNF software architecture is described in the ETSI GS NFV-SWA 001 specification [NFVSWA], developing the concept of a VNF Component (VNFC), characterized as an internal component of a VNF that can be mapped to a single virtualization container (e.g. VM) to provide a subset of the VNF's functionality (see figure below). These VNFCs are software components defined by the VNF Provider when designing and developing the software that provides the VNF as whole functionality. In other words, we can say that the composition of VNFCs and their interfaces to compose a VNF is provider-specific (Fig. 6.7).

Figure 6.8 shows an example of VNF instance made up of 4 VNFC instances, which are of 3 different types, each one with its own requirements on the OS and execution environment (please notice that there are 2 instances of the first VNFC "A", in order to reflect the general case). The VNF Descriptor (VNFD) shown below is a specification template provided by the VNF Provider for describing virtual resource requirements of the VNF, and it is used by the NFV MANO functions to determine how to execute VNF lifecycle operations (such as

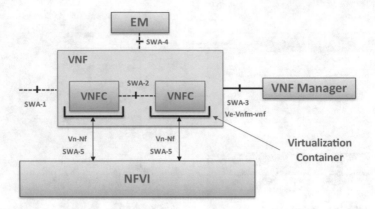

Fig. 6.7 VNF components

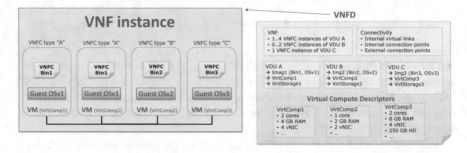

Fig. 6.8 Example of VNFD elements

instantiation). Unlike the NFV MANO interfaces, the interfaces defined by SWA in the first phase of NFV have not been further detailed in the later phases of NFV.

In the context of this book we don't plan to describe all the work done by NFV so far, but only what is needed for the purpose of learning MEC technologies. Currently the ISG NFV (as of end of 2020) is counting 114 companies among members and participants and continues to develop new specifications[1] that meet the needs of the industry, with maintenance cycles for its already published specifications, completing the Release 4. In particular, the ETSI NFV Release 4 aims to specify around the following technical areas: NFVI evolution (e.g. on enhancements to support lightweight virtualization technologies), NFV automation (e.g. by improving lifecycle management and orchestration), evolution of the NFV MANO framework (e.g. by optimizing internal NFV MANO capabilities exposure and

---

[1]The full set of published ETSI ISG NFV specifications can be found at this link: https://docbox. etsi.org/ISG/NFV/Open/Publications_pdf/Specs-Reports.

usage, and by defining a service-based framework) and operationalization (e.g. integration and use of NFV with other management and network frameworks).

In this perspective, one important feature that has been carried over into Release 4 from previous Release 3 is "MEC in NFV" (see also [NVF-R4]) and is covering the following main areas:

- Enhancement support for MEC in NFV deployments.
- Support coordination of NFV MANO with consumers (in particular MEC) for graceful termination/stop support.
- Enhancements on the placement and network constraints during resource allocation for network service and VNF instances.

The "counterpart" standardization work done by ETSI ISG MEC is described in Sect. 6.4, while in the next section an overview of the main orchestration frameworks is provided, as they are key for the management of virtualized resources in NFV environments, which is the most typical scenario where MEC is expected to be deployed.

6.3 Open Source Frameworks on Cloud and NFV Technologies

The main orchestration frameworks are described in the tables below:

OPNFV	https://www.opnfv.org/

Project type/main driver: Collaborative project under Linux Foundation
Description: Creation of a reference Open platform for NFV, to facilitate the development and evolution of NFV components across various open source ecosystems. Focus on *system level integration*, deployment, testing. Validation of multivendor solutions
Key facts: 180 Developers, 20 Organizations, 8,685 Commits, 9 Projects Addressing Cloud Native Features
Licensing: Apache 2.0

OpenMANO	https://github.com/nfvlabs/openmano

Project type/main driver: Telefonica
Description: Reference implementation of NFV-O, interfacing with an NFV VIM through RESTful APIs. The project is also including *Openvim* (a reference implementation of an NFV VIM, offering as well a northbound interface, based on REST APIs, Openvim API), and *openmano-gui* (a web-based GUI, Graphical User Interface)
Key facts: The project OpenMANO has been contributed to the open source community project Open Source MANO (OSM), NOW hosted by ETSI
Licensing: Apache 2.0

OSM	https://osm.etsi.org/

Project type/main driver: *ETSI-hosted project*
Description: Develop an Open Source NFV Management and Orchestration (MANO) software stack aligned with the evolution of ETSI NFV and providing a regularly updated reference implementation of NFV MANO
Key facts: A wide range of industry players are involved in the OSM community. In August 2018 reached 100 participating organizations. Release 4 of OSM in May 2018
Licensing: OSM uses well-established tools and methods to develop code under the Apache Public License 2.0

CloudNFV	http://www.cloudnfv.com/

Project type/main drivers: Collaborative project. Founding members: 6WIND, CIMI Corp, Dell, EnterpriseWeb, Overture Networks, Qosmos
Description: Open platform for implementing the NFV framework based on cloud computing and SDN technologies in a multivendor environment. Using active virtualization to support manager and orchestrator (*active virtualization* is a data model representing resources, functions and services using the active resource and active contract sub-elements)
Key facts: Launched in August 2013
Licensing: The project is accepting code contributed under an open source license that permits free integration with commercial components (see also http://www.cloudnfv.com/IntegrationGuide.pdf)

OpenBaton	https://openbaton.github.io

Project type/main driver: TU Berlin, Fraunhofer FOCUS
Description: Open source implementation complaint with ETSI NFV MANO, offering SDKs for integrating heterogeneous VIMs and VNFMs. *Orchestrator completely designed and implemented following the ETSI MANO specification*. Integration with different VIMs provided via Plugins (default is OpenStack)
Key facts: Supported by different European publicly funded projects (NUBOMEDIA, Mobile Cloud Networking, and SoftFIRE). Open Baton represents also one of the main components of the 5G Berlin initiative
Licensing: The Open Baton project is available under Apache 2.0 License

ZOOM	https://www.tmforum.org/collaboration/zoom-project/

Project type/main driver: ZOOM project (Zero-touch Orchestration, Operations and Management) is under TM Forum
Description: Working to develop best practices to support both the technology and business transformation brought about by the introduction of Network Function Virtualization (NFV) and Software Defined Networking (SDN). https://www.tmforum.org/zoom/zoom-operations/
Key facts: Synergy and alignment with ETSI MANO. Now also with ETSI ISG ZSM
Licensing: IPR policy document available upon registration https://www.tmforum.org/resources/standard/policy-on-intellectual-property-rights/

Cloudify	https://github.com/cloudify-cosmo

Project type/main driver: Originally created by GigaSpaces Technologies. Now a spinoff company (https://cloudify.co/)
Description: Open source software cloud and NFV orchestration product. Defining how the application interacts with the data center through APIs to execute the *defined blueprint configurations*. These "blueprints" are YAML DSL configuration files that permit to describe (1) the deployment plan in cloud environments (2) the execution plans for the lifecycle of the application for installing, starting, terminating, orchestrating and monitoring the application stack. Creating also an *abstraction layer able to describe machines and their images for a chosen cloud, in a sort of Infrastructure as Code (IaC) paradigm*
Key facts: Several awards and open source recognitions. The company is member of various collaborative projects and open source communities
Licensing: licensed under the Apache License Version 2.0

ONAP	https://www.onap.org/

Project type/main driver: Collaborative project under Linux Foundation
Description: Providing a comprehensive platform for real-time, policy-driven orchestration and automation of PNFs/VNFs. Supporting inventory, catalog, orchestration, analytics, policy, mediation, optimization and exposure functionality for deploying VNFs. Providing both design time and runtime capabilities. Goal to enable software, network, IT and cloud providers and developers to rapidly automate new services and support complete lifecycle management. *Possibility for network operators to synchronously orchestrate PNFs and NFVs. Supported by Akraino project*, where ONAP can be leveraged for on-boarding VNFs where deployed (Note: VNF can also be deployed through the Akraino portal directly, where ONAP is not deployed)
Key facts: Formed in March 2017. More than 50 large companies (network and cloud operators and technology providers). In January 2018, created LFN (LF Networking Fund), a sort of governance entity integrating different projects, which are retaining their technical independence and project roadmaps
Licensing: All new inbound code contributions to the Project must be made using the Apache License v 2.0 (approval of alternative licenses on exceptional basis)
ONAP Project Corporate Contribution License Agreement, available at onap.org/cla

For further deepening on orchestration frameworks, the reader is kindly invited to have a look to Exercise 6.1, helping a better understanding of VNFs and Network Services through OpenMANO. When talking about orchestration engines for containers, Exercise 9.1 will be essentially based on Kubernetes, as leading container orchestration engine (COE) used for the deployment of many MEC platforms, like OpenNESS.

6.4   MEC in NFV Environments

The MEC systems are based on virtualized infrastructure, and one of the main assumptions for ISG MEC standardization work since the beginning was to leverage as much as possible the work done by ISG NFV in this perspective. In particular, MEC utilizes the NFV infrastructure and uses a virtualization platform for running applications at the mobile network edge. The NFV infrastructure may be dedicated to MEC or shared with other network functions or applications, but where possible MEC uses the NFV infrastructure management entities.

Thus, based on the need for the industry to clarify how to deploy MEC in NFV environments, the ISG MEC group published in 2018 a study on "Deployment of Mobile Edge Computing in an NFV environment" [MEC-017], where essentially a "NFV flavor" of the MEC architecture is depicted, showing also the mapping between MEC and NFV elements and highlighting common areas of possible standardization enhancements, needed in both SDOs for the support of these joint scenarios.

The architecture design resulting from this study was then captured in the updated MEC Reference Architecture document [MEC-003], where the above Fig. 6.9 shows the Multi-access edge system reference architecture variant for MEC in NFV. From a functional point of view, we can easily make some first considerations by comparing this scheme with the "traditional" MEC architecture in Fig. 3.3:

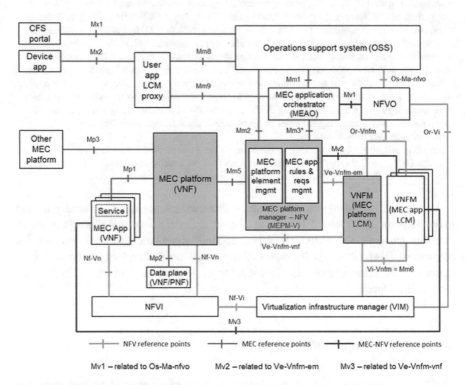

Fig. 6.9 MEC architecture variant of "MEC in NFV". *Source* [MEC-003]—© European Telecommunications Standards Institute 2017. Further use, modification, copy and/or distribution are strictly prohibited

- The MEC platform is deployed as a VNF (note: for that purpose, the procedures defined by ETSI NFV are used);
- the MEC applications appear like VNFs toward the ETSI NFV MANO components (note: this allows re-use of ETSI NFV MANO functionality; it is, however, expected that ETSI MEC might not use the full set of MANO functionality, and requires certain additional functionality);
- the virtualization infrastructure is deployed as a NFVI and its virtualized resources are managed by the VIM;
- The Multi-access Edge Orchestrator (MEO), as appearing in Fig. 3.3, is transformed into a "MEC Application Orchestrator" (MEAO) that uses the NFVO for resource orchestration, and for orchestration of the set of ME app VNFs as one or more NFV Network Services (NSs);
- The MEC Platform Manager (MEPM), as appearing in Fig. 3.3, is transformed into a "MEC Platform Manager—NFV" (MEPM-V) that delegates the LCM part to one or more VNFM(s);
- The Mm3* reference point between MEAO and MEPM-V is based on the Mm3 reference point;
- although not explicitly displayed in this NFV architecture variant, also here the Mp1 reference point between an MEC application and the MEC platform is optional for the MEC application, unless it is an application that provides and/or consumes a MEC service.

As already noticed, in this NFV variant it is assumed that the MEC platform will be realized as a VNF and will be managed according to ETSI NFV procedures. Figure 6.10 shows more in detail the mapping related to the management of the MEC platform as a VNF, where the MEPM-V is acting as the Element Manager (EM) of the MEC platform VNF, and the VNF Manager, according to ETSI NFV, is used to perform LCM of the MEC platform VNF.

Finally, a careful reader might notice that now the Data Plane is a functional block outside of the NFVI (according to ETSI NFV definitions), and consequently also Mp2 is a different reference point separated from Nf-Vn (which is both used to

Fig. 6.10 Management of the MEC platform as a VNF

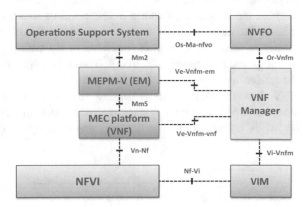

Table. 6.1 ETSI GR MEC 017 final recommendations

ETSI NFV	ETSI MEC
NFV-01: Generic VNF package extension mechanism	MEC-01: Normative architecture
	MEC-02: NS re-use
NFV-02: Enhancements on the Os-Ma-nfvo reference point	MEC-03: Reference points
	MEC-04: VNF Package extensions
NFV-03: Enhancements on the Ve-Vnfm reference point	MEC-05: VNFD for ME app VNF
	MEC-06: ME app VNF package management
NFV-04: Enhancements to the NSD	MEC-07: ME app VNF lifecycle management
NFV-05: VNF migration to support smart relocation	MEC-08: Traffic routing
	MEC-09: Smart relocation

connect the NFVI to the MEC platform and the MEC Applications (which are all acting as VNFs). So, it is likely that Mp2 reference point could need to be further defined by ETSI MEC when it comes to "MEC in NFV" architectural variant.

As anticipated, the study [MEC-017] identified some recommendations, in terms of Normative work suggested to be performed in ETSI NFV and in ETSI MEC. These recommendations are summarized in Table 6.1.[2]

6.4.1   MEC, Virtual RAN, C-RAN, Open RAN

We have widely discussed the progressive process of network virtualization. This process is including also RAN virtualization, i.e. the virtualized implementation of radio access functionalities, which are typically packed into a dedicated base station equipment (e.g. eNB, in case of LTE networks). Essentially, we can see a virtualized base station as a VNF implementation of traditional PNFs of the same functional block, since the "only" difference is how this base station is implemented, but externally (from a 3GPP point of view) the virtual eNB always terminates the radio protocol stack with the same S1 interface as the "physical" eNB. Similarly, other core network functions of the cellular network can be virtualized and implemented as suitable VNFs. These concepts, however, should not confuse the reader, who might be led to think that MEC Applications (also seen as VNFs in NFV deployments) are implementing virtual RAN functionalities. This is not the case, because eNB protocol stack is by definition implementing all low layers up to IP, while instead MEC Applications are working by definition at Layer 7 of the

---

[2]Note for NFV-05 recommendation: at the time of writing this book, VNF migration is (still) not supported in NFV up to Rel.4 so this one won't be covered.

ISO/OSI protocol stack (application level, indeed). Obviously, some cross-layer synergies are possible, e.g. implementing smart algorithms or "radio-aware" applications, but from a functional perspective these VNFs are clearly separated (even if in principle they can be even deployed as co-located and run on the same NFV infrastructure). The scheme below summarizes the main differences between MEC and RAN Virtualization, as good reference for the reader.

■ Focused on porting network functions to virtual environments

■ Enables the migration from a proprietary appliance-based setup to a standard, hardware and cloud-based infrastructure

■ Virtual functions can be connected or chained together to create communication services

RAN Virtualization

■ Supporting the creation of an open environment in the RAN, allowing 3rd-party application/service integration (application-level enablers and APIs)

■ Creates a new value chain and an energized ecosystem, based on innovation and business value

■ Enables a myriad of new use cases across multiple sectors

MEC

■ MEC can reuse the NFV virtualization infrastructure and the NFV infrastructure management to the largest extent possible.

■ The scope of MEC is focused and its business objective differs from that of NFV.

Notes

Additionally, the well-known of the Cloud-RAN architecture introduced by China Mobile Research Institute [CRAN-WP] is now a consolidated trend recognized by many network infrastructure manufacturers. Essentially, C-RAN (Cloud-RAN), sometimes referred to as Centralized-RAN, is an architecture for cellular networks consisting is a centralized, cloud computing-based architecture for radio access networks that supports 2G, 3G, 4G and future wireless communication standards. In more recent developments, the industry [TIP-ORAN] has moved forward by considering with increasing attention the concept of Open RAN, which is about disaggregated RAN functionality built using open interface specifications between elements. In an Open RAN environment, the RAN is disaggregated into three main building blocks:

- the Radio Unit (RU)
- the Distributed Unit (DU)
- the Centralized Unit (CU)

The Open RAN architecture can be implemented in vendor-neutral hardware and software-defined technology based on open interfaces and community-developed standards, and it is currently being specified by O-RAN Alliance (https://www.o-ran.org/). Also in this case, the reader should be aware that these are complementary concepts, which can exist independently from MEC. Nonetheless, the ETSI white paper [MEC-CRAN] describes well the possible synergies between MEC and the possibility to enable and expose RAN Services at application level in MEC (e.g. by consuming information from the MEC RNI API).

6.4.2   *Virtualization Aspects in MEC*

As we anticipated, the ETSI GR MEC 017 study provided also recommendations highlighting common areas of possible standardization enhancements, needed both in ISG MEC and ISG NFV for the support of these deployment scenarios. While some of these enhancements are already available in NFV Release 3, and the current Release 4 of ISG NFV is working on further enhancements, ISG MEC has published a study on "MEC support for alternative virtualization technologies" [MEC-027], with the purpose of identifying the additional support that needs to be provided by MEC when MEC applications run on alternative virtualization technologies (AVTs), such as containers.

In the context of NFV,[3] several virtualization technologies are possible, including hypervisor-based solutions, OS containers, higher-level containers, nesting of virtualization technologies, mixing of virtualization technologies and mixing and nesting of virtualization technologies. The MEC study provides an overview of the impact on MEC, by analyzing the possible options for deploying MEC applications. In particular, as depicted in the above Fig. 6.11, OS container solutions are more lightweight than hypervisor-based VMs and enable faster application instantiation. Consequently, they may be an attractive option for MEC applications, with respect to a hypervisor-based approach, especially in edge-network scenarios.

A careful reader might have recognized that the above figure is indicating the specific term "*OS Containers*". These are is different from the so-called "application containers" (or "high-level containers"), which consist in a level of virtualization technologies more dealing with software code and its development, deployment and runtime environment.

In the *application containers*, the level of abstraction is on the runtime environment, where source code written in a certain programming or scripting language is deployed onto the NFVI. The MEC 27 deliverable recognizes them as an alternative solution to OS containers, because application containers are particularly attractive when there is a need for deploying an application in form of source code, for example, in DevOps environments.

Furthermore, also mixed or nested virtualization technologies are possible, and they are captured as well by the MEC 027 study. The document finally identified also some recommendations, in terms of Normative work suggested for ISG MEC, which are summarized in Table. 6.2.

---

[3]For more context and details, see the ETSI GS NFV-EVE 004 specification.

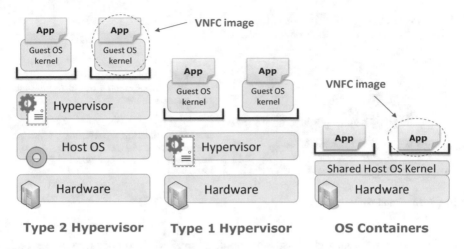

Fig. 6.11 Hypervisor versus OS container solutions for MEC

Table. 6.2 ETSI GR MEC 027 final recommendations

AVT impact on MEC framework	AVT impact on management APIs
Change the MEC application definition as follows: *"MEC application runs as a virtualized application, such as a virtual machine (VM) or a containerized application, on top of the virtualization infrastructure provided by the MEC host, and can interact with the MEC platform to consume and provide MEC services (described in clause 8)"*	• Update ETSI GS MEC 010–2 to expand definition of application lifecycle management operations to cover AVTs • Update ETSI GS MEC 003 to reflect options to support different AVTs

6.5   Cloud Native Computing

This section deserves a small description of the recent concept of "cloud native computing", which is about *Cloud Native* computing is an approach in software development that utilizes cloud computing to *"build and run scalable applications in modern, dynamic environments such as public, private, and hybrid clouds"* (ref. Wikipedia). In practice, developing for cloud native means building apps that can be deployed anywhere in the cloud, with any vendor on any type of server instance (see figure below). Instead of a monolithic application that encompasses every feature in a single package, an application is broken down into a series of loosely connected smaller apps called microservices (Fig. 6.12).

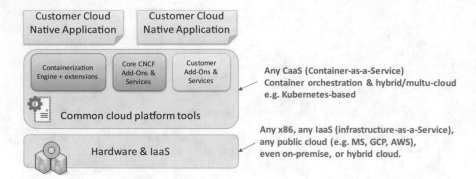

Fig. 6.12 Cloud native applications

This kind of approach, based on application SW decomposition in several microservices, is also very suitable for the most recent trends on agile development, where a separation of individual functions, like new customer sign ups, billing, authentication, analytics or recommendations, results in an easier development, update and deployment of new features.

Terminology: The Cloud Native Computing Foundation (www.cncf.io), an organization that aims to create and drive the adoption of the cloud native programming paradigm, defines cloud native as "Cloud native computing uses an open source software stack to be: • **Containerized**. Each part (applications, processes, etc.) is packaged in its own container. This facilitates reproducibility, transparency and resource isolation • **Dynamically orchestrated**. Containers are actively scheduled and managed to optimize resource utilization • **Microservices-oriented**. Applications are segmented into microservices. This significantly increases the overall agility and maintainability of applications".

The CNCF definition of cloud native includes three specific characteristics: containerization, dynamic orchestration and microservices-oriented applications. The figure below depicts an example of how a possible cloud native application can be realized, by means of decomposition into several microservices (instead of based on a monolithic architecture that would bundle every piece of the application into one package) (Fig. 6.13).

This cloud native approach permits to have applications broken down into a series of microservices, with advantages in terms of faster and scalable development, and allows for flexibility when it comes to allocating resources to specific parts of the app. This approach is also currently used by some MEC platforms (like OpenNESS) that are described in Chap. 9.

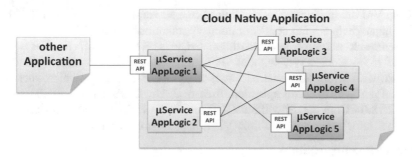

Fig. 6.13 Example of microservice architecture (app SW decomposition)

 Exercises/examples—Chapter 6

Exercise 6.1 *Goal of the exercise*: install a basic edge infrastructure and instantiate your first VNF. The exercise is a step by step tutorial that will let you learn how to manage VNFs and Network Services in an easy and scalable way through OpenMANO, using OpenStack as the VIM, simulating an edge data center.

Many frameworks are present in literature (e.g. OpenNFV, CloudNFV, OpenBaton, OpenMANO, Cloudify, T-NOVA, ONAP, …). The purpose of this exercise is to practice with one of these frameworks, i.e. OpenMANO, and install a basic edge infrastructure. The following setup is used for this exercise:

- Windows 10 Host with Virtual Box.
- Two Virtual Machines with a bridged interface and an internal interface: one with OpenStack installed and the other with OpenMANO.

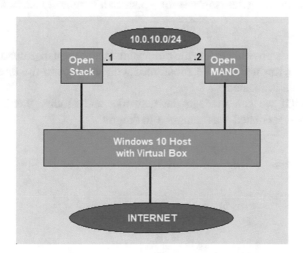

Both VM are configured as an Ubuntu core server, installed via mini ISO 18.04 64-bit https://help.ubuntu.com/community/Installation/MinimalCD

OpenStack machine specifications: 6vCPUs, 8 GB RAM, 100 GB storage, Ubuntu 18.04 Core server without GUI. Installed SW: Microstack, a standalone sandbox, with all the essential packages from OpenStack. https://microstack.run/

OpenMANO machine specifications: 6vCPUs, 2 GB RAM, 50 GB storage, Ubuntu 18.04 Core server without GUI. Installed sw: OpenMANO release NINE, an orchestrator supporting multiple VIM. https://osm.etsi.org/

OpenStack configurations

Once installed, OpenStack should be accessible from any browser at the address used for the bridged interface (192.168.1.X).

From the GUI it's possible to download the "admin_openrc.sh" file that is used to authorize every openstack terminal command. To download it's necessary to access the GUI, under "Project -> API Access", and to click on "Download OpenStack RC File".

The file should be copied on the Ubuntu VM with an SCP issued from a terminal. "scp <path_to_rc_file> <vm_user> @ <vm_bridged_address> :/ <destination_path> ". For example:

scp "D:\users\user\Desktop\admin_openrc.sh" user@192.168.1.10:/home/ user/

Once copied it's possible to use the command "source <path_to>/admin_openrc. sh" where <path_to> needs to be substituted with the path to the directory where admin_opernrc.sh is located.

From the GUI we can manage the networks and modify them. The default "external" network created was renamed to "mgmt".

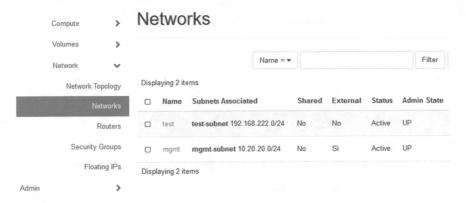

When installed on Virtual Box, the VM instantiated on OpenStack have no access to Internet, so it's necessary to use the following commands to enable it:

source admin_openrc.sh
openstack subnet set --dhcp mgmt-subnet
openstack subnet set --dns-nameserver 8.8.8.8 mgmt-subnet
openstack network set --share mgmt
sudo iptables -t nat -A POSTROUTING -s 10.20.20.1/24 ! -d 10.20.20.1/24 -j MASQUERADE
sudo sysctl net.ipv4.ip_forward=1

Note: the 10.20.20.1 address is used because it's the address of the virtual interface created by MicroStack to communicate with the instantiated VMs on the default public network.

OpenMANO configurations

Once installed, OpenMANO should be accessible from any browser at the address used for the bridged interface (192.168.1.X).

To link OpenStack and OpenMANO one command is needed from the OpenMANO VM:

osm vim-create --name <name> --user <openstack_user> --password <open-stack_password> --auth_url <openstack_auth_address> --tenant <openstack_tenant> --account_type openstack

For example:

osm vim-create --name openstack --user admin --password admin --auth_url http://10.0.10.1:5000/v3/ --tenant admin --account_type openstack

And the result should appear on the OpenMANO dashboard:

VIM Accounts

⊙ PROCESSING ⊘ ENABLED ⊗ ERROR

Name	Identifier	Type	Operational Status
Name	🔍 Identifier	🔍 Select	Select
openstack	5f2f86fa-4712-487d-9a99-f5101c72af0f	openstack	✅

VNF Instantiation

To instantiate a VNF on OpenStack via OpenMANO three steps are needed:

1. Creation of the VNF Descriptor.
2. Creation of the NS Descriptor.
3. Instantiation of the NS.

The descriptors used were downloaded from this path https://osm-download.etsi. org/ftp/Packages/examples/ and modified for this example.

VNFD: hackfest_basic_vnf.tar.gz
NSD: hackfest_basic_ns.tar.gz

A preliminary step is needed on OpenStack since OpenMANO needs to reference an image already loaded at the VIM site. The image can be any image of a VNF.

To load an image on OpenStack the following command can be used:

openstack image create –file=<path_to_img> --container-format=bare --disk-format=qcow2 <img_name>

For example:

openstack image create --file=./xenial-server-cloudimg-amd64-disk1.img --container-format=bare --disk-format=qcow2 **ubuntu16.04**

The xenial-server-clouding-amd64-disk1.img was downloaded from https:// cloud-images.ubuntu.com/xenial/current/xenial-server-cloudimg-amd64-disk1.img

Once the image is loaded on OpenStack it should appear under "Compute - > Images".

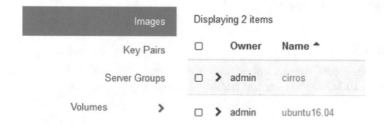

Creation of the VNF Descriptor

The VNFD can be loaded on OSM with the following command from the OpenMANO VM:

osm nfpkg-create hackfest_basic_vnf.tar.gz

The VNF Descriptor contains all the definitions of the VDU that are part of the VNF, the image used, the hardware required and the interfaces needed.
Sample VNFD:

```
vnfd:
  description: A basic VNF descriptor w/ one VDU
  df:
  - id: default-df
    instantiation-level:
    - id: default-instantiation-level
      vdu-level:
      - number-of-instances: 1
        vdu-id: hackfest_basic-VM
      vdu-profile:
      - id: hackfest_basic-VM
        min-number-of-instances: 1
    ext-cpd:
    - id: vnf-cp0-ext
      int-cpd:
        cpd: vdu-eth0-int
        vdu-id: hackfest_basic-VM
  id: hackfest_basic-vnf
  mgmt-cp: vnf-cp0-ext
  product-name: hackfest_basic-vnf
  sw-image-desc:
  - id: ubuntu16.04
    image: ubuntu16.04
    name: ubuntu16.04
  vdu:
  - alternative-sw-image-desc:
    - ubuntu/images/hvm-ssd/ubuntu-artful-17.10-amd64-server-20180509
    id: hackfest_basic-VM
    int-cpd:
    - id: vdu-eth0-int
      virtual-network-interface-requirement:
      - name: vdu-eth0
        virtual-interface:
          type: PARAVIRT
    name: hackfest_basic-VM
    sw-image-desc: ubuntu16.04
    virtual-compute-desc: hackfest_basic-VM-compute
    virtual-storage-desc:
    - hackfest_basic-VM-storage
  version: '1.0'
  virtual-compute-desc:
  - id: hackfest_basic-VM-compute
    virtual-cpu:
      num-virtual-cpu: "1"
    virtual-memory:
      size: "1.0"
  virtual-storage-desc:
  - id: hackfest_basic-VM-storage
    size-of-storage: "10"
```

In this sample VNFD the image loaded on OpenStack is referenced (Ubuntu16.04), as well as the number of interfaces needed for this VM when instantiated and the flavor requested.

This VNF is composed of only one VM with the following requirements:

Interfaces: one virtual ethernet interface.

Image: Ubuntu16.04.

Flavor: 1 vCPU, 1 GB RAM, 10 GB storage.

It's possible to add multiple ethernet interfaces and multiple VDU/VM.

Note: the flavor is created when the ns is instantiated.

Creation of the NSD Descriptor

The NSD can be loaded on OSM with the following command from the OpenMANO VM:

osm nspkg-create hackfest_basic_ns.tar.gz

The NS Descriptor contains the references to the VNFs and the virtual link needed.

```
nsd:
  nsd:
  - description: Simple NS with a single VNF and a single VL
    df:
    - id: default-df
      vnf-profile:
      - id: '1'
        virtual-link-connectivity:
        - constituent-cpd-id:
          - constituent-base-element-id: '1'
            constituent-cpd-id: vnf-cp0-ext
          virtual-link-profile-id: mgmt
        vnfd-id: hackfest_basic-vnf
    id: hackfest_basic-ns
    name: hackfest_basic-ns
    version: '1.0'
    virtual-link-desc:
    - id: mgmt
      mgmt-network: true
    vnfd-id:
    - hackfest_basic-vnf
```

The virtual link can be internal, between the VNFs, or external, as in this example, and should reference the IDs specified inside the VNF Descriptors.

Please note that here the "virtual-link-profile-id" is set to "mgmt" to reference the network on OpenStack. In this way it's possible to choose where the VM will be connected.

Instantiation of the NS

The NS can be instantiated on OSM with the following command from the OpenMANO VM:

osm ns-create --ns_name <ns-instance-name> --nsd_name hackfest_basic-ns --vim_account <vim-target-name>

For example:

osm ns-create --ns_name Test --nsd_name hackfest_basic-ns --vim_account openstack

where:

   --ns_name is the name chosen for the NS instance (can be anything).
   --nsd_name is the name of the NSD as loaded on OSM.
   --vim_account is the name of the linked VIM as loaded on OSM.

When the instantiation is done the following result should appear on OpenMANO:

NS Instances

Name	Identifier	Nsd name	Operational Status	Config Status	Detailed Status
Test	bc4f5ba9-7d52-4cfc-a95d-add9f52d358b	hackfest_basic-ns	✓	✓	Done

While, on OpenStack, should appear:

Displaying 1 item

Instance Name	Image Name	IP Address	Flavor	Key Pair	Status	Availability Zone	Task	Power State	Age
Test-1-hackfest_basic-VM-0	ubuntu16.04	10.20.20.207	hackfest_basic-VU-fv	-	Active	nova	None	Running	0 minuti

Displaying 1 item

The instance name is generated automatically by combining the name of the NS, the id of the NS member, the name vdu-id and an incremental value.

The flavor used was created automatically on OpenStack when the NS was instantiated.

Exercise 6.2   Goal of the exercise: *Setup a development environment by harnessing managed cloud services (e.g. Google Cloud Platform, GCP) to bring up Virtual Machine(s) (VM) on which one can deploy OpenNESS and prototype applications.*

Note: OpenNESS toolkit will be described more in detail in Chap. 9. For the time being, the reader should consider the following preliminary information:

• Open Network Edge Services Software (OpenNESS) is a platform to orchestrate and manage edge applications and Network Functions across diverse network platforms on the edge. It employs a microservices architecture and offers services to edge applications running on the platform.

- OpenNESS is built on top of Kubernetes, whereby an OpenNESS deployment consists of the OpenNESS Kubernetes Control Plane and OpenNESS Edge Node(s) (worker(s)). This enables quick development or migration of existing Cloud Native applications to the edge.
- OpenNESS can also draw upon enhancements across multiple layers of the orchestration stack to provide the best performance on the underlying hardware through optimal placement of applications and services. OpenNESS may be deployed on hardware both On-Premises or across networked compute resources on the Network Edge.
- The OpenNESS Experience Kit is an Ansible playbook that is used to install OpenNESS and its supporting components. The available OpeNESS microservices, k8s extensions and optimizations are conveniently factored into deployment *flavors* which bring together the required components that broadly cover a set of capabilities.

In this exercise, we will install the *minimal flavor* of OpenNESS which serves as a good starting point for application development. For the sake of simplicity, we will deploy a single-node OpenNESS k8s cluster where the OpenNESS k8s Control Plane and Edge node functions reside on the same physical machine.

Proposed solution:

Installing OpenNESS

To realize the benefits of OpenNESS and the Edge Computing paradigm would require a moderately powerful hardware platform on the edge that is capable of handling compute and memory intensive workloads that are offloaded to it. A suggested minimum configuration for this exercise is a multi-core architecture (>= 4 CPUs) and > = 16 GB of RAM. A more comprehensive study of OpenNESS's hardware optimizations and candidate use cases would lead to other recommendations. OpenNESS has been tested and validated on CentOS 7.

A simple way to get started is to harness managed cloud services to bring up Virtual Machine(s) (VM) on which you may deploy OpenNESS and prototype applications. There are many cloud services on the web offering service tiers at different price points. In this exercise, we will utilize the *Google Cloud Platform* which as of this writing offers free usage credits to new users that are usable over 90 days.

Google Cloud Platform (GCP)—Create a CentOS 7 Virtual Machine

After you sign-up and login, access Compute Engine-> VM instances on the side navigation menu. Select Create instance (+). Choose an instance name (e.g. "openness"). Under Machine configuration-> Machine type, select e2-standard-4 (4 vCPU, 16 GB memory). Select "Change" under Boot disk. Look for "CentOS" under Public Images and pick "CentOS 7" as the Version. Enter 40 (GB) as the disk size and hit "Select".

Check "Allow HTTP traffic" and "Allow HTTPS traffic" under Firewall rules. Leave all other settings at their defaults and hit "Create". The VM will soon be up and running and accessible through SSH via the VM instances page. Click "SSH" to login to the VM.

Installation Prework

Configure root access

Once logged into the VM, set a root password and switch to the root user. We will install and run OpenNESS as root.

[id@openness ~]$ sudo passwd
Changing password for user root.
New password:
Retype new password:
passwd: all authentication tokens updated successfully.
[id@openness -]$ su
Password:
[root@openness id]# cd
[root@openness ~]#

Configure local SSH access.

Ansible uses SSH to interact with the OpenNESS host (even if the same machine) while carrying out the installation procedure. We must therefore configure SSH access to the root user.

Generate a key pair using the defaults and do not provide a passphrase.

[root@openness ~]# ssh-keygen
Generating public/private rsa key pair.
Enter file in which to save the key (/root/.ssh/id_rsa):
Enter passphrase (empty for no passphrase):
Enter same passphrase again:
Your identification has been saved in /root/.ssh/id_rsa.
Your public key has been saved in /root/.ssh/id_rsa.pub.
...

Set the appropriate permissions on the public key file and add it to the list of *authorized_keys*.

[root@openness ~]# cd .ssh
[root@openness .ssh]# chmod 600 .ssh/id_rsa.pub
[root@openness .ssh]# cp id_rsa.pub authorized_keys

Update the SSH config to enable the two options below for a password-less login to the root user account.

[root@openness ~]# vi /etc/ssh/sshd_config

```
PermitRootLogin yes
PubkeyAuthentication yes
```

Finally, restart the SSH daemon.
[root@openness ~]# systemctl restart sshd

You may then verify SSH access (to self) as below.
[root@openness ~]# ssh root@openness
The authenticity of host 'openness (10.128.0.25)' can't be established.
Are you sure you want to continue connecting (yes/no)? yes
Warning: Permanently added 'openness,10.128.0.25' (ECDSA) to the list of known hosts.
Last login: Mon Dec 14 20:39:50 2020

Update /etc./hosts

The OpenNESS playbooks need to access the Control Plane and Edge nodes at their individual hostnames. As we have both functions running on the same physical host, we simply set a hostname (e.g. opennessedge) in addition to the current hostname (e.g. openness). Further, we must enable both hostnames to resolve to the localhost IPv4 and IPv6 addresses as shown below.

127.0.0.1   localhost localhost.localdomain localhost4 localhost4.local-domain4 openness opennessedge
::1   localhost localhost.localdomain localhost6 localhost6.localdomain6 openness opennessedge
10.128.0.25 openness.us-central1-a.c.invertible-vine-292718.internal open-ness opennessedge # Added by Google
169.254.169.254 metadata.google.internal # Added by Google

OpenNESS playbooks

Clone the OpenNESS Experience Kit Github repository to obtain the OpenNESS Ansible playbook. But first, install git.

[root@openness ~]# yum install git
...
Complete!
[root@openness ~]# git clone https://github.com/open-ness/openness-experience-kits.git
Cloning into 'openness-experience-kits'...
…
Resolving deltas: 100% (4123/4123), done.
[root@openness ~]# cd openness-experience-kits/
Checkout the 20.09.01 release tag.
[root@openness openness-experience-kits]# git checkout openness-20.09.01
…
HEAD is now at 603268f... Merge pull request #69 from open-ness/openness-20.09.01
[root@openness openness-experience-kits]# ls
action_plugins deploy_ne.sh inventory.ini   network_edge.yml single_node_network_edge.yml
ansible.cfg flavors   LICENSE   README.md   tasks   cleanup_ne.sh
group_vars logs   roles cnab host_vars network_edge_cleanup.yml
scripts

Configure OpenNESS inventory

The OpenNESS inventory file (*inventory.ini*) stores the hostname(s) and IP address (es) of the OpenNESS Control Plane and Edge node(s). Ansible uses these settings to kick off the installation. As we are deploying a single-node OpenNESS cluster, this file must point to the current VM instance under the "controller" and "edgenode" groups via their designated hostnames. The IP address of the VM is 10.128.0.25 (check using *ifconfig*). The resulting inventory.ini should look as below:

[root@openness openness-experience-kits]# vi inventory.ini
[all]
openness ansible_ssh_user=root ansible_host=10.128.0.25
opennessedge ansible_ssh_user=root ansible_host=10.128.0.25
[controller_group]
openness
[edgenode_group]
opennessedge
[edgenode_vca_group]

Execute the playbook

[root@openness openness-experience-kits]# ./deploy_ne.sh -f minimal single

The installation process will take some time to complete. Due to the distributed natured of the deployment with several interdependent microservices, there could be timing issues related to the startup of a recently installed service for which the playbook quits waiting. Upon such failures, a re-run of the above command after a brief gap, possibly with a VM restart usually resolves things.

There may be still other failures involving currently stale configurations/ URLs in the 20.09.01 tag of the OpenNESS Experience Kit. Resolutions for those could be found by looking up GitHub issues in the "edgeservices" and "converged-edge-experience-kits" repositories.

Following a complete installation, you should see messages like below:

PLAY   RECAP
****
****
    openness   : ok=315 changed=206 unreachable=0   failed=0
skipped=184 rescued=0   ignored=15
    opennessedge   : ok=120 changed=47   unreachable=0   failed=0
skipped=127 rescued=0   ignored=1

Running OpenNESS Sample Applications

It is possible to run any containerized application on OpenNESS with minimal modification that directly services end user traffic. Such an application can be provisioned by the OpenNESS Controller but may not utilize any other capabilities of the platform.

Alternatively, applications may expose special services and features on the edge (e.g. MEC Services and APIs). OpenNESS provides the infrastructure to facilitate this via the OpenNESS Edge Application Agent (EAA). Such applications can be broadly classified as Producers or Consumers. Producers advertise an edge service to other edge applications running on the platform and do not directly service end user requests. Consumers subscribe to edge services offered by Producers and may also directly serve end user requests depending on the application's architecture. In this section, we will run applications that are available in the OpenNESS *edgeapps* repository.

Update Kubernetes NetworkPolicy

In the OpenNESS deployment configuration, by default all ingress traffic is blocked to pods and any deployed services are not reachable. To allow ingress traffic, we must apply a *NetworkPolicy* that permits this traffic to pass through. For this, we create a spec file *allow_ingress.yaml* as follows:

```
[root@openness ~]# cat << EOF > allow_ingress.yaml
> apiVersion: networking.k8s.io/v1
> kind: NetworkPolicy
> metadata:
>   name: eaa-prod-cons-policy
>   namespace: default
> spec:
>   podSelector: {}
>   policyTypes:
>   - Ingress
>   ingress:
>   - from:
>     - ipBlock:
>         cidr: 10.16.0.0/16
>     ports:
>     - protocol: TCP
>       port: 80
>     - protocol: TCP
>       port: 443
> EOF
```

Apply this policy to the OpenNESS k8s cluster

```
[root@openness ~]# kubectl apply -f allow_ingress.yaml
```

networkpolicy.networking.k8s.io/eaa-prod-cons-policy created

Clone the edgeapps repository

[root@openness ~]# git clone https://github.com/open-ness/edgeapps.git
Cloning into 'edgeapps'...
…
[root@openness ~]# cd edgeapps/applications/
Checkout the 20.09.01 release tag.
[root@openness openness-experience-kits]# git checkout openness-20.09.01
…
[root@openness applications]# ls
actian-zen cdn-transcode   fpga-sample-app qwilt   sample-app   telemetry-sample-app
cdn-caching eis-experience-kit openvino   README.md smart-city-app
vas-sample-app

Build the sample producer and consumer

The sample producer advertises a service and remains active for roughly 3 min during which it issues notifications to any consumers that are subscribed to the service. Following this period, it deactivates the service and may be relaunched.

[root@openness applications]# cd sample-app/
[root@openness sample-app]# ls
common docker-compose.yml go.mod go.sum Makefile README.md simpleEaaConsumer simpleEaaProducer
Compile the sample producer and consumer
[root@openness sample-app]# make
…
Build docker images for the sample producer and consumer
[root@openness sample-app]# make build-docker
…
Successfully tagged producer:1.0
Verify that the Docker images are available
[root@openness sample-app]# docker images | grep producer
producer   1.0   75d89dc5a226   About a
minute ago 209MB
[root@openness sample-app]# docker images | grep consumer
consumer   1.0   eb0a1d2b9042   About
a minute ago 209MB

Construct k8s pod specification files for the sample producer and consumer

```
[root@openness ~]# cat << EOF > producer.yaml
> apiVersion: apps/v1
> kind: Deployment
> metadata:
>   name: producer
> spec:
>   replicas: 1
>   selector:
>     matchLabels:
>       app: producer
>   template:
>     metadata:
>       labels:
>         app: producer
>     spec:
>       tolerations:
>       - key: node-role.kube-ovn/master
>         effect: NoSchedule
>       containers:
>       - name: producer
>         image: producer:1.0
>         ImagePullPolicy: Never
>         ports:
>         - containerPort: 80
>         - containerPort: 443
> EOF
```

```
[root@openness ~]# cat << EOF > consumer.yaml
> apiVersion: apps/v1
> kind: Deployment
> metadata:
>   name: consumer
> spec:
>   replicas: 1
>   selector:
>     matchLabels:
>       app: consumer
>   template:
>     metadata:
>       labels:
>         app: consumer
>     spec:
>       tolerations:
>       - key: node-role.kube-ovn/master
>         effect: NoSchedule
>       containers:
>       - name: consumer
>         image: consumer:1.0
>         imagePullPolicy: Never
>         ports:
>         - containerPort: 80
>         - containerPort: 443
> EOF
```

Deploy the producer and consumer pods

Producer:
[root@openness ~]# kubectl create -f producer.yaml
deployment.apps/producer created
Consumer:
[root@openness ~]# kubectl create -f consumer.yaml
deployment.apps/consumer created

A listing of the running pods should mention the just deployed producer and consumer.

[root@openness ~]# kubectl get pods
NAME   READY STATUS   RESTARTS AGE
consumer-6f5978c579-drbs5 1/1   Running 0   10s
producer-685fcbc569-svf79 1/1   Running 0   50s

We can now inspect logs from the producer. The producer authenticates with the Edge Application Agent, activates its service and starts issuing notifications per its application logic.

```
[root@openness ~]# kubectl logs -f producer-685fcbc569-svf79
2020/12/14 23:41:41 {[{0xc0002ac200 The Example Producer eaa.openness
[{ExampleNotification 1.0.0 Description for
 Event #1 by Example Producer}]}]}
2020/12/14 23:41:44 Sending notification
2020/12/14 23:41:50 Sending notification
....
```

On inspecting the logs from the consumer, we will see the notifications sent by the producer. Like the producer, the consumer authenticates with the Edge Application Agent, discovers services and subscribes to the service offered by the producer, establishing a WebSocket to receive notifications from that service.

[root@openness ~]# kubectl logs -f consumer-6f5978c579-drbs5
Received notification:
 Name: ExampleNotification
 Version: 1.0.0
 Payload: {"msg":"Mon Dec 14 23:42:23 UTC 2020"}
 URN: {ExampleProducerAppID ExampleNamespace}
Received notification:
...

In summary, we have thus far successfully deployed an OpenNESS cluster and tested the sample producer and consumer applications. These applications are registered to the OpenNESS platform, execute on the cluster and do not interact with the outside world.

Establish outside-in connectivity to the edge application

This section describes a method for servicing end user traffic originating from User Equipment (UE) in an OpenNESS edge application. We start by removing the existing producer and consumer deployments on k8s.

```
[root@openness ~]# kubectl delete deployments.apps producer consumer
deployment.apps "producer" deleted
deployment.apps "consumer" deleted
```

Delete the consumer image from the local Docker registry

```
[root@openness ~]# docker images | grep consumer
consumer                              1.0            eb0a1d2b9042      About
an hour ago  209MB
[root@openness ~]# docker image rm eb0a1d2b9042
Untagged: consumer:1.0
Deleted:
sha256:eb0a1d2b90426559d1f5f1cb0ca191b11f79fc76fd64eb663e44c431dc2130
6a
    ...
```

Update the consumer sample

We must define a resource that we will access externally and for that we make minor modifications to the consumer application. We define a HTTP resource (port 80) at **/lastnot** to return (via the GET method) the payload of the last notification received from the producer's edge service.

For this we instrument the getMessages() function in *sample-app/ simpleEaaConsumer/consumer.go* which processes all received notifications. The "diff" below summarizes the code changes.

```
--- a/applications/sample-app/simpleEaaConsumer/consumer.go
+++ b/applications/sample-app/simpleEaaConsumer/consumer.go
@@ -121,7 +121,18 @@ func connectConsumer(client *http.Client) (*web-
socket.Conn, error) {
   // getMsgFromConn retrieves a message from a connection and parses
   // it to a notification struct
   func getMessages(conn *websocket.Conn) {
-
+      var lastPay string
+      notHandler := func(w http.ResponseWriter, req *http.Request) {
+          if lastPay != "" {
+              fmt.Fprintf(w, lastPay)
+          } else {
+              fmt.Fprintf(w, "No Notifications yet!\n")
+          }
+      }
+      http.HandleFunc("/lastnot", notHandler)
+      go func() {
+          http.ListenAndServe(":80", nil)
+      }()
       err := conn.SetReadDeadline(time.Now().Add(time.Second *
           time.Duration(common.Cfg.ConsumerTimeout)))
       if err != nil {
@@ -143,6 +154,7 @@ func getMessages(conn *websocket.Conn) {
                   fmt.Printf("Received notification:\n   Name: %v\n   Version:
%v\n"+
                       " Payload: %v\n  URN: %v\n",
                       resp.Name, resp.Version, string(resp.Payload), resp.URN)
+                  lastPay = string(resp.Payload)
               }
           }
       }
```

Rebuild the consumer application and Docker image

[root@openness sample-app]# make
[root@openness sample-app]# make build-docker

The updated consumer application is now available in the Docker registry to deploy through k8s.

Define k8s Service for external access to the pod
In order to establish external connectivity to port 80 in the consumer application for receiving HTTP requests, we must define a k8s Service resource and expose a *NodePort* that will consistently map a port on the physical machine to port 80 in the consumer application pod.

```
[root@openness ~]# cat << EOF > service.yaml
> apiVersion: v1
> kind: Service
> metadata:
>   name: consumer-service
> spec:
>   type: NodePort
>   selector:
>     app: consumer
>   ports:
>     - port: 80
>       nodePort: 30007
> EOF
```

We select port 30,007 as the external port on which to receive requests to be routed via the k8s infrastructure to the consumer application pod. This is also a good time to ensure that the router attached to the physical machine (GCP infrastructure in this exercise) permits requests over port 30,007.

GCP's firewall settings can be reached by hitting the ellipsis (next to the SSH link) on the VM instances page, select "View network details" and then click the "Firewall" link in the navigation menu on the "VPC Network" page. Edit the "default-allow-http" policy and add *tcp* port 30,007. The policy would immediately take effect upon saving the updated rule.

Note the external IP address of the VM on the GCP VM instances page (e.g. 34.122.124.30). Following the above configurations, the HTTP resource we've defined would be accessible at http://34.122.124.30:30007/lastnot.

Deploy the producer and consumer applications as before

```
[root@openness ~]# kubectl create -f producer.yaml
deployment.apps/producer created
[root@openness ~]# kubectl create -f consumer.yaml
deployment.apps/consumer created
```

Deploy the k8s service for the consumer application
```
[root@openness ~]# kubectl apply -f service.yaml
service/consumer-service created
```
Check that the applications are running
```
[root@openness ~]# kubectl get pods
NAME                    READY  STATUS   RESTARTS  AGE
consumer-6f5978c579-k2js5 1/1   Running  0        22s
producer-685fcbc569-ppzrq 1/1   Running  0        27s
```

Check that the k8s consumer-service is active
```
[root@openness ~]# kubectl get services
NAME            TYPE       CLUSTER-IP     EXTERNAL-IP  PORT(S)       AGE
consumer-service NodePort   10.109.49.102  <none>       80:30007/TCP
36s
kubernetes      ClusterIP  10.96.0.1      <none>       443/TCP       3h2m
```
Notice that it cites our selected NodePort (30007).

Inspect logs from the consumer pod
```
[root@openness ~]# kubectl logs -f consumer-6f5978c579-k2js5
Received notification:
  Name: ExampleNotification
  Version: 1.0.0
  Payload: {"msg":"Tue Dec 15 01:36:22 UTC 2020"}
  URN: {ExampleProducerAppID ExampleNamespace}
Received notification:
  ...
```
This lists the notifications received from the producer (as before).

Access /lastnot from the UE

You can now use curl on your laptop as representing an external Client (aka UE) to issue a GET request to https://34.122.124.30:30007/lastnot.

$ curl http://34.122.124.30:30007/lastnot
{"msg":"Tue Dec 15 01:37:49 UTC 2020"}
$ curl http://34.122.124.30:30007/lastnot
{"msg":"Tue Dec 15 01:37:55 UTC 2020"}
$ curl http://34.122.124.30:30007/lastnot
{"msg":"Tue Dec 15 01:38:01 UTC 2020"}

Notice that the responses reflect the payloads of the most recent notifications from the producer to the consumer as intended.

We hope you enjoyed this exercise. As next steps, you may further study the source codes of the sample producer/consumer applications to understand the OpenNESS Edge Application Agent APIs for authentication, service activation, service discovery, subscriptions, and notifications which are widely used in OpenNESS applications. We also invite you to explore https://www.openness.org/ in further detail to identify hardware optimizations captured by the project along with other adjacent toolkits (e.g. OpenVINO) for building performant edge applications.

References

[NFV003]   ETSI GR NFV 003 V1.5.1 (2020–01) Network Functions Virtualisation (NFV);
                Terminology for Main Concepts in NFV. https://docbox.etsi.org/ISG/NFV/
                Open/Publications_pdf/Specs-Reports/NFV%20003v1.5.1%20-%20GR%20-%
                20Terminology.pdf
[NFV002]   ETSI GS NFV 002 V1.2.1 (2014–12) Network Functions Virtualisation (NFV);
                Architectural Framework. https://docbox.etsi.org/ISG/NFV/Open/Publications_
                pdf/Specs-Reports/NFV%20002v1.2.1%20-%20GS%20-%20NFV%
                20Architectural%20Framework.pdf
[NVFSWA]   ETSI GS NFV-SWA 001 V1.1.1 (2014–12) Network Functions Virtualisation
                (NFV); Virtual Network Functions Architecture. https://www.etsi.org/deliver/
                etsi_gs/NFV-SWA/001_099/001/01.01.01_60/gs_NFV-SWA001v010101p.pdf
[NVF-R4]   ETSI NFV, NFV(20)000160, NFV Release 4 Definition, v0.2.0 (2020–07)
                https://docbox.etsi.org/ISG/NFV/Open/Other/ReleaseDocumentation/NFV(20)
                000160_NFV_Release_4_Definition_v0_2_0.pdf
[MEC-017]   ETSI GR MEC 017 V1.1.1 (2018–02) Mobile Edge Computing (MEC);
                Deployment of Mobile Edge Computing in an NFV environment. https://www.
                etsi.org/deliver/etsi_gr/MEC/001_099/017/01.01.01_60/gr_MEC017v010101p.
                pdf
[MEC-003]   ETSI GS MEC 003 V2.2.1 (2020–12) Multi-access Edge Computing (MEC);
                Framework and Reference Architecture. https://www.etsi.org/deliver/etsi_gs/
                MEC/001_099/003/02.02.01_60/gs_MEC003v020201p.pdf

[CRAN-WP]   China Mobile Research Institute (2011) C-RAN: The Road Toward Green RAN
[TIP-ORAN]   Telecom Infra Project (TIP), Open RAN. https://telecominfraproject.com/
                openran/
[MEC-CRAN]   ETSI white paper #23, Cloud RAN and MEC: A Perfect Pairing", first edition,
                February 2018. https://www.etsi.org/images/files/ETSIWhitePapers/etsi_wp23_
                MEC_and_CRAN_ed1_FINAL.pdf
[MEC-027]   ETSI GR MEC 027 V2.1.1 (2019–11) Multi-access Edge Computing (MEC);
                Study on MEC support for alternative virtualization technologies. https://www.
                etsi.org/deliver/etsi_gr/MEC/001_099/027/02.01.01_60/gr_MEC027v020101p.
                pdf

Chapter 7
Edge Computing in 5G Networks

We have already acknowledged multiple times in this book that a great advantage of MEC technology is to access agnostic, as the multi-access edge computing architecture (and the related ETSI MEC standard) doesn't depend on the specific underlying network access technology.

Nonetheless, there is a quite unanimous consensus in the industry to consider, in general, edge/MEC deployments coupled with the 5G introduction (often with the progressive introduction of distributed core network), as MEC is considered as a key asset to fully exploit the potential of new cellular networks.

That's why the importance of this chapter, where we will provide an overview of 5G systems, with particular reference to 3GPP standards (from Rel.15 on) and ETSI MEC standards (especially the recent publication of "MEC in 5G" report), in order to provide then an explanation of the recent work on MEC synergized architecture for the harmonization of the two standards. Finally, an overview of the MEC support for network slicing and verticals is concluding the chapter.

7.1 Overview of 5G Networks

5G is the new generation of the global telecommunication network. 5G will support a number of diversified vertical sectors, different types of users and services, both machine- and human-type communications (Fig. 7.1).

Three are the main usage scenarios for 5G as depicted in the above spider chart:

- **Enhanced Mobile Broadband (eMBB)** to deal with hugely increased data rates, high user density and very high traffic capacity for hotspot scenarios as well as seamless coverage and high mobility scenarios with still improved used data rates.

© Springer Nature Switzerland AG 2021
D. Sabella, *Multi-access Edge Computing: Software Development at the Network Edge*, Textbooks in Telecommunication Engineering, https://doi.org/10.1007/978-3-030-79618-1_7

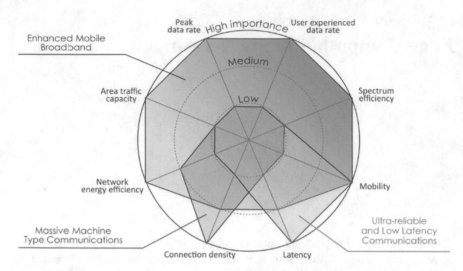

Fig. 7.1 Main 5G usage scenarios. *Source* ETSI: https://www.etsi.org/technologies/mobile/5g

- **Massive Machine-type Communications (mMTC)** for the IoT, requiring low power consumption and low data rates for very large numbers of connected devices.
- **Ultra-reliable and Low-Latency Communications (URLLC)** to cater for safety-critical and mission critical applications.

5G is also expected to create a huge economic value, by contributing (according to GSMA estimations[1]) up to 2.2$ trillion by 2034, on the various sectors, from ICT and Trade, Agriculture and Mining, Manufacturing and Utilities, Professional and Financial Services, and Public Services. In accordance with these forecasts, 5G is also expected to create new and enhanced existing operator monetizable attributes, over the so-called "vertical market segments". Among these monetizable attributes and key components of the new 5G era, there is indeed the *Edge Computing*.

The three above-described usage scenarios (eMBB, URLLC, mMTC) impose a set of heterogeneous KPIs (Key Performance Indicators), to address diverse and demanding new applications and E2E (end-to-end) services. 5G is not limited to a **new radio** design, but encompasses the whole network, including a new **service-based architecture**, and introduces several innovations, like **network softwarization** (see Chap. 6), including **network slicing** and of course **Edge Computing**, as key assets for the fully exploit the potential of new cellular networks.

The above KPIs (see Fig. 7.2) can be translated in a set of requirements (in terms of latency, traffic volume density, throughput, but also reliability, energy consumption, cost, etc.) which are at the basis of the 5G system design.

---

[1]https://www.gsma.com/wp-content/uploads/2019/04/The-5G-Guide_GSMA_2019_04_29_compressed.pdf.

Fig. 7.2 Heterogeneous KPIs
for the main 5G usage
scenarios

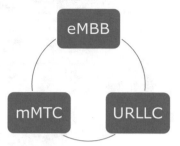

- Low latency
- High spectral efficiency
  / throughput

eMBB

mMTC   URLLC

- Improved link budget
- Low device complexity
- Low device battery life
- High density device depl.

- High reliability
  (Low packet error rate)
- Low latency

It is worth clarifying that, for each usage scenario, these are very *heterogeneous* requirements,[2] and it is simply not reasonable to pretend that 5G systems will address them contemporarily, at the same time. On the contrary, the whole concept of network slicing is supporting this need to meet diverging requirements, coming from different services. In few words, network slicing is about the creation of E2E virtual networks at the service of a class of applications. Slices are obtained through a softwarized management of the network resources; slices can be flexibly and quickly defined and setup, to support diversified vertical services. The interested reader is invited to find more background information in the literature [xx], while the MEC support for network slicing is described in Sect. 7.6.

7.1.1   5G System Architecture

As anticipated, a key component of 5G systems is the introduction of a service-based architecture. Shortly, this is characterized by the following innovations:

- Network control functions expose APIs based on HTTP/2 and RESTful.
- Providing flexibility, simplifying deployment and evolution of the network.
- Harmonizing the entire network control plane with web technologies.

More in detail, the new 5G system architecture has the great advantage to support natively Control and User Plane Separation (CUPS). This isn't a new

---

[2]More details on the 5G performance KPIs can be found in the Final Evaluation Report [Feb. 2020] from the 5G Infrastructure Association on IMT-2020 proposals. Link: https://www.itu.int/en/ITU-R/study-groups/rsg5/rwp5d/imt-2020/Pages/ws-20171004.aspx.

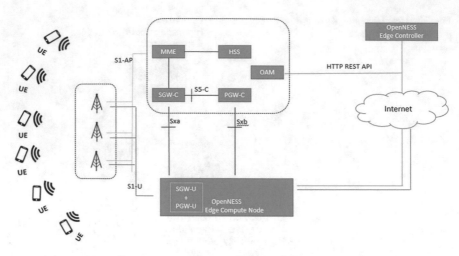

Fig. 7.3 Example of LTE evolved packet core (EPC) with CUPS, as important asset for MEC deployments. *Source* www.openness.org

concept in wireless networks, of course (in fact, CUPS is part of the 3GPP Release 14 standard[3]). Anyway, CUPS is important for 5G because it enables to distribute resources throughout the network, which is key to having an efficient core network. Furthermore, CUPS is essential to 5G networks because it allows operators to separate the evolved packet core (EPC) into a control plane that can sit in a centralized location, for example, the middle of the country, and for the user plane to be placed closer to the application it is supporting (Fig. 7.3).

The key concepts introduced by the 5G system architecture are then:

- Based on NFV and SDN.
- Support for edge computing.
- Service-based interactions between CP network functions.
- Separate the User Plane (UP) functions from the Control Plane (CP) functions.
- Minimize dependencies between the Access Network (AN) and the Core Network (CN).
- Support "stateless" NFs, where the "compute" resource is decoupled from the "storage" resource.
- Support capability exposure.

5G Core is thus "CUPS-native", in the sense that its system architecture support natively this separation between control plane and user plane: Fig. 7.4 is showing the 5G system architecture in "reference point representation", highlighting the interaction existing between the NF services in the network functions described by

---

[3]For CUPS in Rel 14 please refer to 3GPP TS 23.401 and 23.214.

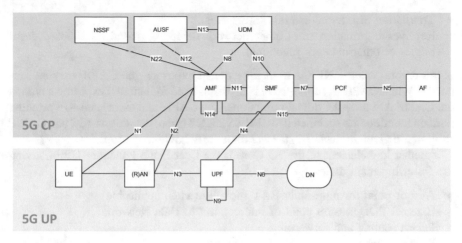

Fig. 7.4 5G system architecture, in "reference point representation" (ref. 3GPP [TS 23.501])

point-to-point reference point (e.g. N11) between any two network functions (e.g. AMF and SMF).

An alternative way of showing the 5G system architecture is the so-called "service-based representation" (Fig. 7.5), where network functions (e.g. AMF) within the Control Plane enable other authorized network functions to access their services. This representation also includes point-to-point reference points where necessary.

A key element of the 5G Core (visible only in this last representation) is the NEF (Network Exposure Function), which is in charge of the following main duties:

- Exposure of capabilities and events.
- Secure provision of information from external application to 3GPP network.

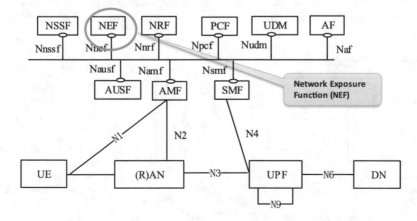

Fig. 7.5 System architecture, in "service-based representation" (ref. 3GPP [TS 23.501])

- Translation of internal-external information.
- Receives information from other network functions (based on exposed capabilities of other network functions).

As a note, when a NEF is used for external exposure, the CAPIF may be supported. When CAPIF is supported, a NEF that is used for external exposure supports the CAPIF API provider domain functions. The CAPIF and associated API provider domain functions are specified in 3GPP TS 23.222 (more details on MEC and CAPIF support are given in Sect. 7.7, talking about MEC 5G integration).

Another key element of the 5G Core is the User Plane Function (UPF), having the following main duties:

- Anchor point for intra-/inter-RAT mobility (when applicable).
- External PDU session point of interconnect to Data Network.
- Packet routing and forwarding.
- Packet inspection.
- Lawful intercept (UP collection).
- Traffic usage reporting.

This is a key element of 5G systems, for MEC deployments, essentially because it is the unique point of access of data traffic from the 5G base station. Section xx will provide a more detailed explanation of the MEC deployment in 5G.

As anticipated, 5G Core is "CUPS-native". In addition, in 5G systems CUPS is extended to RAN. The purpose of this book is not to provide a complete overview of 5G system design (especially the 5G NR, New Radio), as this would deserve a more detailed description. Instead, our goal here is to put the accent on few specific topics which are actually more relevant from a MEC developer perspective. The reader interested also on radio aspects of the 5G radio design is invited to refer to the waste literature in the space [5GRAN20] (Fig. 7.6).

When it comes to actual implementation and 5G deployment options, although several possible 5G configurations have been proposed, two deployment models (or options) have been standardized to meet initial market requirements. These are the Non-standalone (NSA) and Standalone (SA) 5G deployment models. With these,

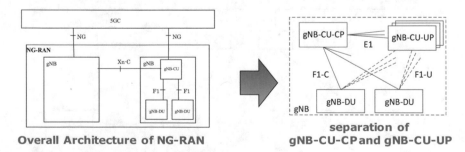

Overall Architecture of NG-RAN   **separation of**
                                                   gNB-CU-CP and gNB-CU-UP

Fig. 7.6 CUPS for 5G RAN (ref. 3GPP [TS 38.401])

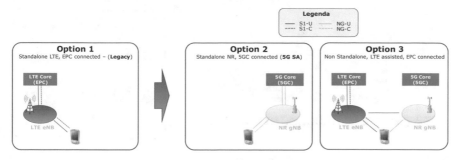

Fig. 7.7 Option 2 (NSA) and Option 3 (SA) for 5G deployments

different operators will have different approaches on *when* and *how* to deploy 5G (Fig. 7.7).

In this book, we will analyze in detail only the above main 5G deployment options, i.e. NSA (Option 2) and SA (Option 3). For a more complete overview of all possible configurations, the reader can have a look at the recent GSMA document on "Operator Requirements for 5G Core Connectivity Options" [GSMA-5G], describing the various deployment options, and including the results of a survey conducted by GSMA with 20 operators (and operator groups) who are planning to launch 5G in the next 2 years to understand which of the options they are considering for their deployment plans.

7.1.2 5G Non-Standalone Deployments

In this mode, 5G NR (New Radio) is used, but still LTE Core (EPC) is used for control (C-Plane) functions, e.g. call origination, call termination, location registration, etc., whereas 5G NR will focus on U-Plane alone. With Option 3, connection to both LTE cell and 5G cell is mandatory (this is the so-called Dual Connectivity, DC. At control plane (dashed line in the figure below), 5G radio control parameters are exchanged via LTE RAT. To have this, functions should be added to eNB. Moreover, UE monitors paging channels as well on LTE RAT (Fig. 7.8).

This scenario is particularly attractive for operators who prefer to migrate their 4G core network in a second phase, and want to first introduce the 5G NR. Thus, Option 3 constitutes a convenient way to optimize their investments on network infrastructure, by progressively adding 5G radio base stations (which are essential to trigger the selling of new 5G terminals).

To be more precise, Option 3 is actually a family of **EN-DC** (E-UTRA—New Radio **Dual Connectivity**), in three variants (3/3a/3x). For all of these dual connectivity variants, both 4G base station (eNB) and 5G NR base station (gNB) are

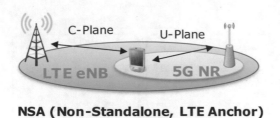

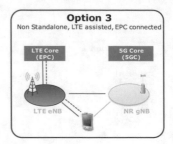

NSA (Non-Standalone, LTE Anchor)

Fig. 7.8   Option 2 (NSA) for 5G deployments

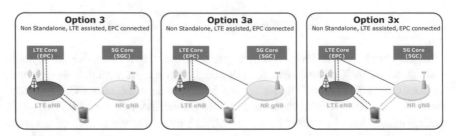

Fig. 7.9   Comparison between the three EN-DC scenarios (Option 3, 3a, 3x)

connected to EPC, where the eNB is acting as Master (MeNB) and gNB working as Secondary (SgNB). Nevertheless, the following differences apply for these variants:

- **Option 3**: eNB terminates S1-U interface, and NR gNB haven't any interface to EPC. The traffic flow is converged at eNB PDCP layer and divided from the eNB to the gNB via X2 interface.
- **Option 3A**: gNB also has S1-U interface to EPC. Traffic flows are split in the core network, and the different service bearers can be carried in LTE or 5G NR.
- **Option 3x**: gNB has S1-U interface to EPC. The traffic flow is converged at gNB PDCP layer and divided from the gNB to the eNB via X2 interface (Fig. 7.9).

7.1.3   5G Standalone Deployments

In this mode, the new 5G Core is actually introduced, and the UE works by 5G RAT alone (LTE RAT is not needed, although multi-mode terminals are of course allowed by standard, and envisaged in the market). In this Option 2, the 5G cell is used for both C-Plane (Control Plane) and U-Plane (User Plane) to take care of both signaling and information transfer. 5G radio control parameters are exchanged via 5G NR RAT, and also UE monitors paging channels on 5G cell (Fig. 7.10).

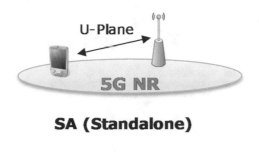
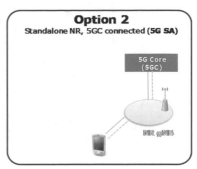

Fig. 7.10 Option 3 (SA) for 5G deployments

This deployment scenario is specifically attractive in areas where there is no legacy LTE system and full-fledged 5G NR access system is required to be deployed.

For early deployment of full-fledged 5G networks, this is the most attractive option where the operators can introduce 5G-only service without 4G interworking.

7.1.4 5G Deployment Phases

By summarizing, as a general statement, we can say that a widely accepted sequence for 5G deployment phases (starting from 4G networks) is to consider first Option 3 (5G NSA) and then Option 2 (5G SA). Of course, this is not a rule, but at least a common assumption shared by many operators (Fig. 7.11).

When it comes to MEC, here below are depicted, respectively, the three hypothetical deployment phases, starting from MEC in 4G networks (where the assumption is that a generic operator might still have its centralized EPC deployed), and then moving it as VNF into 5G networks, first as Option 2 (where still the core network is EPC, but already migrated into a distributed architecture, and possibly virtualized) and then as Option 3 (with 5G Core, and then MEC directly connected after the UPF).

An exemplary (and simple) performance evaluation of these deployment phases can be performed by means of system-level simulation, where the three scenarios are compared fairly and in a controlled environment, with a simulated traffic and users in a multi-cell layout.

A good reference for this comparison can be a recent paper [JSAN2020] using an accurate and comprehensive End-to-End (E2E) system simulation model (exploiting Simu5G (http://simu5g.org/) for radio access and Intel CoFluent for core network and MEC), taking into account user-related metrics, such as response time or MEC latency. Main simulation assumptions for this exemplary evaluation are summarized in the figure below, where a video streaming traffic sent by the MEC

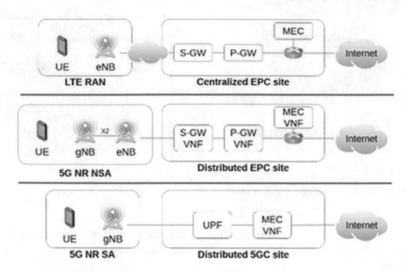

Fig. 7.11 MEC deployment in 4G with centralized EPC (top), 5G NSA (middle) and 5G SA (bottom). *Source* [JSAN-5G]

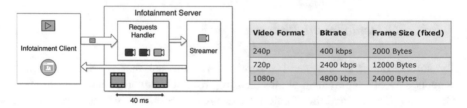

Video Format	Bitrate	Frame Size (fixed)
240p	400 kbps	2000 Bytes
720p	2400 kbps	12000 Bytes
1080p	4800 kbps	24000 Bytes

Fig. 7.12 Parameters of the simulated V2X infotainment service [JSAN2020]

App to mobile subscribers is emulating a typical **infotainment service**, provided in a Cellular Vehicle-to-Everything (C-V2X) environment (Fig. 7.12).

For an accurate evaluation, a multi-cell environment is simulated, with a typical scenario composed of 57 cells, deployed according to a hexagonal tessellation (the figure below shows only the first two tiers of cells for better readability). UEs are deployed according to this urban-grid vehicular scenario, where a growing number of users is simulated (also with different formats of video streams from the MEC App). Main assumptions and parameters are also summarized in the table below (Fig. 7.13).

For this use case evaluation, a first KPI can be the **video frame latency**, defined as the time between the transmission of a video frame from the MEC app and its reception at the UE. This measures the "age" of the information received by the user, and can be further divided into a MEC latency, i.e. from the generation of a video frame at the MEC app to its packets leaving the MEC server, a CN latency, i.e. the time it takes these packets to reach the gNB/eNB, and a RAN latency, i.e.

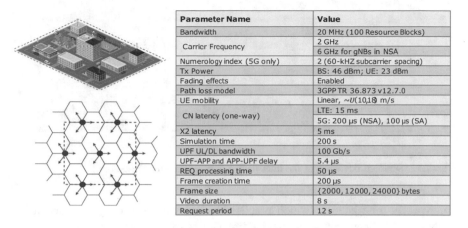

Parameter Name	Value
Bandwidth	20 MHz (100 Resource Blocks)
Carrier Frequency	2 GHz
	6 GHz for gNBs in NSA
Numerology index (5G only)	2 (60-kHZ subcarrier spacing)
Tx Power	BS: 46 dBm; UE: 23 dBm
Fading effects	Enabled
Path loss model	3GPP TR 36.873 v12.7.0
UE mobility	Linear, ~$U(10,18)$ m/s
CN latency (one-way)	LTE: 15 ms
	5G: 200 μs (NSA), 100 μs (SA)
X2 latency	5 ms
Simulation time	200 s
UPF UL/DL bandwidth	100 Gb/s
UPF-APP and APP-UPF delay	5.4 μs
REQ processing time	50 μs
Frame creation time	200 μs
Frame size	{2000, 12000, 24000} bytes
Video duration	8 s
Request period	12 s

Fig. 7.13 Urban grid scenario and main simulation parameters [JSAN2020]

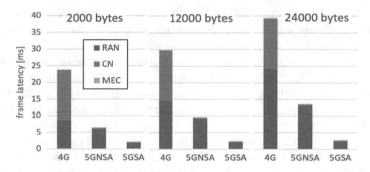

Fig. 7.14 Frame-latency breakdown in a scenario with 24 UEs [JSAN2020]

the time between the arrival of a video frame at the gNB/eNB and its reception at the UE. For more details on MEC performance evaluations and KPIs, the reader is invited to read Chap. 10.

Figure 7.14 shows the trend of the frame latency and its breakdown (for the three deployments considered and in a scenario with 24 UEs). It is quite evident that, in the 4G scenario, both the RAN and the CN introduce a considerable delay; on the other hand, in 5G scenarios, the main latency contribution is given by the RAN. In all the three scenarios, the MEC delay is negligible (in 5G SA deployments, MEC latency starts being a problem only with a bigger number of users in the system).

A careful reader could notice that this first KPI ("initial frame delay") impacts on the starting time of the video playout and can be absorbed by a playout buffer at the receiver side. For that purpose, if we want to assess the actual perceived QoE for this infotainment service, a better indicator of "video freezing occurrence" should be instead another metric. The figure below shows the standard deviation of the

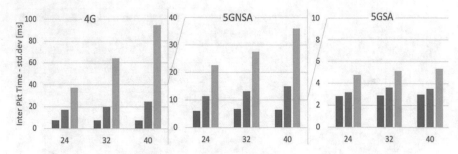

Fig. 7.15 Standard deviation of the inter-packet times at the UE [JSAN2020]

inter-packet time latency (which practically is a measure of the *jitter*), and it is a meaningful indicator of user-perceived quality of experience (Fig. 7.15).

Interestingly, the above plots confirm the same trend of the three deployment phases, but with more insights related to the QoE, and thus the actual user satisfaction. Here, it is quite evident that in the 4G scenarios (where we can still consider the system not congested, from a radio capacity perspective) the jitter is comparable to twice the case of 5G NSA deployments (or even more, in case of 5G SA). These results show then that the 4G network is going to represent the bottleneck for MEC services such as this, and that only deploying 5G will alleviate that.

To summarize, deploying 5G, already in NSA, but especially in the SA option, is going to make a large difference, as it accommodates considerably more users before MEC hosts become a bottleneck. Of course, the previous results are just exemplary of a chosen use case, while real-world measurements of live deployments are expected to provide different numbers; anyway, the comparison is conducted fairly and in a controlled environment (by following state-of-the-art system-level simulation methodologies for these scenarios), thus can provide the user a good benchmark and understanding of the benefits given by 5G deployments.

7.2 Session and Service Continuity in 5G Systems

An important aspect to be discussed before talking more in detail about the edge computing support in 5G is the concept of Session and Service Continuity (SSC), and its three modes of operation. In fact, the support for session and service continuity in 5G system architecture enables to address the various continuity requirements of different applications/services for the UE, and brings significant improvements with respect to 4G systems (the table below summarizes some differences on managing mobility and service continuity, in this perspective).

4G	5G
Mobility, as described at the architecture level in [TS 23.401], was based on a central mobility solution that made it difficult to relocate mobility anchors closer to the end user	In contrast, 5G uses a distributed mobility solution based on multiple anchors providing different IP addresses as the device moves from one area to another
Session continuity is enabled by anchoring a PDN connection (as PDU sessions are referred to in 4G networks) to a P-GW which allocates an IP address to the mobile device: PDN connection and IP address allocation are maintained as long as the device remains attached to the network, even when the device moves around	Different types of session continuity can be provided, and are indicated by a "Session and Service Continuity" (SSC) mode value of 1, 2, or 3 (defined in TS 3GPP.23.501 Sect. 5.6.9)

As anticipated, the 5G system supports different SSC modes, thus providing flexibility in that perspective. Nonetheless, it is worth noting that the SSC mode associated with a PDU session does not change during the lifetime of a PDU Session. The following three modes are specified:

- **With SSC mode 1**, the network preserves the connectivity service provided to the UE. For the case of PDU session of IPv4 or IPv6 type, the IP address is preserved. SSC mode 1 applies to any PDU session type. Additionally, in SSC mode 1, the network may decide to add and remove, dynamically, additional network anchors (and therefore IP addresses) to the PDU session, while always keeping the initial one.[4]
- **With SSC mode 2** (called also break-before-make), the network may release the connectivity service delivered to the UE and release the corresponding PDU Session. For the case of IPv4 or IPv6 type, the network may release IP address (es) that had been allocated to the UE. SSC mode 2 applies to any PDU session type.
- **With SSC mode 3** (known also as make-before-break), the changes to the user plane can be visible to the UE, while the network ensures that the UE suffers no loss of connectivity. A connection through new PDU session anchor point is established before the previous connection is terminated in order to allow for better service continuity. For the case of IPv4 or IPv6 type, the IP address is not preserved in this mode when the PDU session anchor changes. SSC mode 3 only applies to IP PDU session type.

Figure 7.16 shows in a nutshell the main differences between the three SSC modes in 5GS.

---

[4]This would result in a second IP address being allocated on the network interface with which the long-term IP address is associated. This second IP address may be brought down at any time.

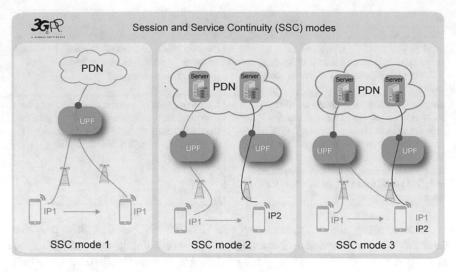

Fig. 7.16 SSC modes in 5GS. *Source* 3GPP

7.3   Network Exposure Function in 5G Systems

Before analyzing the edge computing support in 3GPP, a special attention is deserved by the NEF (network exposure function), which is an entity with a central role in the 5G system architecture (Fig. 7.5). In fact, according to TS 23.501 [TS38-401], the NEF is responsible for service exposure, both for AFs (Application Functions) *inside* and *outside* the 3GPP trusted domain (see also figure below). More in specifics, 3GPP talks about two types of AFs: **trusted AFs** and **non-trusted AFs** (Fig. 7.17).

This is an important clarification in the 3GPP specifications, as it allows in principle the service consumption from exploiting AFs outside the trusted domain, and thus coming from third parties in collaboration with the operator (e.g. application developers or even hyperscalers, in the view of a federation) (see also Chap. 8 for more details on MEC Federations).

7.4   Initial Edge Computing Support in 5G Systems

The support for edge computing in 3GPP specifications can be found since the initial SA requirements for 5G (TS 22.261), which are already covering MEC, even if this term doesn't appear explicitly, but is mentioned more generically as "Service Hosting Environment" (note: the spec is still "stage 1", in 3GPP terminology). In particular, let's have a look at an excerpt of this SA spec:

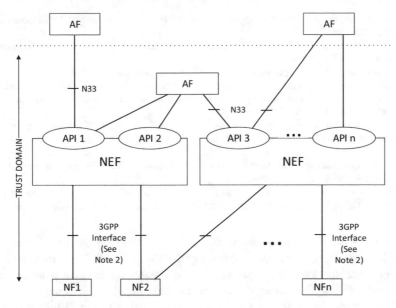

Fig. 7.17 AF inside and outside the 3GPP trusted domain. *Source* [TS 23.501]

3GPP SEES and (e)FMSS features allow the operator to expose network capabilities e.g., QoS policy to 3rd party ISPs/ICPs. With the advent of 5G, new network capabilities need to be exposed to the 3rd party (e.g., to allow the 3rd party to customize a dedicated network slice for diverse use cases; to allow the 3rd party to manage a trusted 3rd party application in a Service Hosting Environment to improve user experience, and efficiently utilize backhaul and application resources).

[…] "Mechanisms for minimizing user plane resources utilization include in-network caching and application in a Service Hosting Environment closer to the end user. These optimization efforts contribute to achieving lower latency and higher reliability"

[…] "The 5G core network shall support charging for services/applications in an operator's Service Hosting Environment"

Interestingly, the goal of operators is not only to enable applications in a Service Hosting Environment closer to the end user, but also to allow the third party to manage a trusted third-party application, which is perfectly in line with MEC assumptions.

Moving to more technical 3GPP work on 5G architecture design ("stage 2" specifications), the support of edge computing is made more explicit and specific (ref. 3GPP TS 23.501). Here below is an excerpt of the related specification:

The 5G Core Network selects a UPF close to the UE and executes the traffic steering from the UPF to the local Data Network via a N6 interface. […]. The functionality supporting for edge computing includes:

- Local Routing: the 5G Core Network selects UPF to route the user traffic to the local Data Network.
- Traffic Steering: the 5G Core Network selects the traffic to be routed to the applications in the local Data Network.
- Session and service continuity to enable UE and application mobility.
- User Plane selection and reselection, e.g. based on input from Application Function.
- An Application Function may influence UPF (re)selection and traffic routing as described in clause 5.6.7.
- Network capability exposure: 5G Core Network and Application Function to provide information to each other via NEF as described in clause 7.4 or directly as described in clause 7.3.
- QoS and Charging: PCF provides rules for QoS Control and Charging for the traffic routed to the local Data Network.
- Support of Local Area Data Network: 5G Core Network provides support to connect to the LADN in a certain area where the applications are deployed as described in clause 5.6.5.

Here, more details on the actual traffic routing and technical enablers for the edge computing support are provided. We can say that this TS 23.501 specification is the main reference for MEC deployments in 5G systems, as it permits us to better understand how MEC architecture can be mapped into 3GPP systems.

In fact, by looking at the 3GPP TS 23.501, according to the definition of User Plane Function(s) (UPFs), they "handle the user plane path of PDU sessions. A UPF that provides the interface to a Data Network supports the functionality of a PDU session anchor".

According to this definition, the logical **UPF** in the 3GPP architecture may correspond to some functionalities defined in ETSI for the **MEC Data Plane** (ETSI GS MEC 003) (Fig. 7.18).

Similarly, according to the 3GPP TS 23.501, an Application Function (AF) in 3GPP contains the following high-level functionalities:

- Application influence on traffic routing (see Sect. 5.6.7),
- Accessing network capability exposure (see Sect. 5.13),
- Interact with the policy framework for policy control (see Sect. 5.14).

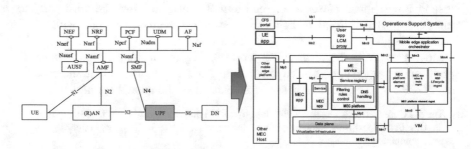

Fig. 7.18 MEC data plane as a UPF in 3GPP

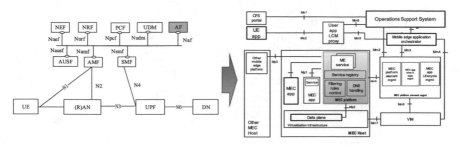

Fig. 7.19 MEC platform as an AF in 3GPP

According to this definition, the logical **AF** in the 3GPP architecture may correspond to some functionalities defined in ETSI for the **MEC Platform** (ETSI GS MEC 003) (Fig. 7.19).

In summary, MEC host can be roughly associated to AF and UPF in 5G arch. In addition, since in 3GPP the user traffic is routed to the local data network, MEC Apps can be mapped in 5G architecture as located in the DN. The figure below shows an example of MEC host deployment, according to this mapping (Fig. 7.20).

5G systems are also expected to run over virtualized environments (see Chap. 6): more detailed schemes and examples of MEC deployment in 5G are also depicted in the figures below. These are showing two cases, respectively: MEC with a 5G virtualized core network (top) and MEC in fully virtualized 5G network, thus co-located with vRAN (bottom) (Fig. 7.21).

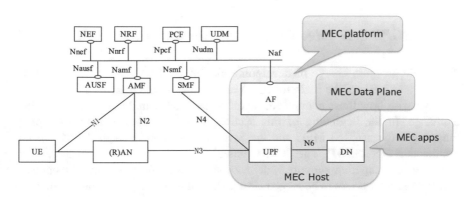

Fig. 7.20 Example of MEC deployment in 5G systems (from Rel 15 on)

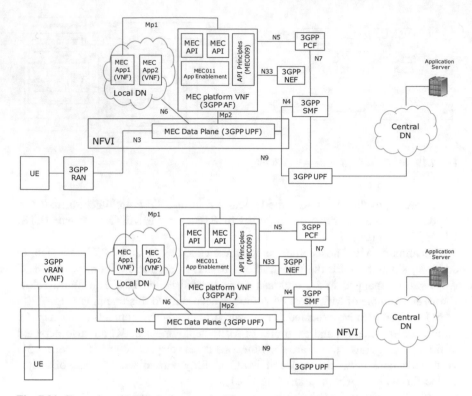

Fig. 7.21 Examples of MEC deployment in 5G systems (top) with virtualized core network, and (bottom) on a fully virtualized 5G network, with MEC co-located with vRAN

7.4.1 3GPP SA2 Edge Computing Support in Rel.15/16

The 3GPP group in charge of developing the "Stage 2"[5] of the 5G system architecture is SA2. Here, the main functions and entities in the network are identified, together with how these entities are linked to each other and the information they exchange. In the beginning of this section, we have anticipated an exemplary mapping between ETSI MEC and the 3GPP entities contained in the 5G system architecture (TS 23.501). The latter is indeed the main SA2 output in the matter, where edge computing support is defined, with the purpose of enabling operator and third-party services to be hosted close to the UE's access point of attachment, so as to achieve an efficient service delivery through the *reduced end-to-end latency and load on the transport network*. Let's analyze more in detail how this edge

---

[5]Note: The output of SA2 is used as input by the groups in charge of the definition of the precise format of messages in Stage 3 (Stage 2 for the Radio Access Network is under RAN's responsibility).

computing support works in 5GS (5G systems) according to the SA2 specification (Release 16).

Here are some key edge computing enablers in 5GS:

- Flexible placement of UPF (user plane function).

  - The data traffic routing is managed by the 5G core network that is selecting a proper UPF close to the UE[6] and executes the traffic steering from the UPF to the local data network via a N6 interface.
  - This enabler is also referred to Fig. 7.18, where it is assumed that the UPF is chosen to serve edge applications close to the user. In addition, as an important added value, the 5G core network may expose network information and capabilities to an edge computing application function.[7]

- Support of multi-homed PDU sessions

  - A UE may establish multiple PDU sessions, to the same data network or to different data networks, via 3GPP and via and non-3GPP access networks at the same time. This support of multi-homed PDU sessions can be realized via either *"Uplink Classifier"* or *"Multi-homed IPv6 PDU Session with Branching Point"*:

    The use of UL CL (Uplink Classifier) is transparent to UE. It uses a single IP address for traffic bound to either local data network (where edge computing server locates) or remote data network.

    The use of multi-homed IPv6 PDU session with BP (Branching Point) is not transparent to UE. UE needs to be provided with routing rules (via IETF defined IPv6 route advertisement message), so it knows which source IPv6 prefix to use for traffic bound to the local network.

- The Session and Service Continuity (SSC) mode 3

  - This SSC3 (named as *"make-before-break"*, with no service continuity interruption) can be used to support 5G edge computing with PSA relocation (IP address change).

The following subsections contain more detailed description of some of these enablers, in order to provide more clarification on how practically data traffic is managed at the edge of 5G networks, since Sect. 7.7 on "MEC 5G integration" will be based also on these enablers.

---

[6]Note: Although defined as a standalone logical function, the UPF can be collocated with the RAN. The absence of LTE's "SGW-like" convergence point facilitates the placement of the UPF.

[7]Note: Talking in ETSI MEC terminology, these can be functionalities that could be on their turn exposed to applications via the MEC APIs. Here, in principle, also according to 3GPP there can be network APIs exposable to edge applications via a generic AF located at the edge.

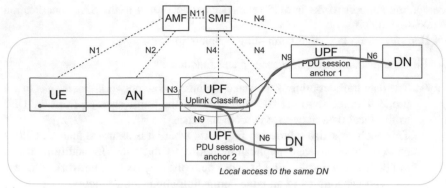

User plane Architecture for the Uplink Classifier

Fig. 7.22 Multi-homed PDU session with UL CL. *Source* 3GPP [TS23.501]

Multi-homed PDU Session with UL CL

The UL CL (Uplink Classifier) is essentially a functionality contained in the UPF that allows steering local traffic to local services (e.g. local CDN server) and the rest of the traffic toward central services (Fig. 7.22).

The above figure shows an UPF with UL CL functionality that provides forwarding of UL traffic toward different UPFs (acting as PDU session anchors, or PSA), who are indeed terminating the UL traffic toward the respective DNs (Data Networks).[8] With this mechanism, the UE uses the same IP address to access either network and is not aware which DN it is communicating with. The UL CL (controlled by SMF over N4 reference point, as showed in the figure) applies filtering rules (e.g. to examine the destination IP address of IP packets sent by the UE) and determines how the packet should be routed. The UL CL supports also connectivity to a local data network (e.g. tunneling) as well as billing/charging, Lawful Interception (LI) and bitrate enforcement.

As an important note, the above figure shows the general case of different UPFs, but of course when it comes to actual deployment options, specific cases may occur. In particular, depending on the need, the UPF (UL CL) and UPF (PSA2) may be also physically collocated.

Multi-homed IPv6 PDU Session with BP

A PDU session may be associated with multiple IPv6 prefixes: this is referred to as *multi-homed PDU session* (here, the UE uses different IPv6 addresses to access the

---

[8]Dually, also the DL traffic from different DNs and coming from different PSAs is merged by the UPF Uplink Classifier back to the UE.

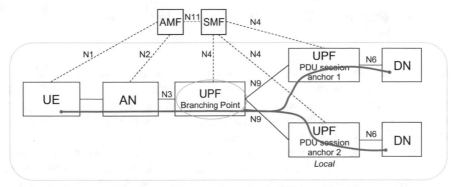

User plane Architecture, when using the UPF Branching Point

Fig. 7.23 Multi-homed IPv6 PDU session with BP. *Source* 3GPP [TS23.501]

network and reach different DNs). In this scenario (figure below), a "common" UPF is referred to as "branching point"[9] that steers the UL traffic toward one or the other IP anchor based on the source prefix of the packet (Fig. 7.23).

The above figure shows the multi-homed IPv6 PDU session with BP that can be used not only to support make-before-break service continuity (i.e. the SSC mode described in the previous section), but also to support concurrent access to a local service (e.g. local CDN server) and the Internet with a different IPv6 prefix.

As an important note for the reader, we should not confuse SSC modes with the deployment of UL CL or BP anchors. In fact, both UL CL and BP can be applied to any SSC mode, as they are enablers to realize multi-homed PDU sessions (for more details, the reader can refer to clause 5.6.9.2 of TS 23.501). Another difference between UL CL and BP is that BP only applies to IPv6 because it requires IPv6 multi-homing support, while instead UL CL can be applied to IPv4, IPv6, or IPv4v6. Moreover, we have also the following difference between UL CL and BP:

- When UL CL is deployed, the uplink packet is determined by UL CL whether it should be routed to the local DN based on the destination IP address in the IP header.
- Instead, when BP is deployed, the uplink packet is determined by BP whether it should be routed to local DN based on the source IP address in the IP header.

---

[9]A branching point for a given PDU session may be inserted or removed by the SMF on the fly.

7.5   Further Edge Computing Support in 5G Rel.17 Systems

As we've seen, Release 15 is the first 3GPP release specifying 5G system, where a support of edge computing is also defined. After further enhancements in Release 16 on some 5G functionalities, in Release 17 3GPP made a significant progress specifically on the edge computing area (figure below shows the Rel.17 timeplan approved in the last December 2020 meeting) (Fig. 7.24).

Figure 7.25 shows a generic edge computing network, as seen from a 3GPP perspective, where an edge data network (owned/managed by a ECSP—edge computing service provider) is communicating with the PLMN operator mobile network, and connected via UPF. As we can see, this figure is coherent with the MEC deployment in 5G described in Sect. 7.4 (and based on Release 15 specifications). In addition, it introduces also the terms EAS (edge application server) and EES (edge enabler server), which are entities introduced in Release 17 by SA6 (see

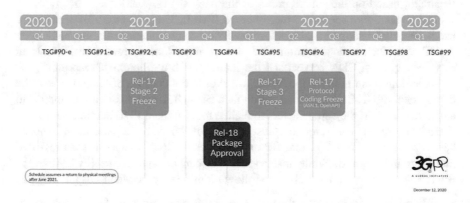

Fig. 7.24   3GPP Release 17 plan approved in December 2020. *Source* www.3gpp.org

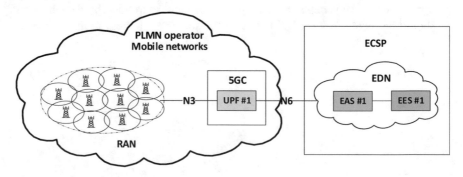

Fig. 7.25   Edge data networks and the PLMN operator mobile networks

Sect. 7.5.1), and basically corresponds, respectively, to MEC applications and MEC platforms, in the ETSI MEC standard.

The following subsections will describe more in detail the 3GPP activities in that perspective: in particular, while so far SA2 defined the system architecture for Release 15 and 16, we now describe the Release 17 work from SA5 for network management and SA6 for application architecture. As an important *caveat*, the reader should be aware that what we write in this book is reflecting the current status of Rel.17 standardization, where a freeze of Rel.17 is planned for the end of 2Q 2021, thus the description in the next subsections, although quite stable, might be subjected to small changes. Furthermore, it is also not excluded that further "stage 3" implementations (across all 2021 and beyond) may reveal any need for CR back in the stage 2 specs. Nonetheless, for the purpose of a general understanding about MEC in 5G, the reader is provided here a quite comprehensive and consolidated overview. Unfortunately, for an actual software implementation of 3GPP APIs, developers should wait for the protocol coding freeze (including OpenAPI representations, planned for 3Q 2021), while an actual availability of 5G products implementing those 3GPP APIs will be expected to come later on, of course.

7.5.1   SA6

The SA6 work on EDGEAPP (which architecture is depicted in the figure below) is basically introducing new functional entities, procedures and information flows necessary for enabling edge applications over 3GPP networks (Fig. 7.26).

The above application layer architecture is conceived to fulfill the architecture requirements and procedures to enable the deployment of edge applications. In

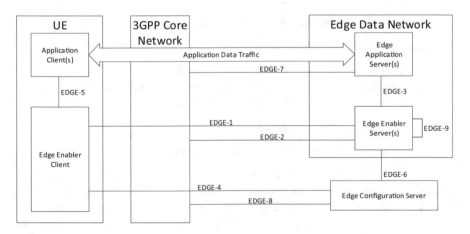

Fig. 7.26 EDGEAPP architecture. *Source* 3GPP [TS 23.558]

particular, other than the well-known client and server application endpoints (here named, respectively, as application client and edge application server), we have three new functional entities:

- **Edge Enabler Client (EEC)**, who enables discovery of edge application servers and provisioning of configuration data.
- **Edge Enabler Server (EES)**, providing to EEC the information related to the edge application servers, and also exposing capabilities of 3GPP network to edge application servers.
- **Edge Configuration Server (ECS)**, providing supporting functions needed for the EEC to connect with an EES.

More in detail, even if the *Synergized Mobile Edge Cloud Architecture* will be presented in Sect. 7.6, we can anticipate that the EES duties are essentially like those of a MEC platform, as defined by ETSI standard (and described in Chap. 3). In fact, the edge enabler server is in charge of the following functionalities:

- Provide information for EAS discovery to the EECs

  - Enablers co-located EASs to register and provide profile information.
  - Distribute EAS information to the ECS.
  - Caters EEC requests for EAS discovery.
  - Supports dynamic notifications to subscribing entities, about EAS updates.
  - Supports Edge-aware Application Clients (ACs).
  - Supports EAS discovery based on desired profiles.
  - Provisioning of EAS source IP address (replacing DNS).
  - Maintains EEC-related context for enhanced services.

- Provide edge services through APIs to co-located EASs

  - APIs for EAS registration,
  - APIs for EAS profile sharing,
  - APIs for providing 3GPP core services to EASs,
  - APIs for UE identification and UE-related events.

- Support application context transfer

  - APIs for subscription for UE-mobility events,
  - APIs assisting the discovery of target EASs (interaction with other EESs),
  - APIs for influencing 3GPP core user plain path settings.

- Support general service discovery

  - Enables EASs to advertise services via CAPIF.

On the other hand, the EAS is basically equivalent to an MEC application, as defined by ETSI standard. In addition, the edge application server(s) (acting as trusted or untrusted AF) may directly access the 3GPP core network capabilities as specified in 3GPP (ref. [TS23-501, TS23-682]). The EES here can re-expose the network capabilities of the 3GPP core network to the edge application server(s) as per the CAPIF architecture specified in 3GPP TS 23.222 (ref. [TS23-222]).

For more detailed information about the EDGEAPP architecture, the reader is invited to look at the 3GPP TS 23.558 (ref. [TS23-558]).

7.5.2   SA5: Management Aspects of Edge Computing

Edge computing is supported in 5GC to enable applications with reduced end-to-end latency requirements hosted closer to the UE's access point of attachment. The goal of *edge computing manage*ment is to deploy the edge computing applications in the local data networks in a manner that the requirements are met.

The current SA5 work on management aspects is captured in the 3GPP TR 28.814 (ref. [TS28-814]), and it first clarifies the different roles of the various players involved. In fact, edge computing environments are heterogeneous, i.e. composed of the interaction of different players, each of them with different roles and duties, thus with an impact from a management perspective. More in detail, Fig. 7.26 shows the roles and relationship of the various actors involved in the deployment of edge computing services (see annex B in TS 23.558). In particular, we have

- The **Application Service Provider (ASP)**, responsible for the creation of Edge Application Servers (EAS) and Application Clients (AC), representing the edge computing application running in the server and UE client, respectively.
- The **Edge Computing Service Provider (ECSP)**, responsible for the deployment of edge data networks (EDN) that contain EAS and EES.
- The **PLMN operator**, responsible for the deployment of 5G network functions, such as 5GC and 5G NR (Fig. 7.27).

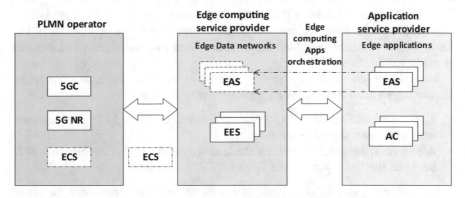

Fig. 7.27 Relationship of service providers in the edge computing network deployment. *Source* 3GPP [TR 28.814]

Based on this split of responsibilities, we can make the following considerations:

- The ASP can have service agreement with one or more ECSP(s) and may request the ECSP to deploy one or more EAS in the EDN.
- Upon receipt of ASP's request, the ECSP should deploy the EAS(s), and then register the EAS(s) to the EES in the EDN.
- The ECSP can have service agreement with one or more PLMN operators and may request the PLMN operators to connect EAS and EES with 5GC network functions.
- The Edge Configuration Server (ECS) may reside in PLMN operator or ECSP.

The edge computing management also covers the lifecycle management of EDN and of the related components (including EAS, EES and ECS), performance assurance and fault supervision of these components.

In any case, the reader should be aware that for the time being, this SA5 study [TS28-814] is still at an early stage, while further development is expected in the future, especially in the view of aligning 3GPP management system with NFV and MEC management, also in accordance with the above-described EDGEAPP architecture. Additionally, SA5 is also working on charging aspects of edge computing in connection with 5GS.

7.6   MEC Synergized Architecture

The recent 3GPP SA6 alignment work with ETSI MEC architecture (described in the Annex C of TS 23.558) deserves a special mention, in the context of this book. This alignment between the two standards is essentially consisting in an exemplary mapping that helps the reader to understand how the two respective architectures are in synergy with each other, and what is the relationship between the various functional blocks defined by the two SDOs. This work is also reported and more widely described in a recent white paper [ETSIwp36], as joint effort of both ETSI MEC and 3GPP SA6 officials and many companies involved in both bodies, published with the intention to show to the industry the value proposition of different standards streams and how those standards may be combined when it comes to deployments (Fig. 7.28).

In particular, the resulting mapping, referred in the White Paper as "*Synergized Mobile Edge Cloud Architecture*", highlights two aspects:

- The equivalence between EAS (edge application server) and MEC applications, who are consuming edge services.
- A basic correspondence between the underlying EES (edge enabler server) in SA6 and the EMC platform in ETSI MEC, as essentially.

On the second aspect, the following table (as an own elaboration from the author of this book) describes more in depth and compares the features and capabilities of

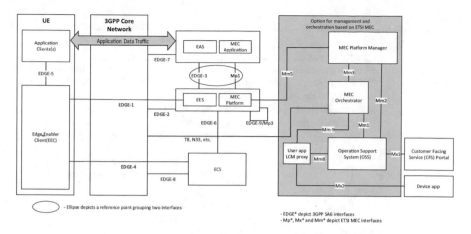

Fig. 7.28 Synergized architecture leveraging ETSI MEC and 3GPP standards. *Source* [TS23.558] [ETSIwp36]

the MEC platform (in ETSI) and the Edge Enabler Server (EES, in 3GPP SA6), to possibly identify similarities and gaps.

Feature/ capability	EES (3GPP SA6)	MEC platform (ETSI MEC)	Comments
General purpose	Provide info for EAS discovery to the EEC provides edge services through APIs to co-located EASs Supports application context transfer supports general service discovery	Provide App look-up procedure through UALCMP Provides MEC services through APIs to co-located or remote MEC Apps Supports application Context transfer supports MEC service discovery (via service registry)	Generally, the two standard groups are defining the same basic platform features
Edge app discovery	EEC/EES interaction Also DNS based is possible ([ETSIwp36], Sect. 3.3.1)	DNS based	While the App discovery in ETSI MEC is only based on DNS, the SA6 standard foresees also other device-based method
Enablement APIs	APIs for EAS registration, APIs for EAS profile sharing,	MEC application support API; MEC service management API; UE App Interface API	Similar set of APIs

(continued)

(continued)

Feature/ capability	EES (3GPP SA6)	MEC platform (ETSI MEC)	Comments
Service APIs	APIs for providing 3GPP Core services to EASs, APIs for UE identification and UE-related events	MEC service APIs (RNI API, Location API, UE Identity API, Fixed Access Info API, Traffic Management APIs, WLAN Info API, V2X Info Service API)	Mainly, the two sets of APIs (related to edge services exposed to the Apps) are complementary. This can be a great added value, especially for products compliant with both standards
Application context transfer	APIs for subscription for UE-mobility events APIs assisting the discovery of target EASs (interaction with other EESs) APIs for influencing 3GPP core user plain path settings	App Mobility Service API (GS 021), application lifecycle, rules and requirements management (GS 010–2)	
General discovery	Enables EASs to advertise services via CAPIF	ETSI MEC aligned with CAPIF (MEC031)	Good alignment

As a preliminary analysis from the above table, at high level the two standard groups (ETSI MEC and 3GPP SA6) are defining the same basic platform features. Moreover, the two sets of APIs (related to edge services exposed to the Apps) are complementary. This can be a great added value, especially for products compliant with both standards.

More in general, the goal of these harmonization efforts is expressing a need from the industry to ensure that both SDOs allow (among the other options) also the possible implementation of a single platform as product compliant with both standards (otherwise imposing to vendors to produce two different products would result in increase of costs, and on an obstacle for the kick-off of the edge computing market). For that reason, in order to open the market, vendors should be allowed (as an option) to possibly implement a single product compliant with two harmonized standards (as claimed, in general, by the white paper).

Since some functions and reference points (e.g. Mp3/EDGE9) are still at early stage of deployment, currently the two standard groups are still working, thus a complete analysis still is impossible (at least until the Rel. 17 will be completed, and Stage 3 definitions will be ready from both sides). Anyway, based on the work done so far by ETSI MEC and 3GPP SA6, we can already see a certain intention from the industry to align, in order to avoid duplication of work and ensure a certain alignment on edge computing standards. In this perspective, the study item MEC 031 on "MEC 5G integration" is a quite useful example of alignment, from ETSI ISG MEC side.

7.7 MEC 5G Integration

The GR MEC 031 is addressing the MEC system interactions with the 5G system, including the correspondence of the current MEC procedures to procedures available in 3GPP 5G system specification, options for the functional split between MEC and 5G Common API Framework (CAPIF), realization of MEC as 5G application function(s). In particular, on one hand, the document clarifies once again the correspondence of MEC as an Application Function (AF), and the Data Plane as an UPF (see figure below). The service exposure, according to 3GPP, is made possible via the NEF that may support Common API Framework (CAPIF) functionality, and more specifically the CAPIF API provider domain functions, for external exposure (according to TS 23.222) (Fig. 7.29).

On the other hand, the report analyzes in detail the MEC system and the Common API Framework (CAPIF). As already anticipated, the NEF is the 5G NF in charge of securely exposing the network capabilities and events to AFs and other kind consumers (TS 23 501), and both trusted AFs and non-trusted AFs. Also, an NEF may support Common API Framework (CAPIF) functionality, and more specifically the CAPIF API provider domain functions, for *external* exposure (TS 23 222). This may help in bridging the two aspects, i.e. service exposure via NEF and CAPIF functionality, so that we can finally summarize the following considerations, for a general mapping between MEC and 5G:

- An MEC application (equivalent to a EAS) is a service consumer, and can be seen as an API invoker in CAPIF, which can be inside or even outside the 3GPP trusted domain; this aspect is enabling third parties to instantiate applications outside the operator's network and (upon agreement with MNOs) consume services via the APIs exposed by NEF and CAPIF.
- Similarly, also a MEC platform consuming a service (acting as AF in 3GPP) can be seen as an API invoker in CAPIF, and can be inside or even outside the 3GPP trusted domain.

Moreover, the ETSI GR MEC 031 highlights how some functionalities of a MEC platform can be associated with the CAPIF Core Function (CCF) in 3GPP, making possible an integration between MEC and CAPIF (for more details on the possible options, the reader is invited to have a deeper look at the MEC 031, Sect. 4.3).

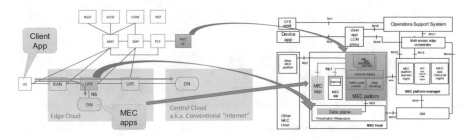

Fig. 7.29 Correspondence between ETSI MEC and 3GPP entities

7.8 MEC Support for Network Slicing and Verticals

Another important technological ingredient of 5G system is network slicing. It was initially defined in the Next-Generation Mobile Network Alliance (NGMN) as "*a concept for running multiple logical networks as independent business operations on a common physical infrastructure*". Essentially, network slicing is transforming a PLMN from a single network to a network where logical partitions are created, with appropriate network isolation, resources, optimized topology and specific configuration to serve various service requirements. When we talk about network slicing in 3GPP, network slice instances can be provisioned to fulfill certain communication service purposes, and to fulfil diverse requirements requested by a particular application. In that perspective, network slicing is not only realized at RAN level (as an evolution of the concept of RAN sharing), but also at core network and transport level (thus, by partitioning all layers of 5G infrastructure). Network slice instances can be also conveniently created/activated/de-activated/modified/terminated based on the service needs (Fig. 7.30).

The above figure shows an example of three different network slices, with different characteristics. In particular, we can make the following considerations:

- In this exemplary deployment, some network functions are in common for slices #1 and #2, including the AMF and the Related Policy Control (PCF) and Network Function Services Repository (NRF).
- While slice #1 provides the UE with data services for Data Network #1, and slice #2 for Data Network #2. Those slices and the data services are independent of each other apart from interaction with common access and mobility control that applies for all services of the user/UE.
- Slice #3: in this example, all network functions serve a single network slice only; thus, we can say that the network slice is entirely dedicated to serving Data Network #3.

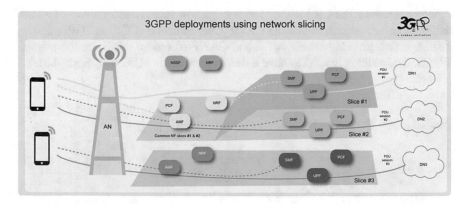

Fig. 7.30 Exemplary 5G deployment of three network slice instances. *Source* 3GPP: https://www.3gpp.org/news-events/1930-sys_architecture

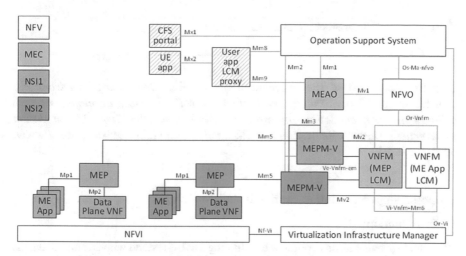

Fig. 7.31 Example of network slicing with MEC. *Source* [MEC-024] © European Telecommunications Standards Institute 2019. Further use, modification, copy and/or distribution are strictly prohibited

Network slicing is particularly interesting and profitable for operators, because it represents a tool for exploiting their 5G infrastructure, for better monetization. In fact, with the recently emerged concept of Network-Slice-as-a-Service (NSaaS), network slice instances can be even offered by an operator (or, in general, a CSP—communication service provider) to its customers in the form of a communication service.[10] Typical examples of these customers, belonging to many vertical market segments, can range from automotive players, industrial automation firms, etc. Additionally, also these customers can play the role of CSP and offer their own services (e.g. communication services) on top of the network slice instance.

Since edge computing is often a key technology for many 5G use cases, the above-described new 5G era monetizable attributes include not only network slicing but also edge computing. In this perspective, MEC support for network slicing is also a relevant topic addressed by ETSI ISG MEC (ref. [MEC-024]), especially in the view of exposing network capabilities to application developers (Fig. 7.31).

For more information about network slicing support, the reader is invited to look at the MEC024 report, while the figure below shows an exemplary case of multiple network slices and association with the different MEC architectural entities belonging to each slice instance.

---

[10]Note: This service allows customers to use and optionally manage the network slice instance.

	Exercises/examples—Chapter 7

Exercise 7.1 Goal of the exercise: *Setup an E2E open source environment, suitable for MEC performance evaluation (or MEC services development) with cellular network emulation, by properly configuring the SimuLTE emulation environment* (https://simulte.com/) *with Intel OpenNESS toolkit* (https://www.openness.org/).

Proposed stepwise solution (in five installation steps).

1. Introduction

We want to provide an environment for building a MEC testbed with an emulated LTE network. The MEC environment is implemented by the Intel OpenNESS software kit, whereas the LTE network emulation is provided by SimuLTE. For the sake of concreteness, we will consider the use case of a video streaming edge application.

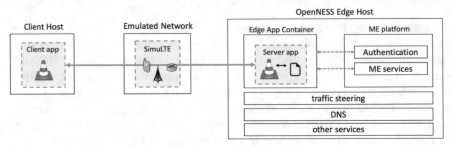

The above figure depicts a high-level representation of the implemented framework, which is composed of a client, a server, and an emulated network in between. The client application may run on either a virtual machine (VM) or a dedicated laptop/desktop and it consists in one instance of the commercial VLC media player. The server application is a Docker container hosted within the OpenNESS edge node and it consists in a customized version of the VLC software. Details of the application are described later in the document. The above container may interact with the MEC platform for exploiting MEC services and receives data plane traffic according to the configured traffic policies in the Edge Host. The 4G network is emulated using SimuLTE, run in real-time mode.

2. Implementation of the edge application

The server-side application in our framework is a video streaming application that

- takes a predefined video file as input;

- applies real-time transcoding, i.e. converts "on the fly" the video to H264 format using the $\times$ 264 encoder;
- makes the output of the transcoding process available for network streaming via HTTP on the well-known TCP port 8080.

The commercial software VLC provides natively all the above features. In particular, we referred to the development branch of VLC 4.0, available on GitHub.[11] Moreover, it is open source, and hence its source code can be modified so as to obtain new functionalities.

In order to make the application run on the OpenNESS environment, we build a Docker container running Linux Ubuntu 18.04 as operating system and including all the necessary dependencies for compiling VLC. Once the container is started, it runs the following command:

```
vlc/vlc -vvv bbb_640x360.mp4 --loop --repeat

--sout                                                  'trans-
code{vcodec=h264,venc=x264,vb=1000}:standard{access=http,mux=
ps,dst=:8080}' --intf telnet --telnet-password vlc
```

The command launches an instance of VLC taking the movie called bbb_640 $\times$ 360.mp4 as input. Options --loop and --repeat allow the video to start from the beginning once it gets the end of the video. Option --sout specifies the chain of command to be executed for streaming the video toward the network. The transcode module specifies that the video will be transcoded to H264 format using the $\times$ 264 encoder, and the initial bitrate is set to 1000 kbps. The standard module specifies how the transcode module's output is made accessible. In this case, the stream can be accessed using HTTP protocol on port 8080.

We remark that for our use case employing the RTSP protocol is tricky, if feasible at all. With RTSP, video and audio streams are sent by the server using a set of UDP ports which are assigned dynamically at the time of the request from the client, hence one cannot know them a priori. As we will describe later, one needs to specify which TCP/UDP ports to enable when configuring the application on the OpenNESS controller, but the current version of OpenNESS requires one to specify ports one by one, i.e. it does not allow one to specify the whole range of ports or a subset of them, making the configuration of RTSP not practical.

The two last options allow us to handle the video streaming remotely, by connecting to the server via telnet and on port 4212. We use this interface to dynamically change the bitrate of the video streaming, by typing the command vb < new_bitrate_value > in the telnet prompt.

---

[11]https://github.com/videolan/vlc.

This edge application does not need to exploit services from the MEC platform, and hence the interactions (e.g. authentication, subscription to MEC services) with the MEC platform are not implemented.

In order to onboard the VLC Edge App on the OpenNESS Edge Node, the Docker image must be saved in a compressed tar.gz archive, e.g. "*vlc-edge-app.tar.gz*".

3. Edge application on-boarding

In this exercise, we refer to OpenNESS, 20.06 version. Make sure that OpenNESS for Network Edge is fully installed and set up as shown in the figure below. Detailed installation instructions are available here: https://www.openness.org/docs/doc/getting-started/network-edge/controller-edge-node-setup.

Note that the Traffic Generator Host and the OpenNESS Edge Node are physically connected through dedicated network interfaces for the data plane (i.e. different from the interfaces used for OpenNESS management). Before going ahead, make sure that the Traffic Generator Host is able to ping the OpenNESS Edge Node and its Edge Apps (e.g. the OpenVINO example app coming with OpenNESS). The blue box in the figure shows the configuration of the IP routing table on the Traffic Generator Host.

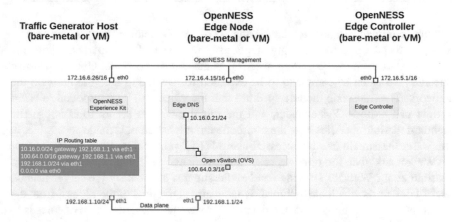

In order to onboard the VLC Edge App on the OpenNESS Edge Node, the archive "*vlc-edge-app.tar.gz*" must be available on the OpenNESS Edge Node.

On the OpenNESS Edge Node

1. Load the image to Docker registry by typing the command:

```
docker load -i vlc-edge-app.tar.gz
```

2. Check that the image was built successfully and is available in the local Docker image registry:

```
docker images | grep vlc-server
```

3. Since the VLC Edge App is a *consumer app*, assign it new name and tag to highlight that, by typing:

```
docker tag vlc-server vlcserver-cons-app:1.0
```

where "vlcserver-cons-app" is the new image name and "1.0" is the assigned tag.

4. Check that the name and tag have been changed by typing:

```
docker images | grep vlcserver-cons-app
```

On the OpenNESS Edge Controller

1. Edit the application deployment script:

```
vi vlcserver-cons-app.yaml
```

```
apiVersion: v1
kind: Pod
metadata:
    name: vlcserver-cons-app
    namespace: default
    labels:
      name: vlcserver-cons-app
spec:
    containers:
    - name: vlcserver-cons-app
      image: vlcserver-cons-app:1.0
      resources:
        limits:
          ephemeral-storage: "2Gi"
      imagePullPolicy: Never
```

2. Once the script has been saved, create the application pod:

```
kubectl apply -f vlcserver-cons-app.yaml
```

3. Check that the pod is running (it might take some time to get it to the "Running" state):

```
kubectl get pods | grep vlcserver-cons-app
```

```
[root@controller ~]# kubectl get pods | grep vlcserver-cons-app
vlcserver-cons-app                1/1       Running      0          44m
```

4. Create a network policy for the VLC Edge App:

```
vi vlcserver_policy.yml
```

```
apiVersion: networking.k8s.io/v1
kind: NetworkPolicy
metadata:
  name: vlcserver-policy
  namespace: default
spec:
  podSelector:
    matchLabels:
      name: vlcserver-cons-app
  policyTypes:
  - Ingress
  ingress:
  - from:
    - ipBlock:
        cidr: 192.168.1.0/24
    ports:
    - protocol: TCP
      port: 80
    - protocol: TCP
      port: 8080
    - protocol: TCP
      port: 443
    - protocol: TCP
      port: 4212
```

5. Once the script has been saved, apply network policy:

```
kubectl apply -f vlcserver_policy.yml
```

6. The last step is to obtain the IP address of the pod running the VLC Edge App, by typing:

```
kubectl exec -it vlcserver-cons-app ip a
```

```
[root@controller ~]# kubectl exec -it vlcserver-cons-app ip a
kubectl exec [POD] [COMMAND] is DEPRECATED and will be removed in a future version. Use kubectl kubectl exec [POD] -- [COMMAND]
instead.
1: lo: <LOOPBACK,UP,LOWER_UP> mtu 65536 qdisc noqueue state UNKNOWN group default qlen 1000
    link/loopback 00:00:00:00:00:00 brd 00:00:00:00:00:00
    inet 127.0.0.1/8 scope host lo
       valid_lft forever preferred_lft forever
59: eth0@if60: <BROADCAST,MULTICAST,UP,LOWER_UP> mtu 1300 qdisc noqueue state UP group default
    link/ether 32:5b:or:19:00:24 brd ff:ff:ff:ff:ff:ff link-netnsid 0
    inet 10.16.0.35/16 brd 10.16.255.255 scope global eth0
       valid_lft forever preferred_lft forever
```

Take note of the IP address as it will be needed by SimuLTE and the client application (10.16.0.35 in this example).

4. Configuring SimuLTE as a network emulator

At this point, we need to let data originated by the client host flow through the network emulated by SimuLTE and then reach the server-side application running in the OpenNESS edge host, and vice versa. Since both SimuLTE and OpenNESS are able to exchange IP packets with the external world, the problem of making them interoperable is a networking problem, i.e. forwarding IP packets from SimuLTE to the OpenNESS domain and vice versa. In order to let SimuLTE work as network emulator, the environment on the host running SimuLTE must be configured accordingly.

We assume the deployment depicted in the following simplified schema, where one host runs SimuLTE and has two network interfaces. Its *eth0* interface is attached to the OpenNESS Edge Host, whereas its *eth1* is attached to another host (the client) on another private local network.

```
   |--------|                                           |---------------|
|--------------------|
   | client | < ---- > [if: eth1]| SimuLTE's host |[if: eth0] < ----
> | OpenNESS Edge Host |
   |--------|                                           |---------------|
|--------------------|
```

We assume that SimuLTE runs an instance of the emulated network. In particular, we use the network called "EmulatedNetwork" included into the file "*simulations/emulation/EmulatedNetwork.ned*". Data packets coming from the server (i.e. OpenNESS Edge Host) are received by eth0 interface and injected into the "router" element of the running instance of SimuLTE. From there, data are forwarded through the emulated LTE network toward the UE. When data packets reach the UE, they are sent to the client host through the eth1 interface.

In order to set up the above scenario, the following steps are required to configure the operating system on this host (these instructions refer to Ubuntu 16.04 OS).

(A) Setting up a (private) local network between the client and the host running the simulation (this host).

   1. Physically connect the client to the SimuLTE's host, e.g. via Ethernet cable.
   2. Within the SimuLTE's host, identify the name of the interface the client is connected to, by running the command ifconfig. This will be eth1 in our example.
   3. If down, activate the interface, by running "ifup eth1".
   4. Assign an IP address to eth1

      – Run "sudo nano /etc./network/interfaces".
      – Add an entry like the following:

```
auto eth1
iface eth1 inet static
address 192.168.55.1
netmask 255.255.255.0
network 192.168.55.0
broadcast 192.168.55.255
```

Alternatively, if you do not want to set up the address manually, set up a DHCP server on interface eth1 and assign address automatically to the client. In any case, take note of the IP address assigned to it.

5. If needed, "reboot" the interface by running: `ifdown eth1; ifup eth1` or restart networking service: `sudo systemctl restart networking`.
6. If DHCP was not used, on the client configure the network so as the client has an address belonging to the same network (e.g. 192.168.55.2) and set the gateway as the IP address of this host (e.g. 192.168.55.1). Add route toward the gateway.

(B) Provide Internet access to the client through SimuLTE's host:

1. Enable IP forwarding, by running "nano /etc./sysctl.conf" and edit the file as follows:

   net.ipv4.ip_forward = 1.
2. Add NAT rule:

```
    sudo    iptables    -t    nat    -A    POSTROUTING    -s
192.168.55.0/24 -o eth0 -j MASQUERADE
        sudo iptables-save
        sudo iptables -L
```

3. Add route for packets destined to the client's network:

```
    sudo    route    add    -net    192.168.55.0    netmask
255.255.255.0 eth1
```

(C) In order to use SimuLTE as network, configure the OS firewall:

1. Run `sudo gedit /etc/ufw/before.rules` and edit the file as follows:

```
            # block traffic to the server on eth1
            -A ufw-before-forward -i eth1 -d 10.16.0.35 -j
DROP
            # block traffic from server on eth0
            -A ufw-before-forward -i eth0 -s 10.16.0.35 -j
DROP
```

where 10.16.0.35 is the IP address where the server makes the stream available in this example (the address we got during the configuration of the Edge App).

2. Reload firewall, by running "sudo ufw reload"

(D) Get SimuLTE from GitHub and build it:

1. Clone the *emulation* branch from the SimuLTE repository from GitHub, next to the "inet" folder.

```
      git clone --branch emulation https://github.com/inet-
framework/simulte.git
```

2. Start the IDE, and ensure that the "inet" project is open and correctly built.
3. Import the project using: File | Import | General | Existing projects into Workspace. Then select the workspace dir as the root directory, and be sure NOT to check the "Copy projects into workspace" box. Click Finish.
4. Right-click on the "inet" folder | Properties | OMNeT + +| Project Features. Tick the "Network emmulation support" box. Click OK.
5. You can build the project by pressing CTRL-B (Project | Build all).
6. To check if the installation went fine, try running an example from the IDE: select the simulation example's folder under "simulations", and click "Run" on the toolbar.

(E) Configure the SimuLTE environment (use the "simulations/emulation" folder as starting point):

1. In the omnetpp.ini file, set the UE's "extHostAddress" parameter to the IP address of the real host's interface (the one connected to the Internet, i.e. eth0). Get the IP address by running the command "ifconfig"

   – `*.ue.extHostAddress` = "192.168.55.2".

2. Configure the routing tables of all the network devices in the simulated network. To do so, edit the.mrt files, included in the folder "routing`.

   – Edit them so as to enable a path from the router to the UE, when the destination IP address is the one of the real host's interface (e.g. `192.168.55.1`).
   – In the omnetpp.ini file, set the "routingTable.routingFiles" parameters to the path of the.mrt files.

3. In the omnetpp.ini file, set the "device" and "filterString" parameter for both the router and the UE.

   – The "device" parameter is the name of the interface which you want to capture the packets from.
   – The "filterString" parameter is filter expression specifying which packets need to be captured.

   Note: The omnetpp.ini file includes one exemplary configuration, e.g. [Config TCP].

5. **Launching the testbed**

In the SimuLTE host, open a terminal (by clicking on the Terminal icon on the left bar) and move to the simulations/emulation folder within your installation of SimuLTE, and run the simulation *with root privileges*. Now, the emulation has started (see the log of the OMNeT++ simulation in the terminal). The next step is to start the VLC client application.

On the Client host, open the GUI of VLC player. Then, click on "Media" → "Open Network Stream…".

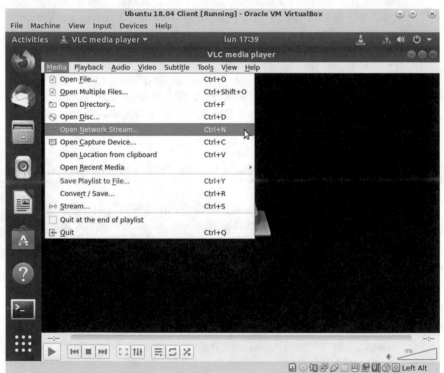

Enter the URL of the video to be streamed `http://<ip_address>:`
`<port>`, where `<ip_address>` is the IP address of the container running the
VLC Edge App (10.16.0.35, in this example) and `<port>` is the port where the
stream has been made available (8080, in this example).

Tick the "*Show more options*" box and select the desired caching time (this is
needed for avoiding playout interruptions). In this example, that is set to 30 s.

Then, click Play.

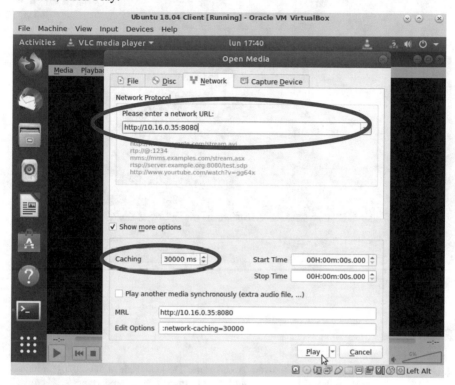

References

[5GRAN20]   5G Radio Access Network Architecture: The Dark Side of 5G (2020) Sirotkin S
              (ed). Wiley-IEEE Press, 448 Pages. https://ieeexplore.ieee.org/servlet/opac?
              bknumber=9289917. ISBN: 978-1-119-55088-4
[MEC-027]   ETSI GR MEC 027 V2.1.1 (2019–11) Multi-access edge computing (MEC);
              study on MEC support for alternative virtualization technologies. https://www.
              etsi.org/deliver/etsi_gr/MEC/001_099/027/02.01.01_60/gr_MEC027v020101p.
              pdf
[TS23-501]   3GPP TS 23.501 V16.7.0 (2020-12) System architecture for the 5G system
              (5GS); stage 2 (Release 16). https://www.3gpp.org/ftp//Specs/archive/23_series/
              23.501/23501-g70.zip
[TS23-682]   3GPP TS 23.682 V16.8.0 (2020-09) Architecture enhancements to facilitate
              communications with packet data networks and applications (Release 16).
              https://www.3gpp.org/ftp//Specs/archive/23_series/23.682/23682-g80.zip
[TS38-401]   3GPP TS 38.401 V15.9.0 (2020-10) NG-RAN; architecture description. https://
              www.3gpp.org/ftp//Specs/archive/38_series/38.401/38401-f90.zip
[TS23-222]   3GPP TS 23.222 V17.3.0 (2020-12) Common API framework for 3GPP
              northbound APIs. https://www.3gpp.org/ftp//Specs/archive/23_series/23.222/
              23222-h30.zip

[TS23-558]   3GPP TS 23.558 V1.2.0 (2020-12) Architecture for enabling edge applications.
                https://www.3gpp.org/ftp//Specs/archive/23_series/23.558/23558-120.zip
[TS28-814]   3GPP TR 28.814 V0.2.0 (2020-10) Study on enhancements of edge computing
                management (Release 17). https://www.3gpp.org/ftp//Specs/archive/28_series/
                28.814/28814-020.zip
[ETSIwp36]   ETSI White paper n.36 (2020) Harmonizing standards for edge computing—A
                synergized architecture leveraging ETSI ISG MEC and 3GPP specifications.
                https://www.etsi.org/images/files/ETSIWhitePapers/ETSI_wp36_Harmonizing-
                standards-for-edge-computing.pdf
[GSMA-5G]   GSMA (2019) Operator requirements for 5G core connectivity options, version
                10.   https://www.gsma.com/futurenetworks/wiki/operator-requirements-for-5g-
                core-connectivity-options/
[JSAN2020]   Virdis A, Nardini G, Stea G, Sabella D (2020) End-to-end performance
                evaluation of MEC deployments in 5G scenarios. JSAN, J Sens Actuator Netw
                9(4). https://doi.org/10.3390/jsan9040057
[MEC-031]   ETSI GR MEC 031 V2.1.1 (2020-10) Multi-access edge computing (MEC)—
                MEC 5G integration. https://www.etsi.org/deliver/etsi_gr/MEC/001_099/031/
                02.01.01_60/gr_MEC031v020101p.pdf

Chapter 8
MEC Federation and Mobility Aspects

In the previous chapters, we discussed MEC in virtualized environments (Chap. 6) and in 5G systems (Chap. 7). In order to complete this third part of the book, it's now worth talking about MEC Federation. This concept is essentially related to a collaboration between various systems (and stakeholders), including edge clouds provided by MNOs, but also CSPs (cloud service providers), neutral hosts, on-premise cloud providers, etc. All these systems are, in principle, applicable to MEC architecture: In fact, MEC stands for multi-access edge computing, thus specifying edge clouds using any kind of access technologies (5G, Wi-Fi, fixed access, etc....). Then, in the first part of the present chapter, we will describe the main aspects related to MEC Federation, the related requirements from the industry (e.g. GSMA OPG) for the implementation of multi-cloud federated environments and the recent standardization efforts in ETSI MEC, to address these requirements in that perspective. In the second part of the chapter, MEC application mobility is also discussed, as an important aspect for the realization of MEC systems in real world.

8.1 Background of MEC Federation

The concept of MEC Federation takes its roots from the concept of Cloud Federation, which is on it turns an evolution of the general cloud computing model [NIST-CC], but then applied to multiple (edge) cloud systems.

There are several literature references that can help in understanding the fundamental basis of the federation model. An important reference can be provided by the recent publication of the "NIST Cloud Federation Reference Architecture" [NIST-FED]. According to the NIST definition, a federation is:

> an organization of self-governing entities that have common policies, administrative controls, and enforcement abilities governing the use of shared resources among members. A virtual administrative domain wherein multiple participating organizations/sites can

© Springer Nature Switzerland AG 2021   245
D. Sabella, *Multi-access Edge Computing: Software Development at the Network Edge*, Textbooks in Telecommunication Engineering, https://doi.org/10.1007/978-3-030-79618-1_8

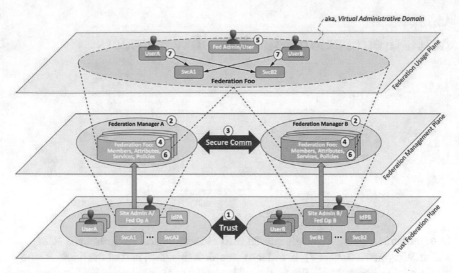

Fig. 8.1 Three-plane illustration of the Cloud Federation Ref. Arch [NIST-FED][1]

define, agree upon and enforce resource discovery, access and usage policies for the sharing of a subset of their resources.

As a consequence, the ultimate goal of a federation is *"the use of shared resources among members"*, and in order to do that, a definition of means for *"resource discovery, access and usage policies"* is needed. The NIST document is thus illustrating its Cloud Federation Reference Architecture in a sort of layered three planes representation of trust, security and resource sharing and usage (figure below is showing the interactions of the actors in these three levels). Here, as a preliminary step, two sites A and B (respectively, in red and blue) need to exchange and share some of their resources in order to reach a common business goal, and thus establish a trust relationship (Trust Federation Plane). Thus, they deploy their own Federation Managers, which are entities providing the necessary federation functions and establishing a secure communication between the two sites (Federation Management Plane). Once this federation is created, a User A belonging to a certain Administrative Domain A can discover and consume services and resources belonging to Administrative Domain B. Since these resources are now part of the federation, we talk about Virtual Administrative Domain (in purple in the Fig. 8.1), where essentially resource/service owners from a certain site (A or B) could make services available to any user in the federation by registering their service endpoints and defining their associated discovery and access policies. These users, services, policies, authorizations, etc. could change dynamically over the course of the federation's lifetime.

---

[1]Reprinted courtesy of the National Institute of Standards and Technology, U.S. Department of Commerce.

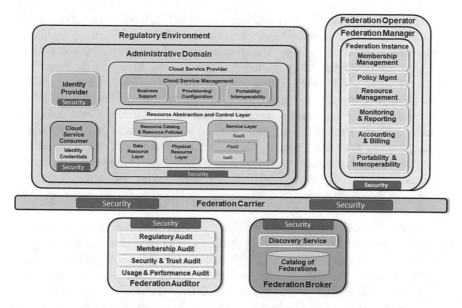

Fig. 8.2 The NIST Cloud Federation Reference Architecture actors [NIST-FED][2]

The above NIST definition is very useful as it identifies some essential characteristics of a cloud federation:

- a virtual **collaboration context**, where the federation is actually not necessarily "owned" by any one user or organization;
- participating entities have a **membership** in the federation and **identity credentials** that are linked to each member;
- users, sites and organizations can participate in a federation by choosing to **share** some of their **resources** and **metadata** and making them discoverable and accessible to other federation members;
- participating members **agree upon the common goals and governance** of their federation, based on well-known roles, attributes and policies.

Cloud federations are thus composed of entities that are agreeing on a certain governance model, with the joint purpose to share cloud resources to reach a common goal (market). Those entities can be widely geographically dispersed and exist under jurisdictions that frequently span multiple national and local domains. We talk then about Administrative Domains (AD) and Regulatory Environments (RE), as shown in Fig. 8.2. The figure shows also the main actors of the NIST

---

[2]Reprinted courtesy of the National Institute of Standards and Technology, U.S. Department of Commerce.

Cloud Federation Reference Architecture: Cloud Service Consumer, Cloud Service Provider, Federation Manager, Federation Operator, Federation Auditor, Federation Carrier and Federation Broker. Those actors are playing different roles in the federation. In particular, users (called here also Cloud Service Consumers) can discover, request and invoke services made available from cloud service providers (even belonging to different entities of the federation), where the underlying resource abstraction and control layer is implementing different cloud computing models, e.g. IaaS, PaaS and SaaS (as described also in the first part of this book).

Here, the central role of security is made evident as part of many interactions between all federation actors. In particular, the Federation Manager has the duty to implement Membership Management, Policy Management and other functionalities, in order to determine which resources can be shared to whom, and with which kind of characteristics.

The NIST model is defined by an entity belonging to the US Department of Commerce, thus mainly focused on US applicability of cloud federations. Recently, also the European Union released its own data strategy [EU-DATA] for the period 2021–2027, where the commission intends to invest €2bn in a European High Impact Project on European data spaces and *federated cloud infrastructures*.

The cloud is in fact a high-priority item in the EU agenda of the data strategy, as it is considered as essential to deploy technologies such as AI, IoT blockchain, or data analytics. This Impact Project will, among other, federate energy-efficient and trustworthy cloud infrastructures and related services. Cloud technologies[3] that have been developed within Horizon 2020 funded research and by market actors will be deployed via the Connecting Europe Facility 2 (for cloud infrastructures interconnection) and Digital Europe (for cloud-to-edge services and cloud marketplaces) Programs. The cloud federation component of the High Impact Project will foster the gradual rebalancing between centralized data infrastructure in the *cloud* and highly distributed and smart *data* processing at the *edge*. In this context, the commission will foster synergies between the work on **European cloud federation** and Member States' initiatives such as GAIA-X.

The GAIA-X project (called also "a federated data infrastructure for Europe"[4]) is an initiative started in 2019 initially from few national governments, and it is progressively getting traction from other EU members states, in collaboration with many industrial players in the field of the IT and telecommunication and cloud service domains. The purpose of the project is to cater for European standards and reference architectures to create EU-based "virtual hyperscale providers". The high-level federation concept of GAIA-X intends, in fact, to represent a strong open infrastructure ecosystem, able to provide services across multiple providers and

[3]https://ec.europa.eu/digital-single-market/en/policies/cloud-computing.
[4]See also the GAIA-X website [GX-Web]

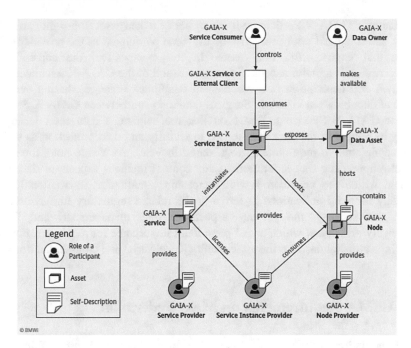

Fig. 8.3 Major relations between GAIA-X Assets and GAIA-X participants. *Source* [GX-TechArc]

nodes of the ecosystem, and as a foundation of digital sovereignty. This federated concept addresses the following challenges:

- Decentralized processing locations.
- Multiple actors and stakeholders.
- Multiple technology stacks.
- Special policy requirements or regulated markets.

More in detail, the technical architecture of GAIA-X could help the reader to have a better idea of the federation [GX-TechArc], which is requiring an appropriate communication infrastructure to enable hybrid cloud and multi-cloud scenarios. Figure 8.3 shows the major relations between GAIA-X Assets and GAIA-X Participants. In this federation picture, GAIA-X Nodes are essentially computational resources in the most general sense, even realized in a hierarchical way (i.e. at country/region/local level). A node here can represent data centers, edge computing, basic hardware, network and infrastructure operation services to more sophisticated, but still generic infrastructure building blocks like virtual machines or containers.

A GAIA-X Data Asset is a dataset that is made available to consumers via a service that exposes the data asset. Not all data is made available in the federation, as consumers and providers can also host private data within GAIA-X. Most importantly, a GAIA-X Participant is the owner of a data asset (top-right part of the figure), and this must not necessarily be the same participant as the provider of the service that exposes the data asset. In this perspective, the capability of self-description is a major aspect of data assets (and of the GAIA-X Architecture as whole). A self-description is a mechanism enabling exchange, sharing and brokerage of data between GAIA-X Services (and also and between GAIA-X Services and non-GAIA-X Services). Based on that mechanism, a data asset is able to specify its own requirements with regard to security and data protection as well as other administrative requirements (e.g. data lifecycle). As a side note, the reader may also appreciate that federation in this context includes aspects of data management which are considered with increasing attention at international level, especially for the implications in terms of the related regulatory framework (e.g. GDPR, for security and privacy aspects of data management) and Digital Sovereignty principles, which are of paramount importance for regulators, not only in public sectors but also for the applicability in industrial and commercial domains.

8.2 GSMA Requirements on MEC Federation

While international institutions (as described in the above) are establishing regulatory frameworks for the federation, also in the industrial domain several stakeholders are already trying and realizing MEC Federations. In particular, the recent GSMA work of the Future Networks Programme,[5] which led the creation of a TEC (Telco Edge Cloud) taskforce of 25 Operators (from Europe, Middle East, Asia-Pacific and the Americas), with the aim to prepare for service launch, and of the GSMA Operator Platform Group (OPG) as a global forum (including 30 Operators and 15 Technology providers) for the definition of the technical framework of the MEC Federation. These are signs of the need for global collaboration between mobile operators, cloud providers and other players in the field of edge computing. This huge effort from many stakeholders led to the publication of two GSMA white papers [OPG-WP, TEC-WP], which are, respectively, defining in detail all technical requirements and main commercial aspects of an MEC Federation.

The GSMA Operator Platform Telco Edge Proposal Whitepaper [OPG-WP] is not defining a standard for that, but mandates a set of functional requirements for the technical realization of an **MEC Federation** (that they call "Telco Edge Cloud concept"), and also poses the basis for a harmonized implementation of the related standards (that most likely could be covered by different SDOs). In particular, the Telco Edge Cloud is defined by GSMA as a "*global distribution system that offers Edge Computing services for customers to develop, deploy and manage edge-based*

---

[5]See also the GSMA Operator Platform Concept Whitepaper, January 2020, https://www.gsma.com/futurenetworks/resources/operator-platform-concept-whitepaper/

solutions over a global footprint, benefiting from the unique capabilities that Telecom operators offer". Main characteristics of this TEC concept are:

- It offers a digital one-stop shop for *edge computing services*, on a **trusted** and **secure environment**.
- It presents a fair and transparent **commercial model** across the value chain.
- TEC is built on *edge technology* based on a combination of **Open Source**, **Cloud** and **Telco standards**.
- It provides a **single and simple interface** (GUI, CLI, API) that facilitates the relationship with multiple **Service Providers** and multiple **Mobile Operators**, interconnected by open and standard federation mechanisms.

The Telco Edge Cloud is composed of multiple OP (operator platform) instances connected together, where each OP instance has the goal to facilitate access to the edge cloud capability of an operator or federation of operators.

GSMA OPG is defining the technical framework and requirements of the platform (called OP—operator platform) that will support the Telco Edge Cloud service. The high-level reference architecture of this MEC federation is depicted in Fig. 8.4, and it essentially consists in a common exposure and capability framework characterized by four types of interfaces:

- **Northbound Interface (NBI)**: It enables service management and fulfilment of enterprise and application providers' use case requirements.
- **East/Westbound Interface (E/WBI)**: The interface between instances of the OP that extend an operator's reach beyond their footprint.
- **Southbound Interface (SBI)**: It is connecting the OP with the specific operator infrastructure that will deliver network services and capabilities to the user.
- **User-Network Interface (UNI)**: It enables the User Client (UC) hosted in the user equipment to communicate with the OP.

A more detailed view of this architecture can be seen in Fig. 8.5, where the different kind of OP implementations are more explicitly depicted, i.e. an edge

Fig. 8.4 High-level reference OP architecture. *Source* GSMA [OPG-WP]

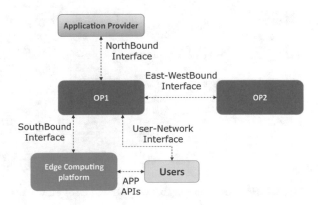

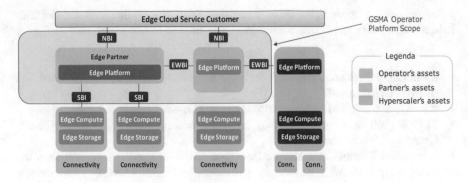

Fig. 8.5 Federation of different MEC stakeholders in the telco edge cloud. *Source* GSMA [OPG-WP]

platform fully managed by operators (in green), or in collaboration with partners (in blue, where they can even act as aggregators) or provided by an hyperscaler (in purple). In all above cases, GSMA is focusing on connectivity block (in green) provided by operators, which in case of 5G network is essentially including both RAN and UPF (as explained in the previous chapter), while the edge hardware resources are provided by the operator or by the hyperscaler. Edge platforms instead can be brought by several kind of stakeholders, hence the need for defining an interoperable EWBI interface in the MEC Federation.

Figure 8.5 clarifies also different possible business roles played by operators, which are also related to the kind of infrastructure owned and managed by them or by partners and other members of the federation.

GSMA OPG/TEC is also building links to other industry bodies from SDOs to associations from other industries. When it comes to ETSI MEC, the GSMA TEC concept and the various roles of the OP architecture can be mapped into the corresponding elements of the MEC framework (Fig. 8.6), while the edge resources

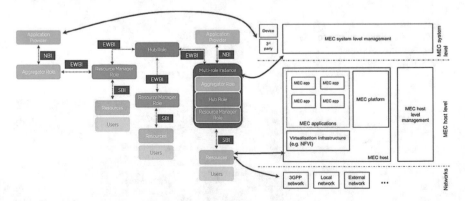

Fig. 8.6 High-level mapping between OP and ETSI MEC framework. *Source* GSMA [OPG-WP]

correspond to the MEC Hosts, other roles of the OP, like Aggregator Role, Hub Role and Resource Manager Role, can be mapped into the MEC system-level management part of the ETSI MEC architecture; furthermore, NBI should connect the application provider, which corresponds to the device in MEC.

Of course, this conceptual mapping doesn't necessarily imply that all ETSI building blocks are already fully supporting MEC Federations (or actually need to do that); anyway, this initial GSMA mapping exercise is, of course, useful for the reader to understand the ongoing standardization efforts in this perspective. In fact, inter-MEC communication work is planned in ETSI ISG MEC as a follow-up of the current MEC 035 work item, and it is believed that this standardization work will be relevant to the area of the EWBI for the purposes of the MEC federation required by GSMA TEC. Next section will describe more in detail this ongoing effort in ETSI to support federation requirements, compatibly with the scope of ISG MEC (since, as anticipated, the whole system is so complex and articulated, that it will need a harmonized implementation of the multiple standards, most likely covered by different SDOs).

The second white paper published by the TEC task force on "Edge Service Description and Commercial Principles" [TEC-WP] is complementing the first document on technical requirements, as it instead deals with the TEC service description, in order to help stakeholders understand the kind of services it will deliver, the different roles of actors involved in the federation and their contribution. Neither this second document is aiming to define any standard for the MEC Federation, but establishes a set of business assumptions agreed in the TEC task force that are complementarily having technical implications for the actual realization of the federation (in particular, this white paper is helping the OPG group complete the specification of requirements, incorporating those needed to accommodate the TEC commercial principles Fig. 8.7).

The underlying reason of the creation of this TEC concept in GSMA is given by the increasing need for operators to better exploit their newly acquired 5G assets

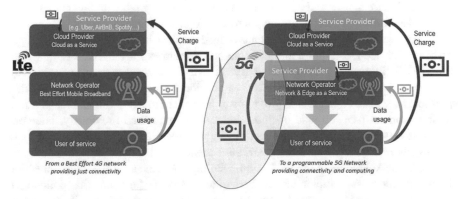

Fig. 8.7 Operator platform: problem statement. *Source* Telefonica's keynote at Droidcon MEC Hackathon 2020

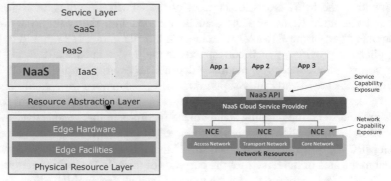

Fig. 8.8 Network-as-a-Service (NaaS) as evolution of cloud computing models (left); network and service capability exposure in the federation (right). *Source* GSMA [TEC-WP]

with respect to previous 4G LTE networks. The above figure well explains this comparison, based on the clear recognition that 5G provides opportunities for operators to *further* monetize their network investment. According to this view, the potential revenue stream for MNOs is not only related to data usage or traditional service charge to service providers, as in LTE networks so far (where MNOs were more focused on providing best effort mobile broadband), but also including a certain amount that users will pay to service providers in partnerships with MNOs. Here network operators will be more focused on offering *network and edge as a service* capability to their partners. This concept of NaaS (network-as-a-service) should be seen not in alternative but in addition to the classical cloud computing models (IaaS, PaaS, SaaS), and it is based on the added value provided by the exposure of network capabilities to application developers and customers. In fact, while the traditional cloud models are all still valid options for the federation, the additional NaaS layer (left side of the figure below) is providing on demand connectivity, between user and edge nodes (via LBO), among edge nodes (managed by load balancing), and from edge nodes to public cloud Fig. 8.8.

In a typical MEC Federation scenario, as described in the GSMA White Paper on Edge Service Description and Commercial Principles for the Telco Edge Cloud [TEC-WP], there are multiple OP partners (MNOs and edge service providers). Some of them are capable (or willing) to provide their edge platform, while other partners may need edge service providers to help them in extending the reach out to their customers (see left side of figure below). As explained previously, interconnections between OP partners are technically realized via EWBI, and they also can eventually make use of an *Edge Interconnection Hub*. This can be a role taken by one of the stakeholders in the federation, who may become an Edge Interconnection Hub Provider and implement a Federation Broker function (as defined in [TEC-WP]) in order to enable one to many interconnections (see right side of Fig. 8.9).

Fig. 8.9 Different OP partners connections in the federation (left); role of Edge Interconnection Hub (right). *Source* GSMA [TEC-WP]

8.3   MEC Federation in ETSI

As briefly introduced in the previous subsection, ETSI ISG MEC has completed a work item [MEC-035], which is a study on inter-MEC communication aiming at identifying recommendations on the corresponding normative work needed for standardization. This normative work may affect the existing specifications such as requirement, use cases, and reference architecture as well as defining a new data model and APIs based on the recommendation from the study.

For the purpose of avoiding market fragmentation and to ensure the end-to-end follow-up of the use cases, ETSI ISG MEC has been considering to coordinate with GSMA OPG and other organizations. Therefore, in MEC 035, ETSI refers GSMA OPG to align with their activities addressing the requirement raised in the group, which is captured as typical MEC Federation scenarios. Based on that, the study defines the information exchange between different MEC systems as follows:

- for system to establish a security trust by authorizing each other;
- for an application provider/customer to deploy its application across multiple MEC systems using a single MNO relationship and integrations;
- for an MEC application in need of consuming an MEC platform service (i.e. remote service consumption) or
- for an MEC application in need of communicating with each other MEC applications (i.e. MEC app-to-app communication).

In order to realize such an information exchange, structuring the hierarchical signaling is the first step of the study. Figure 8.10 is a simplified architecture that provides a conceptual mapping between the OP architecture (described in the previous section) and ETSI MEC architecture, where for the sake of simplicity we showed just the case of a federation composed of two MEC systems. Here, the higher part of each MEC system (which is an OP instance) is the MEC system level and the lower part is the MEC platform level. The NBI (North-Bound Interface) defined in GSMA OP corresponds to the interface(s) coming out from the OSS (e.g. Mx1, Mx2, not shown here). The East/West-Bound Interface (E/WBI) links between system levels of two MEC systems. Note that federation manager is a new

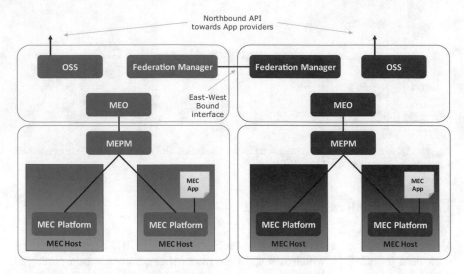

Fig. 8.10 The considered hierarchical functional levels of an MEC federation (simplified scheme based on ETSI ISG MEC [MEC-035])

entity that is still not present in v2 of the ETSI MEC architecture, and indeed is a new requirement of MEC 035 for the introduction in the architecture.

As enablers for the needed hierarchical signaling between the MEC systems, two options are envisaged. The first option is considering the usage of a federation manager (as depicted in Fig. 8.11), which is located in the MEC system level and connected to the MEO. This implies the introduction of two new reference points, called in the figure below *Mfm-fed* (connecting MEO and federation manager) and *Mff-fed* (connecting two federation managers). In accordance with GSMA OP requirements, the federation manager is responsible for the following functionalities: authorization, security, application lifecycle management, resources/platform publishing and discovery, exposure of the catalog of MEC systems, and publishing and discovering dealing all assurance functionalities.

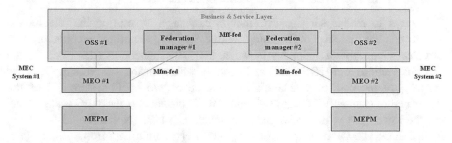

Fig. 8.11 High-level framework for federation manager and reference points. *Source* [MEC-035] —© European Telecommunications Standards Institute 2021. Further use, modification, copy and/or distribution are strictly prohibited

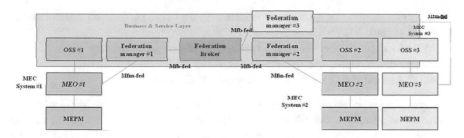

Fig. 8.12 The proposed federation management reference point Mfm-fed connecting an MEC system's MEO with a federation manager. *Source* [MEC-035]—© European Telecommunications Standards Institute 2021. Further use, modification, copy and/or distribution are strictly prohibited

The second option is based on using a federation broker, which is located between multiple federation managers as depicted in Fig. 8.12. This is an alternative solution, to be considered in presence of multiple MEC systems, in order to avoid the complexity to reach a high number of federation agreements. In this context, an additional reference point is needed (called *Mfb-fed* in the figure below) that connects the federation broker to the various federation managers. In practice, the federation broker is regarded as a hub function among high number of federated MEC systems.

In order to enable service consumption or MEC app-to-app communication, MEC system needs to identify which MEC systems are the member of an already established MEC federation, or which MEC systems are available to form a new MEC federation. This process is called MEC system discovery and is depicted in Fig. 8.13. The MEC 035 document considers the following three steps:

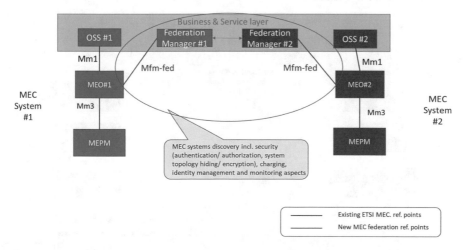

Fig. 8.13 The proposed federation management reference point Mfm-fed connecting an MEC system MEO with a federation manager. *Source* [MEC-035]—© European Telecommunications Standards Institute 2021. Further use, modification, copy and/or distribution are strictly prohibited

1. A service communication query is issued by an MEC application instantiated at MEC system #1.
2. MEO #1 contacts the federation manager to obtain the necessary information of other MEC systems via federation manager.
3. MEO #1 finds out or selects the desired MEC systems.

Once the target MEC system is discovered (or selected), an MEC platform discovery is required. As illustrated in Figs. 8.14 and 8.15, two options can be possible in this process. One is the direct communication between two MEOs, the

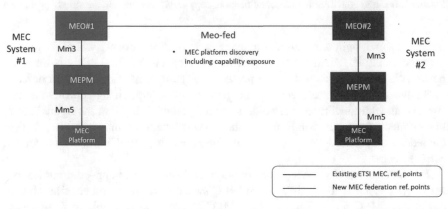

Fig. 8.14 The role of the Meo-fed reference point connecting configured MEOs is to enable inter-MEC system platform discovery including capability exposure. *Source* [MEC-035]—© European Telecommunications Standards Institute 2021. Further use, modification, copy and/or distribution are strictly prohibited

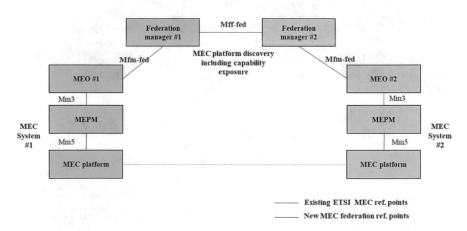

Fig. 8.15 Inter-MEC platform discovery by using the federation manager modules. *Source* [MEC-035]—© European Telecommunications Standards Institute 2021. Further use, modification, copy and/or distribution are strictly prohibited

other is coordination via federation managers. In order to find the appropriate MEC platform, the decision may be made based on the performance (i.e. key performance index values), location or other metrics. The KPI values may include the communication latency between source and target MEC platforms. Therefore, MEOs have to exchange the information in addition to MEC platform identifier. In the case of direct communication between MEOs, since MNOs would not be eager to share details of the internal structure of their managed MEC systems to other MNOs in general, only information essential to the subsequent information exchange. This information exchange should be conducted based on the security policies and agreement between MNOs. In the case of the coordination via federation manager, each MEO shares the required information by means of the exposure of the catalog of MEC systems.

After the MEC platform discovery, remote service consumption and MEC app-to-app communication are finally enabled.

8.4 Edge Resources Exposure: A Case Study

As we have said in the previous sections, the presence of an MEC Federation implies the opportunity to expose edge computing resources to end users. An interesting use case in this perspective is represented by "task application offloading" (i.e. the possibility to offload a computationally demanding task to an MEC host, which can be in the same MEC system or even in another MEC system of the MEC Federation). Moreover, considering today's co-existence of multiple Radio Access Technologies (RATs) such as LTE, 5G New Radio (NR) and Wi-Fi, this section aims at providing a case study by modeling the general problem of multi-RAT application computation offloading for increased energy efficiency at the network and offloading device side.

8.4.1 Application Offloading Use Case in MEC

The evolution of communication systems poses increasing challenges from energy perspective, as today's devices, such as smartphones, are equipped with radio modems for multiple RATs; this is translated into computationally complex signal processing functionalities, such as channel estimation or beamforming. On top of that, application-related tasks (e.g. security keys generation) are becoming increasingly challenging at the terminal side, as seen from an energy consumption standpoint. However, unfortunately, battery capacity does not evolve with the same pace.

Application computation offloading is one of the typical use cases enabled by MEC technology, as listed in ETSI GS MEC 002 [MEC-002]. As shown in Fig. 8.16, an MEC host executes compute-intensive functionalities with higher

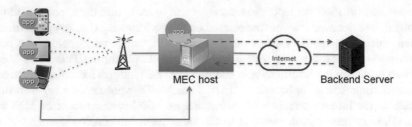

Fig. 8.16 Application computation offloading using MEC. *Source* [MEC-002]—© European Telecommunications Standards Institute 2017. Further use, modification, copy and/or distribution are strictly prohibited

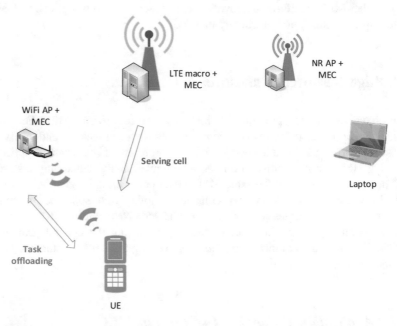

Fig. 8.17 Example of task offloading opportunity. *Source* [MEC-002]—© European Telecommunications Standards Institute 2017. Further use, modification, copy and/or distribution are strictly prohibited

performance, as compared to mobile devices, hence resulting in improved user experience. This enables a consumer to use low-complexity devices by offloading computing execution load to MEC infrastructure. The task offloading "opportunity", an example of which is depicted in Fig. 8.17, depends on the way the trade-off between time and energy needed for computations, on one hand, and the time and energy needed to transmit task instructions (i.e. the input/output files of the offloaded task), on the other hand, will be addressed. As a result, each offloading opportunity may be evaluated by its affordability, depending on the specific application considered and the capabilities (processing, energy autonomy, radio communication) of the potentially offloading device. Application examples are the following:

- Video processing: characterized by huge data to process, but also by huge data to transfer;
- Antivirus (see also work done in EU project [EUTROPIC], https://www.ict-tropic.upc.edu): this scenario is, instead, characterized by a smaller amount of data to be transferred.

At the end, the task can be conveniently offloaded to a server, with energy benefits for the terminal.

In further detail, the incurred challenge for this use case is to address computationally (and, therefore, energy) demanding tasks by exploiting different available RAT resources and differentiating the data traffic to be steered to each RAT node, accordingly. The offloading scheme should take into consideration the required application offloading communication capacity, channel state, backhaul state, task parameters (i.e. number of input/ output bits and CPU cycles per input bit), along with the expected energy consumption of both computation and communication. The RAT interworking, including the allocation of radio resources, should be performed taking into account both the available device energy and application performance requirements, such as the total latency measured between the time instant a task is generated and the time instant the task's output is received by the device.

According to a multi-RAT-aware task offloading framework, the terminal will evaluate: (i) the presence of different RATs (e.g. Wi-Fi, LTE and NR) and their characteristics, including channel state and backhaul state and (ii) application parameters describing the task's processing intensity (CPU cycles per input bit), and the number of exchanged data with the edge server, i.e. task input/output bits. There are two possible ways to accomplish this:

1. **Decentralized approach**: The terminal, by means of communication to each RAT node under coverage, such as an e Node B (eNB), a Wi-Fi Access Point (AP) or a 5G base station, will evaluate trade-offs and decide upon the best offloading solution.
2. **Centralized approach**: The terminal issues an offloading request to the MEC system co-deployed with radio-access infrastructure, that, in its turn, will evaluate trade-offs and assign the appropriate MEC host(s) to process the requested tasks.

An illustration of the system setup appears in Fig. 8.18. It is assumed that a single MEC host is collocated with each RAT access node.

For the purpose of task offloading optimization, the offloading decision entity will need to collect the following information:

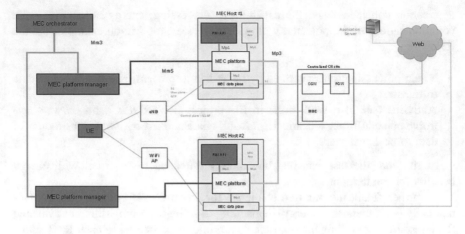

Fig. 8.18 Focused multi-RAT system setup

- Characteristics of the different APs and RATs, including channel state and backhaul state.
- Application parameters, i.e. workload size (measures in CPU cycles) and input/output message payload characteristics.
- Pre-assessment of energy consumption and latency for the different offloading opportunities considering both computation and communication aspects of task offloading and execution.

Based on the above information, the MEC system (centralized approach) or the terminal itself (decentralized approach) will evaluate trade-offs and identify the best offloading solution, by sending a request to offload certain tasks X, Y, Z to edge servers co-located with RAT#1, RAT#2 or RAT#3.

At the end of this chapter, Exercise 8.1 will provide a suitable step-by-step calculation, in order to exemplify these two multi-RAT task offloading approaches (centralized and decentralized).

8.5   MEC Mobility Aspects

When it comes to MEC systems, one of the key assumptions is, of course, the mobility of users connected from client side. This is, of course, both the case of MEC deployed with 5G and with Wi-Fi access networks. And it is also especially true in an MEC Federation, and with an increasing percentage of Internet access from mobile terminals [CISCO-MIDX]. As a consequence, the MEC system should put measures in place to correctly manage the *mobility*, from an MEC perspective. Nevertheless, the reader should not confuse radio mobility issues with MEC and application mobility issues, which are placed at different levels. In fact, if from one

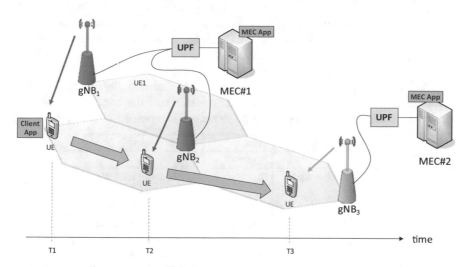

Fig. 8.19 Radio mobility versus MEC mobility

hand the user terminal mobility may cause change of serving radio base station (e.g. gNB), in cellular networks, this radio handover could not necessarily trigger a consequent change of the MEC host (and MEC application) associated with that user client (and client application) (Fig. 8.19).

The above figure is depicting a typical case where radio mobility and MEC mobility are not coinciding, essentially because the deployment mapping of MEC hosts and radio base stations is not necessarily 1:1. In fact, in this simple example, the MEC#1 is associated to gNB_1 and gNB_2, thus in the handover between these base station (T1T2) is not triggering any change from the MEC side (i.e. we talk about *intra-MEC host mobility*). This is instead happening in the second time interval (T2T3), when the UE is moving under the radio coverage of gNB_3, which is connected with MEC#3 (then we have *inter-MEC host mobility*).

In the first time interval (T1T2), still the MEC system doesn't need to relocate service (i.e. the MEC application instance being served to UE and/or UE context) to keep service continuity. Instead, the second interval scenario may result in interruption of service to the UE, thus the MEC system needs to relocate the service to UE from source MEC host (MEC#1) to target MEC host (MEC#2), in order to provide service continuity to UE.

The above figure is showing only a simplified example, suitable to explain the difference between radio/network mobility and MEC mobility. General cases are more articulated. As an additional aspect to consider, depending on the nature of the MEC application (*stateful* or *stateless*), the UE state information transfer is needed or not, respectively. User context is, in fact, a user-specific application state, which is to be maintained/transferred for stateful MEC applications during mobility. Moreover, in the most general cases, the MEC application can be *dedicated* to the UE or *shared* (i.e. serving multiple UEs). In the latter case, it is evident how mobility should also consider constraints due to the presence of other users served by the same MEC host, and this can condition the application instance relocation. The application mobility capability information (e.g. *UserContextTrasnfer Capability* according to ETSI GSM MEC 021 specification [MEC-021]) may be included in the application descriptor (*AppD*) to indicate the stateful/stateless characteristic, the support of user context transfer and the application mobility service dependency. The summary of all the cases is provided in the table below.

Service mobility	App scope	State	Events
Intra-MEC host	Any	Any	No
Inter-MEC host	Dedicated	Stateless	App instance relocation
		Stateful	State transfer and/or App instance relocation
	Shared	Stateless	App instance relocation (conditional)
		Stateful	State transfer and/or App instance relocation (conditional)

The above table is, of course, considering as underlying assumption an intra-MEC system application mobility, while inter-MEC system mobility could include more steps, especially form authentication/authentication/security/charging point of view. At time of writing, ETSI MEC has started working on MEC Federation aspects (see previous sections of the present chapter), but more normative work still has to come. Thus, we cannot exclude that more complex cases may occur, when analyzing inter-MEC system mobility.

Figure 8.20 shows the actors involved in MEC application mobility (considering the intra-MEC system case). MEC application mobility includes the support for the transfer of the user context (i.e. user-specific application state) from the source MEC host to the target MEC host, and the ability of the MEC system to instantiate the application in the target MEC host as a consequence of a UE-mobility event.

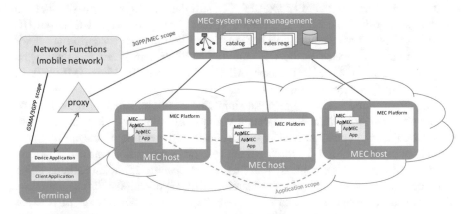

Fig. 8.20 Actors in MEC application mobility (intra-MEC system scope)

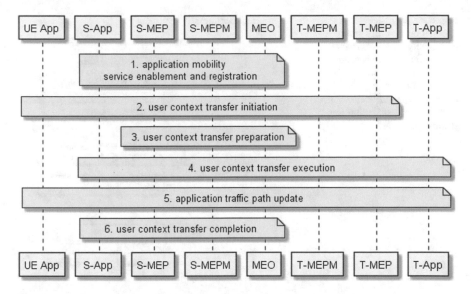

Fig. 8.21 High-level application mobility service information flow

The whole procedure for MEC application mobility is very detailed, and can be summarized in the sub-procedures shown in Fig. 8.21 (note: according to the cases listed in the above, some of them may or may not be present in the actual mobility scenario).

The main sub-procedures of the above sequence diagram are:

1. **Application mobility enablement and registration**: This sub-procedure illustrates the general procedure on enabling the application mobility service and allowing the application instances to register to the required application mobility services.
2. **User context transfer initiation**: This sub-procedure illustrates various detecting and triggering mechanisms for transferring the user context to the target application instance.
3. **User context transfer preparation**: This is an optional sub-procedure for MEC-assisted user context transfer, and used for MEC system to prepare for the transfer.
4. **User context transfer execution**: This sub-procedure illustrates how the user context is transferred to and synchronized on the application instance running on the target MEC host.
5. **Application traffic path update**: This sub-procedure illustrates how MEC system reconfigures the data plane to redirect the traffic to the application instance on the target MEC host.
6. **User context transfer completion**: This sub-procedure illustrates how MEC system to clean up the user context and/or application instance at source MEC host after the user context has been transferred.

Fig. 8.22 High-level application mobility service information flow. *Source* [MEC-021]—© European Telecommunications Standards Institute 2020. Further use, modification, copy and/or distribution are strictly prohibited

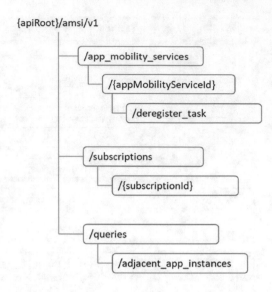

Moreover, mobility aspects related to MEC services like RNIS on the source MEC host and the target MEC host may be involved in the application mobility, and should be seen as part of the individual sub-procedures.

ETSI MEC specifies an AMS (application mobility service) API, in which resource URI structure is illustrated in Fig. 8.22. Many operations are allowed to support MEC E2E mobility, from the typical management of the AMS resources (e.g. create, read, delete, update, register, deregister), to the usual pub/sub-commands for managing notifications (see other APIs in Chaps. 4 and 5). For more detailed information on how to use this API, the reader is invited to have a look at the ETSI GS MEC 021 specification.

8.5.1 Example of MEC Application and E2E Mobility

As in Chap. 6, we have described MEC in 5G, its worth now provides a simple example on how MEC mobility can practically happen when MEC is deployed in 5G systems, and with a fully virtualized network. In what follows, the example provided is corresponding to a typical automotive use case requiring ultra-low latency mobility scenario (Fig. 8.23).

The simple mobility scenario is, of course, triggered by changing radio conditions due to the fast movement of the car under the coverage of different gNBs.

Let's analyze what happens at the different time instants between t_0 and t_4:

- t_0—the UE is attached to RRH1, the vRAN1 is associated to MEC platform # 1 and the MEC app instantiated at Local DN # 1;

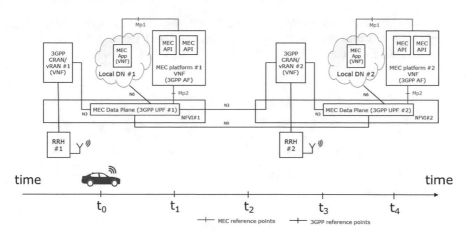

Fig. 8.23 MEC mobility in a typical automotive scenario

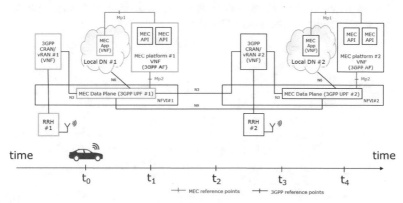

- t_1—the UE is attached to RRH # 2, while the vRAN # 2 is still associated to MEC platform #1 and the MEC app instantiated at Local DN # 1 (as a first step, the RAN handover is performed);

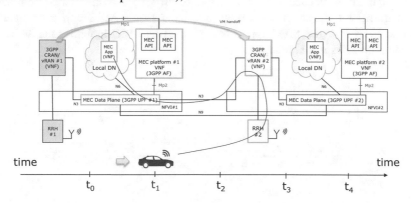

- t_2—the UE is attached to RRH # 2, and the vRAN # 2 is still associated to MEC platform # 1 and the MEC app instantiated at Local DN # 1 VM handoff is performed from UPF # 1 to UPF # 2;

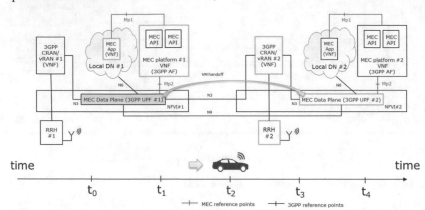

Note: As depicted below, at time instant t_2, both UPF # 1 and UPF # 2 will maintain connectivity to the Local DN # 1.

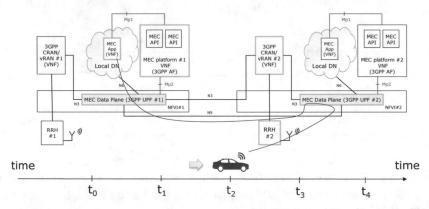

- t_3—the UE is attached to RRH # 2, and the vRAN # 2 is still associated to MEC platform # 1; however, it is now associated to the MEC app instantiated at Local DN # 2 (the MEC app is first affected by MEC mobility; the MEC app communicates through Mp3);

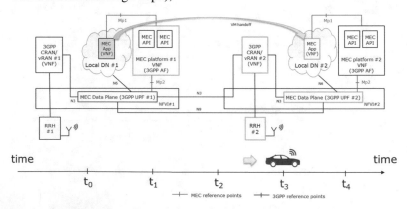

Note: The below figure shows the associations after the MEC app migration and after the handoff of VMs from MEC platform # 1 to MEC platform # 2.

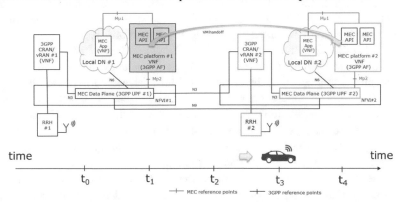

- t_4—the UE is attached to RRH # 2 and the vRAN # 2, associated to MEC platform # 2 and the MEC app instantiated at Local DN # 2. Completion of VM migration procedure;

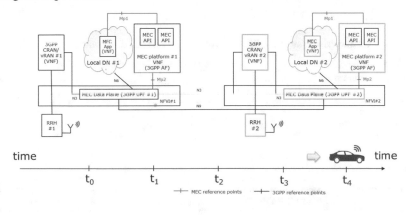

Exercises/examples—Chapter 8

Exercise 8.1 *Goal of the exercise*: Understand the two multi-RAT task offloading approaches (centralized and decentralized) and the criteria that influence the decision entity to offload a computationally burdensome task from a device with limited energy and processing capabilities to an MEC host.

- **Decentralized mechanism**

In Tables 8.1, 8.2, 8.3 and 8.4 and we describe the input parameters needed for the evaluation of the final solution and provide indicative numerical values of these parameters. It should be noted that Table 8.4 is partially based on the inputs of Tables 8.1, 8.2, 8.3 as the latency of the input/output data transfer depends on the quality of the radio links (Table 8.1). Computation latency depends on the current per-unit processing speed of the device/MEC host (measured in msec/ GFLOP, see Table 8.3), while computation energy consumption can be calculated by multiplying the task computational load (in Table 8.2, with the per-unit energy consumption of the task executing device/ MEC host (measured in mJ/ GFLOP, see Table 8.3). Finally, the energy consumption per transmission is obtained by multiplying the input (resp. output) data volume per transmission with the consumed energy needed to transmit a single input (resp. output) bit (measured in mJ/ bit).

Assuming that the input parameter values are provided to or measured by the terminal, the terminal will evaluate trade-offs and reach the best task delegation solution, by sending a request to offload a given generated processing task to MEC hosts co-located with RAT#1 or RAT#2 APs, in case local execution is deemed as

Table 8.1 Characteristics of the APs and RATs to be considered for offloading

MEC host candidates for task offloading	Characteristics of the radio connection		
	Type of RAT (Wi-Fi, LTE, 5G)	Average data rate [Mbps]	Average round-trip time [ms]
MEC Host #1	LTE	10	60
MEC Host #2	Wi-Fi	15	80

Table 8.2 Parameters of the processing task to be potentially offloaded

Task frequency [runs/min]	Task computational load [GFLOPS]	Input data volume [Mbyte/run]	Output data volume [Mbyte/run]
60	1	2	0,5

Table 8.3 Per-unit parameters for processing speed and computation energy consumption and their values (single service of "run")

Task hosting device	Current processing speed [msec/GFLOP]	Per computation energy consumption [mJ/ GFLOP]	Per unit payload reception/ transmission energy consumption [mJ/ Mbyte]
UE	900	300	n.a
MEC Host #1	50	50	(10/ 40)
MEC Host #2	20	100	(5/ 20)

Table 8.4 Pre-assessment of (both radio transmission and processing) latency and energy consumption for the different offloading opportunities

Task hosting device	Latency per transmission (input/ output) [ms]	Energy consumption per transmission (input / output) [mJ]	Computation latency [ms]	Computation energy consumption [mJ]
UE	n.a	n.a	900	300
MEC Host #1	(30 / 30)	(20 / 20)	50	50
MEC Host #2	(40 / 40)	(10 / 10)	20	100

non-viable. In detail, for each emerging task, the terminal will first collect the measures of candidate MEC hosts and their collocated APs, in order to then build the input parameter tables, and decide upon the best MEC host for task offloading, if this is needed. The decision is made by means of comparison of the latency/ energy budget for every possible solution, including the one of local task execution. The involved signaling is shown in Fig. 8.24.

With regard to output, an output table will be derived by summing the latency and energy consumption components of Table 8.4. For our example, output parameter values that will drive the decision appear in Table 8.5

In this case (see Table 8.5), MEC Host #1 is providing a better energy efficiency, and a slightly worse latency than MEC Host #2. The choice of the best target

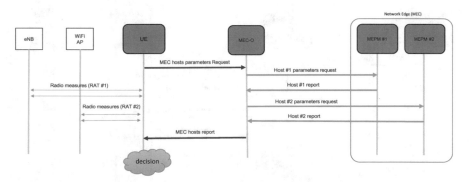

Fig. 8.24 Sequence diagram explaining the decentralized mechanism to decide upon possible task offloading

Table 8.5 Output performance parameter values useful for task offloading decision

Task hosting device	Total latency budget [ms]	Total energy consumption budget [mJ]
UE	900	300
MEC Host #1	110	90
MEC Host #2	100	120

workload host can be made by considering different possible policies, such as (i) total latency budget minimization, (ii) selection of a shortlist of target devices based on a total latency threshold and subsequent selection from the shortlist by minimizing the total energy consumption or (iii) selection of a shortlist of target devices based on a total energy consumption threshold and subsequent selection from the shortlist by minimizing the total latency.

- **Centralized mechanism**

According to this solution variant, the decision-maker is the MEC system that receives from the UE an offloading evaluation request, and collects all parameter values needed to provide a final response to the UE on the optimal task execution solution (i.e. either local execution at the device or delegation to an MEC host). The advantage of this centralized mechanism is that the MEC Orchestrator (MEO) already has full knowledge of MEC host-processing capabilities. Nevertheless, the MEO also needs to obtain information of radio parameters from the RATs focusing on the requesting UE. Such radio parameter values may be acquired by either radio measurement data available through RNI APIs or directly by the device itself. Figure 8.25 explains the involved signaling which is needed for the MEC system's MEO to draw a decision (e.g. offloading to MEC host #2).

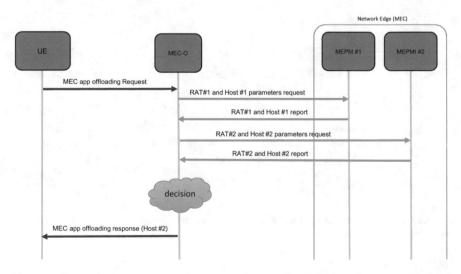

Fig. 8.25 Sequence diagram explaining the centralized mechanism to decide upon possible task offloading

Exercise 8.2   *Goal of the exercise*: Get a closer understanding on MEC application mobility policies, based both on radio and computation aspects.

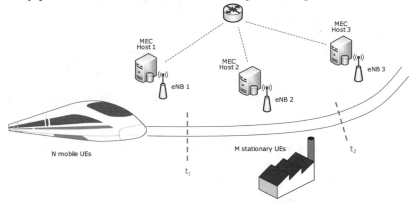

Exercise scenario

A train hosting N mobile UEs travels along a railway, which is covered by three eNBs, called eNB_j, with $1 \leq j \leq 3$. Each of those eNB has a MEC host co-located (MEC_j), which can host applications for UEs connected to any eNB (please notice that in this case there is a 1:1 mapping between radio base stations and MEC hosts, which is a very particular case considered for this exercise; more complex topologies can occur in real-world cases, and in these cases the model below can be suitably updated). M stationary UEs are attached to eNB_2, and they run MEC applications that are hosted by MEC_2. Let's call $d_{i,j}$ the delay incurred in the network transit between MEC_i to eNB_j, and vice versa. Let $d_{i,j} = |i{-}j|D$, i.e. the delay is null if the MEC host that is running an application is co-located with the eNB to which the UE is attached, and non-null otherwise. Assume that the radio-access delay is constant and negligible in the three cells (obviously, this assumption is not realistic, but made up for sake of simplicity, and to focus on MEC application mobility aspects). Each application run at an MEC host incurs a computation delay which depends *linearly* on the number of applications concurrently hosted at that MEC host. If A_i is the number of applications hosted at MEC_i, then the computation delay for each of these is $c_i = \alpha A_i$.

The N mobile UEs are initially attached to eNB_1, and their applications are hosted at MEC_1. At time t_1 and t_2, the N UEs make handovers to eNB_2 and eNB_3, respectively (assume handovers are instantaneous). Assume that the cost of application relocation is null.

1. Write down a formula for the round-trip delay of the N mobile applications in the intervals $[0; t_1]$, $[t_1; t_2]$, $[t_2; t_3]$, under the following scenarios:

   a. The mobile apps are always relocated to the nearest MEC host
   b. The mobile apps are relocated to the *least loaded* MEC host (ties are broken according to distance)

2. Find a scenario (i.e. sets of values for the above parameters, N, M, α, D) such that the delay-optimal policy for the mobile apps is

   a. always to relocate to the nearest MEC host
   b. never to relocate
   c. to stay on MEC_1 until t_2, and then relocate to MEC_3

Solution

1. In $[0; t_1]$, all the N UEs are attached to eNB_1, and their applications are hosted at MEC_1. Therefore, their round-trip delay is $RTT = 2\, d_{1,1} + \alpha\, N = 2\, |1-1|D + \alpha\, N = \alpha\, N$
2. In $[t_1; t_2]$, the round-trip delay of the mobile applications is:
   a. $RTT = 2\ \ d_{2,2} + \alpha\, (N+M) = \alpha\, (N+M)$
   b. $RTT = 2\, d_{1,2} + \alpha\, N = 2\, D + \alpha\, N$

   In $[t_2; t_3]$, the round-trip delay of the mobile applications is

   $RTT = 2\, d_{3,3} + \alpha\, N = \alpha\, N$ in both cases.

(2) Assume that the policy is "always relocate to the nearest MEC Host". This guarantees the smallest round-trip delay in intervals $[0; t_1]$ and $[t_2; t_3]$, since α sN is the minimum delay under the above model. In order for that policy to be optimal also in $[t_1; t_2]$, the following inequality must hold:

$$\alpha \cdot (N+M) \le 2\ \cdot D + \alpha \cdot N$$

After a few straightforward algebraic manipulations, the above yields

$$(\alpha \cdot M)\, /\, D \le 2$$

Assume now that the policy is "never relocate". According to this policy, the delay in $[t_2; t_3]$ would be $RTT = 2\, d_{1,3} + 4D + \alpha \cdot N$. Since relocating to MEC_3 would instead yield $RTT = 2\, d_{3,3} + \alpha\, N = \alpha \cdot N$, this policy can never be optimal.

Assume that the policy is "stay on MEC_1 until t_2, and then relocate to MEC_3". This is optimal in intervals $[0; t_1]$ and $[t_2; t_3]$. For it to be optimal in $[t_1; t_2]$ as well, the opposite inequality must hold, i.e.

$$(\alpha \cdot M)\, /\, D \ge 2$$

| **Quiz—Part 3** |
| (Chapters 6, 7 and 8) |

(1) **CUPS is**

    a. present since Rel.8 networks, but adopted only in Rel. 14, as a CU/DU split in the eNB

    b. a concept only present in 5G core networks

    c. consisting in control and user plane separation, introduced only from Rel.15 onward

    d. part of the Rel. 14 standard

(2) 5G is an important technology because:

    a. 5G will satisfy contemporarily the needs of different scenarios (eMBB, URLLC, mMTC)

    b. 5G will try to satisfy the needs of different scenarios (eMBB, URLLC, mMTC), and not contemporarily

    c. 5G will introduce a lot of innovations, mainly related to new radio design, to support low-latency and high-throughput services

    d. 5G will support low-latency and high-throughput services, expanding the set of 4G addressable services

(3) MEC is compatible with 5G system because

    a. Edge computing is part of the 5G system architecture

    b. MEC is part of the 5G system architecture

    c. Edge computing is part of the 5G system architecture, and MEC is leading standard for edge computing

    d. MEC supports 5G system architecture

    e. Edge computing supports 5G system architecture, and MEC is leading standard for edge computing

    f. None of the above

(4) According to the below diagram on technical 5G performance requirements set by the ITU-R:

    a. 5G system should provide 20Gbit/s of peak data rate, and always with less than 1 ms of latency

    b. 5G system should provide 20Gbit/s of peak data rate, i.e. about 20 times of LTE systems

   c. LTE system can provide up to 20Gbit/s of peak data rate, but not less than
      1 Gbit/s

   d. 5G system can provide up to 20Gbit/s of peak data rate, or less than 1 Gbit/s

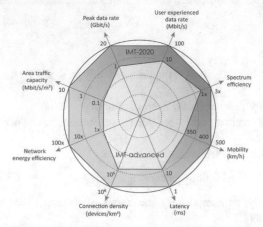

(5) The 5G deployment option depicted in the figure below is

   a. Non-standalone, NR assisted, 5GC connected
   b. Standalone, NR assisted, 5GC connected
   c. Non-standalone, LTE assisted, EPC connected
   d. Standalone, LTE assisted, 5GC connected

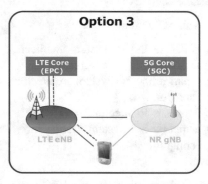

(6) The 5G deployment depicted in the following three figures are, respectively:

   a. Option 3 (NSA), Option 2 (SA), Option 4 (SA with EN-DC)
   b. Option 3 (NSA), Option 2 (NSA), Option 4 (NSA)
   c. Option 3 (NSA), Option 2 (SA), Option 4 (SA)
   d. Option 3 (NSA), Option 2 (SA), Option 4 (NSA)

(7) Key components of 5G systems are

a. New Radio, network softwarization, network slicing, service-based architecture
b. New Radio, radio and core network separation, network slicing, service-based architecture
c. New Radio, control and user plane separation, network slicing, service-based architecture
d. New Radio, network services exposure, access network softwarization, service-based architecture

(8) Consider the SSC modes below:

a. SSC mode 2 is also known as "make-before-break"
b. SSC mode 3 is also known as "make-before-brake"
c. SSC mode 3 is also known as "break-before-make"
d. SSC mode 2 is also known as "make-before-fake"
e. None of the above

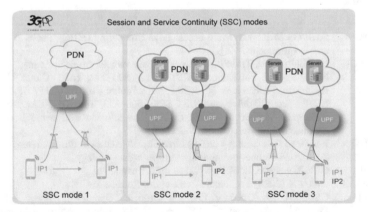

(9) In the 5G Core, the Network Exposure Function (NEF) is in charge of:

a. Exposure of capabilities and events
b. Exposure of entire network slices
c. Mobility and session management control
d. Connecting control plane with user plane functions

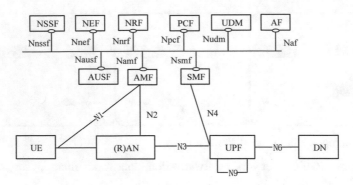

(10) Considering MEC in 5G systems:

    a. the AF is a particular implementation of MEC platform
    b. the UPF in the 3GPP architecture may correspond to some functionalities
       defined in ETSI MEC for a ME application
    c. the MEC application can be mapped with a UPF
    d. the AF in the 3GPP architecture may correspond to some functionalities
       defined in ETSI MEC for a ME platform
    e. None of the above

References

[MEC-002]   ETSI GS MEC 002: Multi-access Edge Computing (MEC); Phase 2: Use Cases and Requirements, v2.1.1, 2018

[EUTROPIC]   FP7 EU project TROPIC, website: https://www.ict-tropic.upc.edu/

[NIST-CC]   NIST Special Publication (NIST SP) - 500–292, NIST Cloud Computing Reference Architecture, 2011, link: https://doi.org/10.6028/NIST.SP.500-292

[NIST-FED]   NIST Special Publication (NIST SP) - 500–332, The NIST Cloud Federation Reference Architecture, 2020, link: https://www.nist.gov/publications/nist-cloud-federation-reference-architecture

[EU-DATA]   European Data Strategy (2020) Link: https://ec.europa.eu/info/strategy/priorities-2019-2024/europe-fit-digital-age/european-data-strategy

[GX-Web]   Gaia-X website, https://www.data-infrastructure.eu/

[GX-TechArc]   GAIA-X (2020) Technical Architecture, Release

[OPG-WP]   GSMA OPG (Operator Platform Group): "Operator Platform Telco Edge Proposal Whitepaper", Version 1.0, Link: https://www.gsma.com/futurenetworks/resources/op-telco-edge-proposal-whitepaper/

[TEC-WP]   GSMA (2020) Telco Edge Cloud: Edge Service Description & Commercial Principles Whitepaper, Link: https://www.gsma.com/futurenetworks/resources/telco-edge-cloud-october-2020-download/

[MEC-021]   ETSI GS MEC 021 V2.1.1 (2020–01) Multi-access Edge Computing (MEC);
             Application Mobility Service API, Link: https://www.etsi.org/deliver/etsi_gs/
             MEC/001_099/021/02.01.01_60/gs_MEC021v020101p.pdf
[MEC-018]   ETSI GR MEC 018 V1.1.1 (2017–10) Mobile Edge Computing (MEC); End to
             End Mobility Aspects, Link: https://www.etsi.org/deliver/etsi_gr/mec/001_099/
             018/01.01.01_60/gr_mec018v010101p.pdf
[MEC-035]   ETSI GR MEC 035 V3.1.1 (2021-06): "Multi-access Edge Computing (MEC);
             Study on Inter-MEC systems and MEC-Cloud systems coordination, Link:
             https://www.etsi.org/deliver/etsi_gr/MEC/001_099/035/03.01.01_60/gr_
             mec035v030101p.pdf

Part IV
Edge Computing Software Development

Chapter 9
Software Development for Edge Computing

In Part 2 of this book we have described the MEC standard, introducing an edge architecture between client and server applications. Then, Part 3 then was focused on deployment aspects (e.g. MEC in virtualized environments, MEC in 5G networks, MEC federations), mentioning also the recent harmonized architecture leveraging ETSI and 3GPP standards. A deep understanding of all these aspects is very important to learn how MEC is realized. However, as we have clarified at the beginning of this book, the MEC ecosystem is composed essentially of two main categories of stakeholders: infrastructure owners and application developers. In order to bring MEC to market success, it is important to involve both categories of stakeholders. For this purpose, we must also recognize that often the goal of software developers is not to understand all details of the standards, but to focus pragmatically on what matters to actually develop applications. Thus, this last part of the book focusses on the software development for MEC (often by referring to ETSI ISG MEC standard, for convenience).

From an application point of view, MEC introduced a 3-tier paradigm, with a sort of "*middle point*" between user and remote server. In this paradigm, which can be called as the "*end-user perspective*" of MEC (figure below), the application is split into "terminal", "edge" and "remote" components, rather than a two-tier client/server paradigm found in cloud computing (Fig. 9.1).

With an additional edge tier, the developer has an additional degree of freedom for locating cloud components, either in a remote cloud (as is typical for cloud computing), and in an edge cloud located with respect to the client.

In this paradigm, application components can be factored according to the following considerations:

- Client App

  - Minor changes, and possible reuse of application logic, in order to minimize legacy service re-design for edge (at least on the client side).

© Springer Nature Switzerland AG 2021
D. Sabella, *Multi-access Edge Computing: Software Development at the Network Edge*, Textbooks in Telecommunication Engineering,
https://doi.org/10.1007/978-3-030-79618-1_9

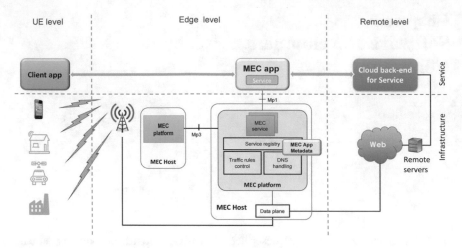

Fig. 9.1 Three-tier MEC application paradigm (the "end-user perspective")

- The Client App may need to authenticate to the Edge and/or Cloud application and negotiate local / remote capabilities based on available resources.

- Edge App

  - Proxy authentication to Remote Cloud.
  - May cache user context for local processing.
  - Needs to synchronize user context with Cloud.
  - May use UE location and other APIs if needed.
  - Use connection quality/bandwidth.

- Remote Cloud App

  - "Owns" service users.
  - Authenticates and maintains user context.
  - Maintains list of edge instances and mapping to served users.

These aspects, in addition to user application mobility-related lifecycle man-
agement operations, cannot be covered having in mind the end-user perspective.
Instead, a *developer perspective* should be considered, which is indeed the focus of
Part 4 of this book.

9.1   MEC: The Application Developer Perspective

The following figure depicts the MEC architecture from the application developer
perspective. With MEC, a UE is able to communicate not only with a remote
application server, but also with another (intermediate) end point (the MEC
application), which leverages a low-latency environment.

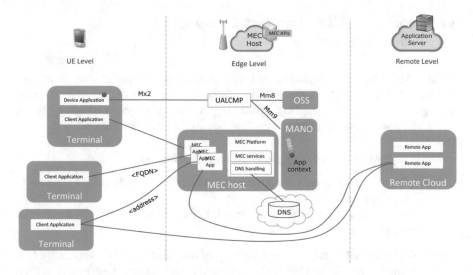

Fig. 9.2 Reference scenario for MEC (the "developer perspective")

In Chap. 4 the client tier was split into a Client App, which interacts with the application components in the edge and remote tiers, and the Device App, which interacts with the MEC system tier. Their roles in the developer perspective will be explained below (Fig. 9.2).

The ETSI GS MEC 016 [MEC-016] is specifying the lifecycle management of the user applications over the UE Application interface (introduced in Chap. 3), which is over the Mx2 reference point between the UE application in the UE and the user application lifecycle management proxy in the mobile edge system.

From a pragmatic point of view, the *simple truth* is that developers don't matter too much about standards and specifications (which are essentially telling us how a MEC system is built), but actually "how to *use* the MEC system", for application development purposes. For this reason, in order to facilitate this understanding and satisfy this need, let's simply describe how things are working from MEC application design point of view. For this purpose, MEC communications can be divided into phases:

Phase 1—MEC application packaging & on-boarding.
Phase 2—MEC application instantiation & communications.
Phase 3—Usage of the MEC platform and services.

These three phases are described in detail in the following subsections.

9.1.1 Phase 1: MEC Application Packaging and On-Boarding

Initially, MEC Apps are packaged by application developers (or in some cases also by MEC operators), typically in a VM or Container. Assuming that an application package exists, the on-boarding procedure is shown in the following figure. The ETSI specifications do not mandate details of the container or VM usage (Fig. 9.3).

These high-level steps for **application on-boarding** are

1. The OSS receives requests for application management and decides whether to grant the requests.
2. Granted requests are forwarded to the MEC Orchestrator (MEO) for further processing.

   • NOTE: the MEO has the responsibility of on-boarding MEC Apps into MEC systems, including checking the integrity and authenticity of the signed packages, validating app rules and requirements and if necessary, etc. and preparing the virtualization infrastructure manager(s) to handle the applications.

3. The MEO assigns an application package ID and provides the MEC Platform Manager (MEPM) with the location of the application image if it is not yet on-boarded in the MEC system.

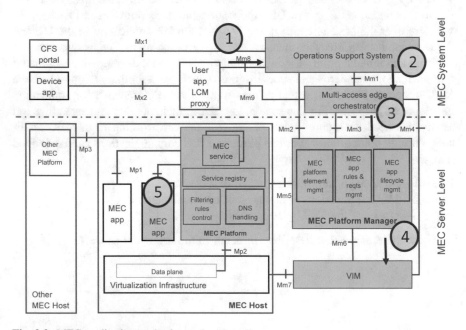

Fig. 9.3 MEC application packaging and on-boarding

4. The MEPM prepares the Virtualized Infrastructure Managers (VIMs), selected by the MEO for application instantiation, by providing the necessary infrastructure configuration information and sending the application images, which are then stored by the VIM.
5. Once on-boarded, the app package is in the "Enabled, Not in use" state.

9.1.2 Phase 2: MEC Application Instantiation and Communications

ETSI MEC specification in principle offers three alternative mechanisms for triggering application instantiation (see figure below) (Fig. 9.4):

1. MEC operator-internal channel.
2. Request from the Customer Facing Service Portal.
3. Request from the Device app via User app LCM proxy.

Regardless of the mechanism used, the MEC system always internally instantiates the application as shown below. At the end of this phase, the application is in the "Instantiated" state (Fig. 9.5).

9.1.3 Phase 3: Usage of the MEC Platform and Services

At this point, a new MEC app instance is created (the MEP indicates service availability, once the MEC app instance indicates that it is operational). In the final phase (after on-boarding and instantiation), communication between the Client App

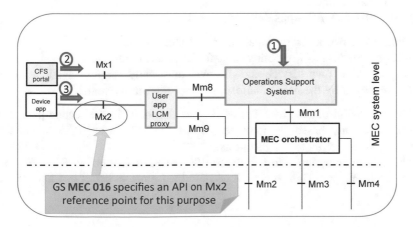

Fig. 9.4 Options for the MEC application instantiation

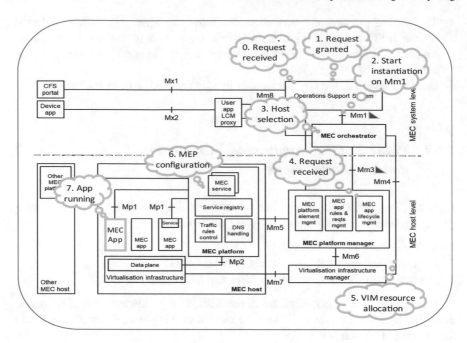

Fig. 9.5 Internal steps for the MEC application instantiation

and the MEC Application occurs, as well as edge service consumption by MEC Apps. Edge service consumption is enabled via MEC APIs, such as the Radio Network Information (RNI) API, which provides authorized applications with radio network related information; the Location API, which provides authorized applications with location-related information; and the Bandwidth Manager API, which allocation of bandwidth to certain traffic routed to and from MEC Apps and prioritizes certain traffic).

In this phase, MEC Apps developers will need to:

- Discover network, users, capabilities and local services.
- Manage traffic, DNS, mobility (e.g. for V2X use-cases, etc.).
- Register their own service and discover third-party services available locally (i.e. exposed by the MEC Platform through the service registry).

These APIs introduced by ETSI MEC (and described in the second part of this book) are mainly based on Representational State Transfer (REST) paradigm, widely used in distributed applications.

There is an ongoing effort in ETSI ISG MEC to define its RESTful to be OpenAPI compliant. This work will be described more in detail in Chap. 10.

9.2   Open Network System Services Software (OpenNESS) Toolkit

We will now look at a particular realization of a MEC system and application, and trace (to some extent) how standards in this area translate to practice.

The Open Network Edge Services Software (OpenNESS) toolkit is designed to foster open collaboration and application innovation at the network and enterprise edge. It is an open source reference toolkit that enables the ecosystem to create and deploy new edge applications and services.

OpenNESS enables the creation of MEC platforms and applications for a variety of domains, such as enterprise, smart city and V2X. It follows what could be called an "edge-native" paradigm, which allows high performance computing elements to be containerized and delivered in configurations that are well suited for the edge computing environment. It supports many different system configurations optimized for various use-cases.

OpenNESS is inspired by edge computing and mobile communication standards. Its general architecture was inspired by the ETSI MEC architecture, although various architectural decisions around security, mobile network integration and cloud native computing have resulted in some differences.

Its approach to integration with mobile networks is based on 3GPP and O-RAN Alliance specifications. It incorporates RESTful interfaces to core network components and can support novel deployment configurations with them that allow for improved performance.

Edge computing, at a high level, can be thought of as cloud computing where the physical location of the cloud matters. An edge computing network can be thought of as a cloud with special constraints, and so cloud native techniques ought to be applicable to edge networks. The special constraints include:

- Footprint and compute density: because an edge node is located in the field, or on-premises at an enterprise, rather than in a data center, the space available for computing resources is constrained. Edge nodes therefore need to provide high compute density, and possibly be specialized for the workloads expected to run in the edge.
- Cardinality and orchestration: the location of edge nodes is chosen to be near to end users. Therefore, the number of nodes in an edge network will be larger than for a cloud. Both resource constraints and cardinality will drive architectures to "scale-out" rather than "scale-up", past a given threshold of required compute capacity.
- Heterogeneity: the three-tier paradigm described earlier implies that the edge tier will offload compute workloads from an end device tier (perhaps for conserving battery power in those devices), and offload real-time compute functions from the remote tier. To achieve the compute density required for these requirements, heterogeneous accelerator hardware may be deployed in edge nodes. Edge applications must be supported in their use of this heterogeneous hardware.

- Access networks: data sources reach edge applications via access networks. For many important use-cases, these are mobile networks, which have control and data plane separation and full sets of protocols and APIs to support routing traffic from an end device to an edge application.

The net effect is that an edge-native system has more "moving parts", more heterogeneity and more constraints in operation than a cloud native system.

9.2.1 OpenNESS System Architecture

The OpenNESS system architecture is depicted in Fig. 9.6[1]. It depicts a "typical" edge node, on which edge applications execute, and a controller, which is used to deploy and manage edge nodes. An edge cluster consists of one or more edge nodes. Multiple edge nodes may be associated with one controller.

The edge nodes and controller contain a variety of functions to support the basic functionality of the system, to deploy accelerators and to manage the lifecycle of edge applications. Not all OpenNESS building blocks are required for every deployment of an OpenNESS platform; the system owner is able to optimize the platform for the use-case.

The OpenNESS system is based on Kubernetes. Thus, many of the building blocks shown here are containers, grouped together in pods so that they can be

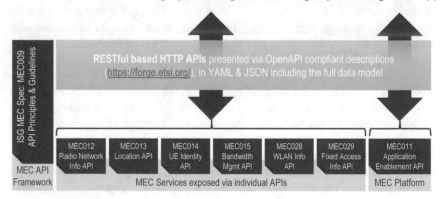

Fig. 9.6 MEC system and the Application Development Community

---

[1]As of June 2021, OpenNESS began a transition to Intel® Smart Edge Open.

deployed together. OpenNESS supports virtual machines for applications, but does so via KubeVirt, rather than natively.

In the edge node cluster, we see CentOS, the host operating system, and Docker, for low level container management. Elsewhere, you can see the standard Kubernetes components, Kube-proxy, Kubelet, Kubectl, etc., that comprise a working Kubernetes cluster.

OpenNESS components appear in both the controller and the edge node. They may be implemented by a variety of methods, including containerized functions, custom resources, or optimizations to the stock Kubernetes distribution. Intel has invested heavily in optimizing Kubernetes for advanced computing resources, and the OpenNESS project takes advantage of these innovations. Some of the optimizations are open source and have been upstreamed to the appropriate communities, and some are proprietary and used in commercial distributions of the toolkit.

System building blocks integrate closely with Kubernetes and provide the fundamental framework for edge computing services and applications. The System building blocks in this figure are shown in two pods in the Edge Node—the System Pod and Container Networking Pod—so that they can be managed together, as well as two building blocks in the Controller.

The Container Networking pod implements the CNI plug-ins that allow the building blocks in the cluster to communicate with each other. The platform uses Kube-OVN as a default CNI but can support other CNIs as well, such as Weave or Calico. Support of multiple CNIs is important because some applications and building blocks have specific CNI requirements.

Because many edge applications require multiple ports, which are not yet supported natively by Kubernetes, OpenNESS uses the Multus CNI to support this functionality.

Other elements of the System pod provide the ability to collect telemetry from the hardware, and to use the telemetry to schedule applications in an optimal manner.

The Platform Building Blocks are deployed in one or more Platform pods. These building blocks are used to support computing and networking accelerators. When an application is deployed, in addition to selecting a destination with enough computing headroom, it can also select a candidate cluster with required special-purpose hardware. These building blocks are specialized for certain families of hardware and work with other building blocks in the controller.

In the controller, the OpenNESS building blocks work with the edge node building blocks, discovering information from the edge node and making orchestration decisions based on that information. The information received includes capabilities and attributes, as well as telemetry information.

Edge applications and services are run in Application Service pods. They are developed by third parties, are made available via application marketplaces and are on-boarded and instantiated from the control plane. These processes correspond to Phases 1 and 2 of the application life cycle described earlier.

Applications may deploy their own software development kits, and consume and produce streams on the data plane between each other, or with the outside world. They use the CNIs and the IP networking capabilities of the edge platform to do this.

The On-Premises and Network Functions block represents the access network, which may be either a cellular network or a "non-3G" network, such as a Wi-Fi or wired network. The mobile network and the edge platform must cooperate to steer traffic from the network to edge applications. OpenNESS provides reference implementations of the Core Network building blocks, as well as building blocks through which applications can configure routing.

So—to summarize this description: the OpenNESS architecture is based on Kubernetes, with a Controller that controls the deployment and operation of building blocks on edge nodes. OpenNESS has System, Platform and Mobile Networking building blocks, which reside in both the Controller and Edge Nodes. They support applications that are deployed as application pods and integrate with external networks.

9.2.2   OpenNESS APIs

The APIs available to applications include RESTful APIs implemented by the OpenNESS system, as well as APIs implemented by Kubernetes, and APIs implemented by runtime libraries for applications.

OpenNESS-specific APIs provide the same functionality as the core APIs specified by the ETSI MEC standards, although they are factored somewhat differently. These are the Edge Application API, Controller API, and Edge Application Authentication API. Jointly, these APIs allow applications to authenticate with the platform, to discover services and other applications and to communicate with services and other applications—all functionality described earlier in Phases 2 and 3 of the life cycle.

Because Kubernetes is used as the underlying container orchestration technology, all components, including applications, operate according to Kubernetes semantics. This includes on-boarding and instantiation of the OpenNESS platform itself, as well as on-boarding and instantiation of applications, covering Phase 1 of the application life cycle.

Within an application container, the application typically runs in a Linux guest operating system and therefore has access to Linux runtime libraries. This includes networking stacks to provide access both to the container network and the external Internet as data planes. The data plane is not specified by ETSI MEC, so 3GPP, IETF and Linux are taken as de jure and de facto standards.

Additionally, edge applications may be linked with application frameworks that provide both runtime libraries and access to hardware accelerators. The most prominent of these frameworks for OpenNESS is the Intel OpenVINO machine learning inferencing framework.

9.2.3   Example OpenNESS Application

This section will discuss an example MEC application that performs video analytics processing from web camera streams. We will describe the application, and how it is implemented using OpenNESS functionality.

This application, "Smart City", uses media streaming and video analytics to recognize people or vehicles from video streams and displays its output in a web interface for visualization by human operators. It uses building blocks that include media streaming and transcoding, as well as hardware to accelerate AI inferencing to detect and classify objects. It uses the OpenNESS APIs to create the video pipelines.

The end-to-end system is depicted in Fig. 9.6. The system takes its inputs from webcams connected to the OpenNESS platform via the public internet. In the example version that we distribute, the cameras are replaced by camera simulators.

The camera input goes into edge platforms that execute the media pipeline. The pipeline is a standalone function that transcodes a media stream and pipes the stream into an analytics pipeline, which looks for the objects that its neural net model is designed to detect. The output of the analytics pipeline goes to an output port, which points to an external function—in this case, a web browser.

Figure 9.7 shows a particular configuration of an OpenNESS platform developed to deploy the application.

As described earlier, since not all building blocks are required for a particular use-case, it is possible to define special "flavors" to support use-cases.

The OpenNESS project has created and validated a number of "flavors" for common use-cases. The flavor shown here is the "Video Analytics Service Mesh" flavor. Figure 9.8 shows an example of Smart City Application architecture, while Fig. 9.9 shows the OpenNESS system deployment for that Smart City App.

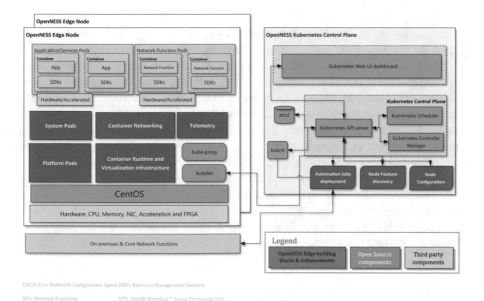

Fig. 9.7 OpenNESS system architecture

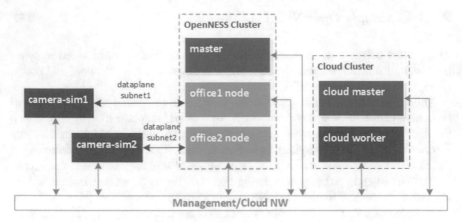

Fig. 9.8 Smart city application architecture

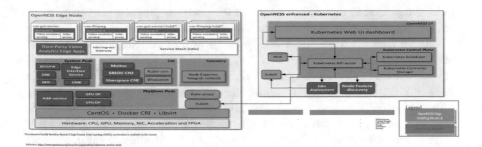

Fig. 9.9 OpenNESS system deployment for smart city app

In this flavor, there are building blocks for traffic steering, for implementing the APIs, for managing the accelerators and for managing the data plane network.

The new, interesting parts of this platform start with the application pods at the top. Each of these is a visual analytics pipeline. Four of them are depicted to show different combinations that can be used—use of either the Gstreamer or Ffmpeg transcoding frameworks to operate the pipeline, and use of either the CPU or an accelerator to run the inferencing engine. Multiple instances of these pods can be run in a system, although a pod that uses an accelerator consumes the entire accelerator while it is running.

Each of these pods has a second container, labeled "Istio-proxy". The application uses a service mesh architecture to wire itself together. Service mesh is a general concept in cloud-native computing; it is not special to OpenNESS. It can be used to in effect build an application-level network to offload the complexity of building a distributed application.

Along with the Istio-proxy "sidecars", the rest of the service mesh is implemented by a service mesh server and a service mesh ingress gateway, shown just below the pipeline pods.

Finally, there is a box labeled "Third-Party Video Analytics Edge App". This is the application that provides the top-level logic of the application.

The OpenNESS controller has the usual functional elements that we have seen earlier, but the pods in the edge node are more specialized.

There is one application pod shown, a "Smart City Edge App". This app receives streaming video from multiple cameras, and performs a variety of visual analytics operations, such as pedestrian and traffic spotting, and outputs inference results into a web interface. It therefore has message queues, camera discovery, storage and display applications running in the pod.

The applications use video analytics functions powered by a Movidius High Density Deep Learning (HDDL) accelerator, produced by IEI. This accelerator is managed by network functions running in the HDDL pod. In order to steer traffic to the accelerator, various other building blocks, shown in the System Pod and the CNI pod, are used. These building blocks ensure that the BIOS and firmware on the server are set appropriately and configure DNS and CNI for proper traffic steering. The applications require proper node affinity (i.e. applications and network interfaces must be on the same NUMA node), and building blocks such as the CPU Manager are used to achieve that.

In Fig. 9.10 we see an individual edge node and the active building blocks in more detail. This node shows three pipeline pods; each of them uses FFMPEG as the pipeline framework, and either the CPU or one of the accelerators to execute the inference engine.

One of the pods in the lower left is labeled "eaa"; this is the component that implements the Edge Application API. Each of the pipeline pods registers as a service (or "producer app") with EAA, as depicted by the various orange lines. In

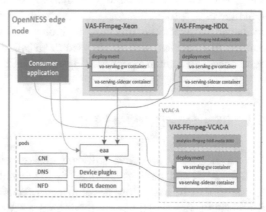

Fig. 9.10 Edge application API usage in smart city app

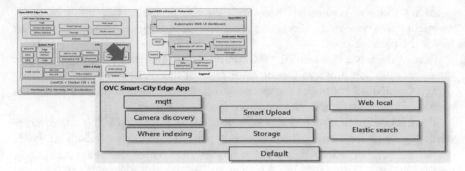

Fig. 9.11 Container deployment in smart city app

particular, it is the "sidecar" container that does this; this is how the application uses the service mesh to assemble itself.

After this is done, a client (or "consumer app") can use EAA to find them, and then can send messages to them.

The Smart City application uses video analytics pipelines to perform data plane processing and uses the Edge Application API to set up pipelines on request by an application.

When the client app subscribes to a particular service, the sidecar to which it subscribed responds with the endpoint of the gateway container, which is the pipeline.

After this wiring is accomplished, the client application interacts with the outside world (such as the human operator, and the cameras) to set up the stream from the camera to the video analytics pod. This is done via setting up routing in the external network, and in the CNIs in the edge platform, and is not shown here.

Once the pipeline is created, and the media is steered, the application simply runs until the operator intervenes.

Figure 9.9 focuses on the various edge applications, in separate containers or VMs, but managed cooperatively, that comprise the Smart City app. It is representative of a "production" application, with functionality to allow web cameras to connect/disconnect dynamically, to include storage and other functions, and to connect to a backend web server used to visualize recognition events (e.g. of cars or pedestrians) to human operators (Figs. 9.10 and 9.11).

9.3   The OpenNESS4J Library

The OpenNESS4J is a Java software library designed to simplify the interaction with the APIs and services provided by the OpenNESS toolkit in order to speed up the development and allow developers to focus on the target business logic and application's functionalities by relying on a set of secure and modular connectors mapping available features. The architectural layers presented in Fig. 9.10

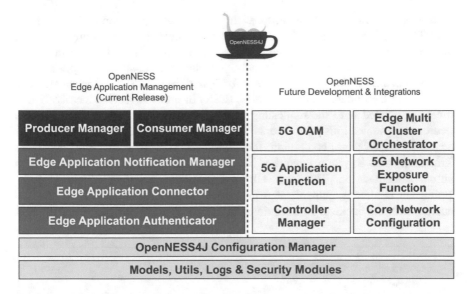

Fig. 9.12 High-level overview of the OpenNESS4J Library considering both the current release and the future developments and integrations

schematically depict existing components and how they are organized in the library. Furthermore, it highlights also the future developments and integration that will be included in the next release in order to map all the APIs and features provided by the toolkit. Currently the library focuses on the OpenNESS producers and consumers Edge Applications (EA) on-boarding process for the Network Edge mapping the authentication flow, service registration and discovery and the internal notifications management system (Fig. 9.12).

The library shares a common set of classes related to data structure models, logs, security and an internal configuration manager responsible to handle a local storage for application's startup parameters, cached certificates, Java Trust and Key Stores. These layers can be reused and extended across all the different library modules in order to support new developments through a shared development framework. The main three library's modules dedicated to EA development are (i) *Edge Application Authenticator*; (ii) *Edge Application Connector*; and (iii) *Edge Application Notification Manager*. The following subsections will present the main characteristics of these components and how they interact to support the definition of producer and consumer OpenNESS applications.

9.3.1   Edge Application Authenticator

The Edge Application Authenticator implements the methods to interact with the OpenNESS Edge Application Authentication (EAA) APIs enabling authentication of apps that intend to securely interact with the platform. The main exposed method is *auth* used to validate the identity of the requesting application and issuing it a valid TLS certificate. The authenticator can be initialized by specifying the url of the target OpenNESS controller and the password to protect the local security files handled by the library and exposes two basic methods: *authenticateApplication()* and *loadExistingAuthorizedApplicationConfiguration()*. The first one creates and executes the Certificate Signing Request (CSR) in order to receive the X.509 certificate associated with the requesting application and required for all subsequent secured communications with OpenNESS APIs. Furthermore, the method decodes the server's response containing the certificate and generates the Java files (TrustStore and KeyStore) required for secure HTTPS communication through mutual authentication. The latter allows instead the developer to check if the application is already authenticated in order to minimize any redundant communication with the backend by checking and loading available existing security files.

```
String BASE_AUTH_URL = "http://eaa.openness/";

String APP_ID = "OpenNessConnectorTester";

String ORG_NAME = "DIPIUniMore";

String NAMESPACE = "test_namespace";

String JAVA_STORE_PASSWORD = "mypwd_test";


//Create EdgeApplicationAuthenticator

EdgeApplicationAuthenticator eaa = new EdgeApplicationAuthenticator(BASE_AUTH_URL,JAVA_STORE_PASSWORD);

//If available load the existing configuration for the target application

Optional<AuthorizedApplicationConfiguration> appConf =

eaa.loadExistingAuthorizedApplicationConfiguration(APP_ID,ORG_NAME);

//If the configuration for the target application is not available authenticate the app on the OpenNESS deployment

if(!storedConf.isPresent())

        appConf = eaa.authenticateApplication(

                NAMESPACE,

                APP_ID,

                ORG_NAME);
```

List. 9.1 OpenNESS4J EdgeApplicationAuthenticator Initialization and application authentication

The code listed above illustrates how to easily create an instance of the EdgeApplicationAuthenticator to authenticate a target application (identified by id, namespace and organization) on the OpenNESS Toolkit. Firstly the developer can use the *loadExistingAuthorizedApplicationConfiguration()* method to check if an authorization configuration is already available in the local storage. If all the required security files are locally available an instance of the *AuthorizedApplication Configuration* class will be returned by the method in order to be used with the *EdgeApplicationConnector*. Otherwise, the developer has to call the *authenticate*

Application() in order to authenticate the application and save all security files in a new configuration. The same authenticator instance can be used within the same software to authenticate multiple applications at the same time thanks to a dedicated and independent management of certificates, files and configurations.

9.3.2 Edge Application Connector

The Edge Application Connector implements the methods to interact with the OpenNESS Edge Application (EA) APIs enabling apps to exploit OpenNESS services such as registration of services, their discovery, subscription to notifications, publication of notifications, and the like. As such, some of the exposed methods are intended for usage by OpenNESS producer applications, that is, applications willing to provide services to other OpenNESS applications, and some other are intended for OpenNESS consumer applications, that is, applications willing to exploit available application services via the OpenNESS toolkit.

```
    String CONTROLLER_URL = "https://eaa.openness/";

    String CONTROLLER_WS_URL = "wss://eaa.openness/";

    String SERVICE_NAME = "demo_service";

String SERVICE_STATUS = "default_status";

    String SERVICE_INFO = "{type:test, param:dummy}";

    String ENDPOINT_URI = String.format("%s/%s", NAMESPACE, APP_ID);

//Create EdgeApplicationConnector

    EdgeApplicationConnector eac = new EdgeApplicationConnector(CONTROLLER_URL, appConf,

CONTROLLER_WS_URL);

    //Load the list of available and registered services (mainly called by the consumer)

    EdgeApplicationServiceList availableServiceList = eac.getAvailableServices();

//Define the notification's structure of the target service

List<EdgeApplicationServiceNotificationDescriptor> notifications = new ArrayList<>();

    notifications.add(new EdgeApplicationServiceNotificationDescriptor(

                    "test notification",

                    "0.0.1",

                    "test description"));

//Define a new service description (producer)

final EdgeApplicationServiceDescriptor service = new EdgeApplicationServiceDescriptor(

                    new EdgeApplicationServiceUrn(applicationId, nameSpace),

                    SERVICE_NAME, ENDPOINT_URI, DEFAULT_STATUS,

                    notifications, SERVICE_INFO);

//Register the new service with the OpenNESS controller

    eac.postService(service);
```

List. 9.2 OpenNESS4J EdgeApplicationConnector Initialization and main available methods

As illustrated in Listing 9.2 main available methods related to service discovery and registration are (i) *getAvailableServices()*: to retrieve the list of all registered services on OpenNESS. This method is typically called by a consumer to find the target services of interest; and (ii) *postService()*: to allow a producer to register a new service associated through a specific descriptor and notification structure. Additional management methods are also available to handle services update and deletion during the operational flows of both producer and consumers.

9.3.3 Edge Application Notification Manager

In order to easily and effectively manage the OpenNESS notification system within both producers and consumers a dedicated set of methods have been integrated into the connector. On one hand consumers can subscribe with HTTPS declaring the interest to a set of notifications (associated to one or more producers) and receive in real-time the updates through active WebSocket channels. On the other hand, the producers can publish their notifications delegating the complexity of handling the list of subscribers and their communication channels to the OpenNESS toolkit.

```
//Retrieve the list of active subscriptions and extract target notification of interest

    SubscriptionList subscriptions = eac.getSubscriptions();

List<EdgeApplicationServiceNotificationDescriptor> notifications = extractTargetNotificationList(subscriptions);

//Define a Custom Notification Handler

    MyNotificationsHandler myNotificationsHandler = new MyNotificationsHandler();

    eac.setupNotificationChannel(NAMESPACE, APP_ID, myNotificationsHandler);

//Subscribe to the target notification list

    eac.postSubscription(notifications, NAMESPACE, APP_ID);
```

List. 9.3 Consumer subscription to a target list of notifications with the creation of a dedicated notification channel

Listing 9.3 shows how the developer can use the *getSubscriptions()* and *postSubscription()* methods in order to dynamically discover and register to a set of notification of interest. The first one returns a list of active subscriptions for the caller while the latter subscribes to a set of notifications (descriptors) associated with a service within the specified namespace. Before posting a new registration, it is mandatory to set up the notification channel through the method *setupNotificationChannel()* specifying the customer handler (created extending the class *AbstractWebSocketHandler*) to be used to receive and process the incoming messages according to the target application business logic.

```
//Define a new Producer Notification

NotificationFromProducer producerNotification = new NotificationFromProducer(

                    "My test notification",

                    "0.0.1",

                    "{id:test, value:2}"

        );
//Post the notification

eac.postNotification(producerNotification);
```

List. 9.4 Producer notification definition and distribution

On the other hand, code in Listing 9.4 describes how a producer can easily define and send a new notification (using an HTTP API) to the OpenNESS toolkit in order to be forward to all the active WebSocket subscribers. The first step is to define the structure of the new notification by specifying the name and the version together with a custom payload associated to any serializable Java object and describing the content of the new message. After that, the developer can use the *postNotification()* method to publish to OpenNESS a new notification compatible with the registered service descriptor.

9.4   MEC and Open Source

In the beginning of this chapter we have talked about developing software based on the ETSI MEC standards. Such standards have a fundamental part to play in ensuring interoperability can be achieved between components. This is particularly important in scenarios where it is desirable that different entities may be offered by different vendors. This is clearly the case in the edge environment where edge applications may be created by many different third-party application developers, as is already the case for the applications provided through app stores for today's mobile and tablet devices.

Open source initiatives also have a critical role in the development of the overall ecosystem. This is since they can provide implementations of the components between which interoperability is targeted and thereby help accelerate widespread standards adoption. Such open source development is outside the scope of the majority of standards development organizations (SDOs). Thereby it is apparent that SDOs and open source bodies can have a highly complementary relationship, which opposes the all too common view that standards and open source have a fraught relationship and are in competition with one another.

With respect to ETSI MEC there are concrete examples of such collaborative relationships with open source bodies. This is exemplified by the cooperation agreement signed between ETSI and LF Edge in 2019. Through this association, an increasing number of ETSI MEC APIs and architectural components are being showcased in a number of the LF Edge's Akraino[1] project blueprints. Specific examples have been captured on the MEC ecosystem wiki,[2] which at this time already lists three MEC solutions offered through **Akraino blueprints**, specifically:

- Connected-Vehicle Blueprint (CVB), which provides a V2X focused MEC platform and platform manager through which services are delivered to applications hosted on connected vehicles according to a set of policies for data dispatch and response. This blueprint is undergoing further development in order to offer an increasing number of connected-vehicle applications and services.
- Enterprise Applications on Lightweight 5G Telco Edge (EALTEdge) blueprint, which is focused on enabling enterprise applications at the telco edge and is compliant to the ETSI MEC standard. To achieve this, a unified portal for both platform management and application developers is offered. Further support is provided through toolchains, a sandbox and SDKs.
- Public Cloud Edge Interface (PCEI) blueprint, which has the purpose of specifying a set of open APIs that enable multi-domain interworking. Such domains are able to provide edge capabilities and applications that in turn require close cooperation between mobile edges, the public cloud and the underlying infrastructure, including data centers and networks.

9.4.1   Akraino API Portal

The above blueprint examples help highlight the convergence that is occurring in the MEC ecosystem, which is expected to accelerate as the industry moves toward commercial MEC deployment. In addition to the blueprints, Akraino has also created an API portal through which information on key Akraino APIs is provided.[3] The site also hosts the API map (see Fig. 9.13), which offers an innovative presentation the APIs grouped into four distinct high-level categories: **edge enabler** (that will include the MEC service APIs); **application enabler** (which includes the MEC application enablement [ETSI GS MEC 011] and Application lifecycle, rules and requirements management [ETSI GS MEC 010–2]); **edge infrastructure**; and **management and orchestration**.

---

[1] https://www.lfedge.org/projects/akraino/.

[2] https://mecwiki.etsi.org/index.php?title=MEC_Ecosystem.

[3] Release 4 is the first time the Akraino project, through the guidance of the API subcommittee, has made it a mandatory requirement for each project to report APIs exposed and consumed as part of the project release review. This is a reflection of the trend in adoption of IT technologies in edge domain. APIs are increasingly used by edge stacks either to expose their capabilities, or to interact with components of various implementations.

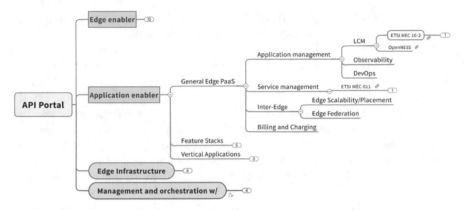

Fig. 9.13 Akraino API Portal (https://apiportal.akraino.org/apimap.html) (copyright ©
Akraino API subcommittee)

The hierarchy of the map is based on the functional diagram from the whitepaper
"Public Cloud Interfacing 5G Edge",[4] written by the Akraino API subcommittee.
Although the whitepaper focuses on 5G Edge, the key functional blocks in the
diagram apply to Telco edge in general.

For each project API set reported, the map traces back to its applicable standards
adoption or upstream projects.

A summary of the Akraino Release 4 project API status is as follows:

- A majority of APIs are focused on application management and service
  management.
- APIs related to inter-edge or edge billing / charging are to be identified.
- There are two mobile network interface API sets (3GPP specified), specifically
  regarding Application Function (AF) influencing and Network Exposure
  Function (NEF).
- There are a rich set of APIs relating to edge infrastructure; however, consoli-
  dation is considered desirable.
- Kubernetes is the most adopted container management platform. Its well-defined
  APIs are the de facto interface specifications for container-based projects.

In the application management and service management spaces within the
application enabler category (see Fig. 9.13), several upstream projects served as
key choices: EdgeGallery (edgegallery.org), TARS (tarscloud.com), OpenShift and
Kubernetes. The latter two are popular container management platforms used
widely in cloud-native implementations. EdgeGallery focuses on the MEC platform
framework of the carrier network's edges. TARS focuses on microservice
ecosystem with high performance and polyglot support. Of these upstream projects,
EdgeGallery is currently the most closely coupled with ETSI MEC. Specifically it

---

[4]https://www.lfedge.org/2020/10/07/akraino-white-paper-cloud-interfacing-at-the-telco-5g-edge/.

implements APIs based on ETSI MEC specifications motivated by its strong MEC platform use-case scenarios, especially interoperability support.

The underlying edge infrastructure is known for its heterogeneous nature and complexities. Ideally, consolidated APIs can be adopted for better interoperability. However, APIs in edge infrastructure category tend to be vendor specific and not standardized, except for adopted Kubernetes open source APIs. For example, the "Radio Edge Cloud (REC)" project uses DANM for container network configuration, general Kubernetes for container CPU pinning and container deployment, and OpenStack for bare metal server deployment configuration. The "Integrated Cloud Native (ICN)" project uses the OpenWrt UCI for unified network configuration, and its own APIs for IPSec tunnel, network controller and binary provision agent.

For realizing ETSI MEC management, highlighted in its reference architecture, ONAP, Kubernetes and OpenShift are the three main upstream projects used in Akraino blueprint projects in the category of management and (edge) orchestration. APIs from these upstream projects are abundant for edge infrastructure, edge node and multi-cluster edge cloud orchestration.

It is also worth mentioning that 5G network capability exposure APIs are introduced in the "5G MEC for Cloud Gaming" project. In this case, the upstream project providing AF influencing and NEF APIs is OpenNESS. The project implements a Traffic Influence API and PFD (Packet Flow Description) management APIs as specified in 3GPP Release 16 onwards.

9.4.2   ETSI MEC DECODE WG and Akraino

In this section, it has been highlighted that ETSI, as an SDO, is not directly involved in open source development. However, within its MEC group, it has been recognized that it is able to provide a key role in supporting the ecosystem. This effort is being led by a dedicated MEC working group (DECODE), which was established in 2018. Addressing issues around enabling the use of MEC APIs in forums such as open source is of key focus. Overall, the group's aim is to accelerate and encourage the development of new and innovative services in the growing edge ecosystem, where Akraino is considered a key partner.

9.5   ServerlessOnEdge

Serverless is an emerging trend in cloud technologies where the notion of server is hidden from the developers, who are provided instead with a minimal abstraction that only allows to execute functions in response either to external events or to the execution of other functions in a chain. The functions are executed within containers on a scalable computing infrastructure, with the promise of infinite

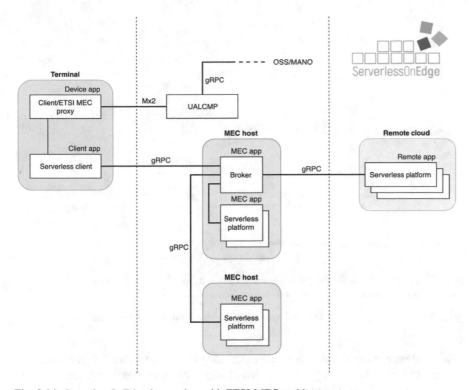

Fig. 9.14 ServerlessOnEdge integration with ETSI MEC architecture

scalability, especially when adopting the so-called Function-as-a-Service (FaaS) model, where functions are stateless. In the cloud, usually there is a single serverless platform that handles all the incoming requests, passing through a load balancer. Serverless is also moving toward the edge, which however does not have just one entry point and aspires at federating multiple co-existing infrastructures, as well as cloud resources.

ServerlessOnEdge[5] addresses specifically this use-case, since it is a decentralized framework for the distribution of function execution to multiple serverless platforms. ServerlessOnEdge can be integrated with ETSI MEC as illustrated in the figure below (Fig. 9.14).

The terminal is a serverless client that wishes to execute stateless functions (also called *lambdas*). Via its embedded device application, it queries the UALCMP of an ETSI MEC domain using the Mx2 interface to retrieve the end point of the MEC app, called Broker in the figure, that will act as its entry point toward the serverless platforms available. The broker is aware of all the federated serverless platforms, which can be local to the same MEC host or running in another MEC host or

---

[5]https://github.com/ccicconetti/serverlessonedge.

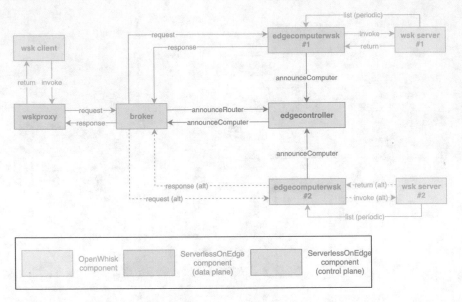

Fig. 9.15 Overview of functional diagram of ServerlessOnEdge with two platforms

executed on remote cloud, through an element called *edgecontroller* (not shown in the figure), also interacting with the MEO and other elements in the management plane of the ETSI MEC domain. The role of the Broker is to dispatch the incoming lambda requests from all its associated clients so as to achieve fairness or meet other QoS requirements depending on the scenario.[6] In general, an ETSI MEC domain may have multiple Brokers and the association between them and the terminals may need to be adapted over time: this can be achieved through a management interface available on the UALCMP. Except for the Mx2 between the device application and UALCMP, all other interfaces are implemented through gRPC,[7] which is a universal framework for remote procedure calls based on HTTP endorsed by the Cloud-Native Computing Foundation (CNCF).

ServerlessOnEdge runs out of the box with serverless clients/platforms using Apache OpenWhisk,[8] which is a versatile FaaS system that manages infrastructure and scaling using Docker containers. A functional diagram with two serverless platforms is illustrated in the figure below (Fig. 9.15).

Integration is enabled thanks to two dedicated components. First, *wskproxy* acts as an HTTP proxy and translates action invocations into lambda function requests

---

[6]C. Cicconetti, M. Conti, A. Passarella. A Decentralized Framework for Serverless Edge Computing in the Internet of Things, IEEE Transactions on Network and Service Management, 2020, https://doi.org/10.1109/TNSM.2020.3023305.

[7]https://grpc.io/.

[8]https://openwhisk.apache.org/.

through gRPC calls; every wskproxy is bound to a given broker. Second, *edge-computerwsk* receives incoming lambda request functions and forwards them to an OpenWhisk server, whose API host URL and authentication token must be known at launch time. In any case, the broker forwards lambda functions calls without any notion on whether they are eventually intended for an OpenWhisk server or not. The control plane, as managed by the *edgecontroller*, is also totally transparent to the OpenWhisk components: actions are advertised, complete with their namespace, as if they were lambda functions executed locally.

9.6   Open Edge Computing—Toward Globally Consistent Edge Services

For the global business success of edge computing, it is important to create a well-structured value chain that provides attractive business opportunities for all companies involved. Therefore, it is important to make sure that the products and services of application developers, edge platform providers and edge infrastructure service providers seamlessly work together and jointly provide value to the end customer. Obviously, edge software development and the related standards and specifications play a key role in that process. This chapter describes the industry initiative "Open Edge Computing Initiative" [OEC001] and its active role in shaping the edge ecosystem on a global scale.

9.6.1   Open Edge Computing Vision

As indicated earlier in this chapter, edge computing requires (at least) three tiers, an UE level (application, IoT system), an Edge level (distributed edge nodes) and a Remote level with the backend cloud service (Fig. 9.16). To create a public edge infrastructure service, it is required to establish a distributed set of edge nodes that are connected to the applications as well as the backend cloud. Depending on the size and quality of the edge infrastructure service this requires significant investments.

On the other hand, as edge computing is still developing, there are a large variety of edge use-cases, technologies and players from very different industries entering into the emerging edge markets. Today, there is still a relatively unstructured set of edge application categories (gaming, AR, drones, automotive, etc.), IT and telecom technologies and business models. The edge technologies and the related markets are fragmented and as of today, it is difficult to address a large portion of the overall edge market. To enable significant investments into edge computing service infrastructure and edge applications, there has to be a globally uniform mechanism for edge computing services.

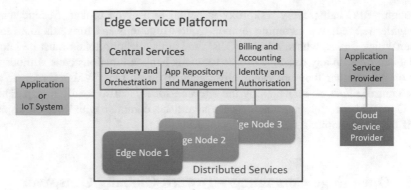

Fig. 9.16 Distributed edge software architecture (3-tier model)

Obviously, agreeing global mechanisms and specifications is typically done by established standardization organizations and open source communities which typically are associated with a specific industry. Edge computing positions itself right at the intersection of telecom and IT. Therefore, the challenge with edge computing is that it requires both, the IT- and the telecom industry to jointly agree the architecture, interfaces and processes around edge computing to create an attractive value chain for edge business.

This fundamental challenge of bringing together IT and telecom industry to shape edge computing led to the formation of Open Edge Computing Initiative in 2014, an industry consortium that shapes and drives the edge ecosystem development on a global scale.

9.6.2 Open Edge Computing Initiative (OEC)

The Open Edge Computing Initiative (OEC) was created to bring together relevant industry players from different industries, who are eager to make edge computing a global business success. By now, the Initiative has active members from both the telecom industry (e.g. Vodafone, Verizon, Deutsche Telekom, InterDigital) and the IT industry (e.g. Microsoft, VMware). The key academic partner of the Initiative is the Carnegie Mellon University (Pittsburgh, USA), one of the globally leading universities in edge research. Furthermore, the Initiative has well-established relationship to several edge-relevant standardization bodies and open source communities. For the latest update on OEC projects and members please refer to OpenEdgeComputing.org.

Together with its member companies the OEC does identify key challenges and hurdles for edge computing business and technology development and addresses these areas in joint workstreams. Since 2014, the OEC has delivered a number of solutions and concepts. Here are some example work results (selection):

- First OpenStack-based implementation of an edge node stack (2015) [OEC002, OEC003].
- Numerous edge-native applications (e.g. OpenRTist, Sandwich Maker, PingPong, 2014 – 2020) [OEC004, OEC005].
- Edge-based virtual desktop solution (2019) [OEC006].
- TPOD: toolset to build edge-native applications quickly without writing code (6/2020) [OEC007].

These work results were created in a joint collaboration between the OEC-team, Carnegie Mellon University and the OEC member companies. To drive and shape the edge ecosystem, all work results are published and the resulting source code is released (refer to openedgecomputing.org/tag/development). Some of the work results have been very well received by the edge community and are widely used, for example, the edge-native application OpenRTIST ([OEC005], see also Exercises in this chapter).

A very important part of the OEC work is the interaction and cooperation with edge-related standardization bodies and open source communities. There is a large variety of edge-related activities and initiatives and it is important to make sure that they align or at least avoid overlap and conflicts. The OEC engages closely, for example, with ETSI MEC, GSMA, Small Cell Forum, Linux Foundation Edge and OpenStack Foundation. Typically, the collaboration does include mutual visits, presentations and workshop participation as well as alignment on new concepts and ideas.

Since quite a few years, the OEC has developed very close cooperation relationships with ETSI MEC and its member companies. For example, the OEC has provided key use-case and the business rational in the area of 3rd party interfaces of the ETSI MEC specifications. Furthermore, OEC is regularly reviewing ETSI MEC specifications and provides constructive feedback.

Edge software development and hands-on experiments in real-world environments are a key element of the OEC work. With the help of its member companies, the OEC has established the Living Edge Lab in Pittsburgh, USA in 2017. This lab is an operating edge infrastructure with small cells distributed in the City of Pittsburgh, that are connected through a fiber network to set of edge nodes. This lab has been further expanded into a flexible and versatile edge infrastructure with up-to-date technologies from leading companies (Verizon, Microsoft Azure, AWS, NVIDIA, etc.). The Living Edge Lab provides an excellent environment for bringing together different edge software and hardware from the IT and the telecom industries and jointly developing new applications, innovations and solutions.

9.6.3 Edge Software Development—Challenges, Projects and Outlook

Edge Software Development is a broad area spanning application development, edge node architectures and APIs as well as the overall edge resource management and orchestration (Fig. 9.16). In all of these areas there are competing solutions and approaches and it is not clear right now which approach will be leading in the market.

From the perspective of an edge application developer this situation is challenging, because it is not clear which interaction model (e.g. application API's) with the edge infrastructure management platform his application should adopt. There are solutions from ETSI MEC compliant platform vendors, independent platform vendors as well as offers from big cloud service providers. As a consequence of this fragmented technology landscape, the development of edge-native applications [OEC008] is progressing very slowly.

Therefore, there is an urgent need to further converge these different edge application API-models and deploy uniform edge mechanisms globally across the IT and the telecom industry to provide attractive business opportunities to the application developer industry. Going forward, the Open Edge Computing Initiative is planning trial projects that progress the convergence of edge approaches from the telecom and the cloud industries.

Another relevant area of edge software development is legacy applications. Many legacy applications use the well-established client–server model (2-tier model). However, as indicated earlier in this chapter for edge computing the execution of an application needs to be distributed into three tiers (client, edge and cloud). Therefore, the latency and compute sensitive parts of an application need to be extracted and executed on the edge. Right now, this extraction process is quite time consuming and often requires to re-write the entire application. A simple methodology for the re-structuring of legacy applications to utilize edge resources would be an attractive business opportunity for edge computing.

There are also edge-specific software development challenges on the edge node. Many edge applications on edge nodes require a specific hardware accelerator type for good performance. However, in a real-world edge environment with many edge nodes from different edge services providers, there will be many different types of hardware accelerators. The challenge is that edge applications should be able to run on any hardware accelerator independent of the vendor, technology or the release type. Here it is important to agree an abstraction layer that allows the application developer to use hardware acceleration resources, independent of the actual type of accelerator available in a specific edge node.

More on the telecom network side of edge computing there is a key challenge around low-latency interworking between telecom operators. Today, the time needed to send packets from an edge node in one operator's network to an edge node in another operator's network is quite high (often above 200 ms round-trip latency). However, edge-native applications often require latencies of below 50 ms. Hence, edge applications with users collaborating in two or more telecom networks will get poor quality of service.

Therefore, telecom operators who are interested in edge computing traffic and services must optimize their interconnect latencies to other telecom operators. Supported by some major telecom companies, the Open Edge Computing Initiative has done a detailed analysis of the problem and published several white papers and technical reports on the matter [OEC009].

Going forward the OEC will continue to engage closely with ETSI MEC to bridge the gap between the telecom and the IT industries and jointly progress the business success of edge computing.

Overall, edge computing has already established itself as an important extension of cloud computing. But there are still many business and technology challenges that will drive the development of edge-specific hardware and software technologies. Therefore, edge computing will continue to be an exciting and attractive new work area and the Open Edge Computing Initiative will continue to shape and drive the edge ecosystem on a global scale.

Exercises/Examples—Chapter 9

Exercise 9.1 Goal of the exercise: Write a MEC application that follows the design steps identified in Exercise 4.1, using the information provided by the Radio Network Information (RNI) MEC Service (RNIS) to compute the percentage of failed handovers occurred for a specific UE in an area covered by the MEC service (e.g. a shopping center). The development of the MEC application must make use of the OpenAPI representation of the RNI service.

Recap of design steps from Exercise 4.1:

Step 1	Query of RAB information for the target UE
	HTTP GET request to the *{apiRoot}/rni/v2/queries/rab_info* resource URI, by filtering on the query parameter *ue_ipv4_address* that identifies the IPv4 address of the target UE
Step 2	**Subscription to cell change notifications for the target UE**
	HTTP POST message to the *{apiRoot}/rni/v2/subscriptions* resource URI, providing the *CellChangeSubscription* details in the message body, as filter parameters to identify the intended types of notifications:
	• *associateId* = UE's IPv4 address
	• *hoStatus* = list of handover status we want to be notified of (i.e., completed, rejected and cancelled ones).
Step 3	**Reception and processing of cell change notifications**
	Implement a REST server that:
	• accepts the subscribed notifications (HTTP POST messages on the resource URI provided in the callback field of the subscription created in step 2).
	• parses the *hoStatus* field of the received notification
	• increment the related counters, which can be used to compute the percentage of failed handovers according to the equation defined at the beginning of Exercise 4.1
	• calculate the Failed handovers (HOfail) by using the formula:
	$\text{HOfail } \% = 100 \cdot \dfrac{\text{HOrej} + \text{HOcanc}}{\text{HOrej} + \text{HOcanc} + \text{HOsucc}}$

Proposed solution: Step 4 (Coding in practice).

Let's now start coding. First of all, we need to build the skeleton for our MEC application. This can be done from scratch, going through the document and implementing manually all the classes of the information models, the HTTP messages, the REST clients and servers, and so on. But this is a very long procedure and quite prone to errors. It is much easier to use the OpenAPI specification, provided by ETSI, as starting point to automatically build the skeleton of our application. Using the swagger-codegen[9] tool it is possible to autogenerate all the classes we need in our preferred programming language. The OpenAPI for all the ETSI MEC documents are publicly available in the **ETSI Forge repository,**[10] including the RNIS OpenAPI.[11]

The swagger-codegen tool can be used in two different modes: (i) locally, by downloading the swagger-codegen-cli tool and launching it in our machine, or (ii) online, using the online swagger editor.[12] In this example, we use the swagger-codegen-cli running locally in our environment. As reference programming language for our MEC application we use Java, but the reader can easily extend the exercise to other languages.

Once the swagger-codegen tool is downloaded (*swagger-codegen-cli-3.0.22.jar* file in our case), we can launch the following command to generate the structure of the MEC application:

```
$ java -jar swagger-codegen-cli-3.0.22.jar \
  generate \
  -l java \
  -i https://forge.etsi.org/gitlab/mec/gs012-rnis-
api/raw/stf593/RniAPI.yaml \
  -o MEC012
```

The code will be generated for a REST client written in java (command option *-l java*), starting from the OpenAPI YAML specification defined with the -i option (*-i* https://forge.etsi.org/gitlab/mec/gs012-rnis-api/raw/stf593/RniAPI.yaml) and in the MEC012 folder (*-o MEC012*). In our case we have selected the client option, since our MEC application acts as consumer of the RNIS API.

The tool autogenerates a code folder similar to the one in Fig. 9.17. The actual java code is located in the src folder. In particular, the information model classes can be found in the io.swagger.client.model package, while the implementation of the REST client is available in the RniApi class in the io.swagger.client.api package.

---

[9]https://swagger.io/tools/swagger-codegen/.

[10]https://forge.etsi.org/rep/mec.

[11]https://forge.etsi.org/rep/mec/gs012-rnis-api.

[12]https://editor.swagger.io/.

```
elian@sasha-virtual:~/MEC012$ ll
total 116
drwxrwxr-x  7 elian elian  4096 dic  9 09:46 ./
drwxr-xr-x 97 elian elian 12288 dic  9 09:41 ../
-rw-rw-r--  1 elian elian  2861 dic  9 09:41 build.gradle
-rw-rw-r--  1 elian elian   851 dic  9 09:41 build.sbt
drwxrwxr-x  2 elian elian  4096 dic  9 09:41 docs/
-rw-rw-r--  1 elian elian   291 dic  9 09:44 .gitignore
-rw-rw-r--  1 elian elian  1664 dic  9 09:44 git_push.sh
drwxrwxr-x  3 elian elian  4096 dic  9 09:41 gradle/
-rw-rw-r--  1 elian elian    50 dic  9 09:41 gradle.properties
-rw-rw-r--  1 elian elian  4971 dic  9 09:44 gradlew
-rw-rw-r--  1 elian elian  2314 dic  9 09:44 gradlew.bat
-rw-rw-r--  1 elian elian  6846 dic  9 09:41 pom.xml
-rw-rw-r--  1 elian elian 24980 dic  9 09:41 README.md
-rw-rw-r--  1 elian elian    40 dic  9 09:41 settings.gradle
drwxrwxr-x  4 elian elian  4096 dic  9 09:41 src/
drwxrwxr-x  2 elian elian  4096 dic  9 09:41 .swagger-codegen/
-rw-rw-r--  1 elian elian  1030 dic  9 09:41 .swagger-codegen-ignore
drwxrwxr-x  5 elian elian  4096 dic  9 09:46 target/
-rw-rw-r--  1 elian elian   357 dic  9 09:44 .travis.yml
elian@sasha-virtual:~/MEC012$ █
```

Fig. 9.17 Autogenerated java code

We can start defining a Java interface (in our case *NotificationsApi*, placed in the io.swagger.client.mp1 package) to be used as baseline for the management of RNIS notifications.

```
public interface NotificationsApi {

  @ApiOperation(value = "Enable ",
                nickname = "startCollection", notes = "")
  @ApiResponses(value = {
     @ApiResponse(code = 201, message = "Started")
  })
  @RequestMapping(value = "/collection/",
            consumes = { "application/json" },
            method = RequestMethod.POST)
  ResponseEntity<?> startCollectionPost();

  @ApiOperation(value = "Receive notification ",
     nickname = "receiveNotification", notes = "")
  @ApiResponses(value = {
     @ApiResponse(code = 204, message = "Received")})
  @RequestMapping(value = "/subscription/hoStatusChange",
     consumes = { "application/json" },
     method = RequestMethod.POST)
  ResponseEntity<?> postNotification(
       @ApiParam(value = "" ,required=true )
       @RequestBody CellChangeNotification body);
}
```

This Java interface is used as baseline to build a REST controller that (i) starts the service, performing the RAB information query, (ii) subscribes to the cell change notifications and (iii) handles these notifications. Then, we need to create a new class, e.g. *NotificationController*, that implements the *NotificationsApi* interface and overrides the *startCollectionPost* and the *postNotification* methods. The former, invoked by an external REST request, will create the subscription, while the latter will process the notifications.

```
@RestController
class NotificationController implements NotificationsApi{

  @Override
    public ResponseEntity<?> startCollectionPost() {
    [...]

  @Override
    public ResponseEntity<?>
    postNotification(@ApiParam(value = "", required=true )
    @RequestBody CellChangeNotification body) {
    [...]
```

In our *NotificationController* we initialize the handovers counters and configure our variable parameters, i.e. the resource URI of the RNI MEC service that our MEC application will interact with and the IPv4 address of the target UE.

```
@RestController
class NotificationController implements NotificationsApi{

 private int failedHandover = 0;
 private int completedHandOver = 0;
 private String basePath = "http://10.30.8.54:8080/rni/v2";
 private String ueIpv4Address = "10.10.1.201";

 [...]
}
```

In the *startCollectionPost* method, the first step is the instantiation of the REST client of our MEC application, which we will use to send the HTTP requests to the RNI MEC service.

```
/* REST client configuration and instantiation */
ApiClient apiClient = new ApiClient();
apiClient.setBasePath(basePath);
RniApi rniApi = new RniApi(apiClient);
```

Now we can use the REST client *rniApi* to query the RAB information, using the method *rabInfoGET* of the *RniApi* class. This method receives as input parameters all the possible filtering attributes of the RAB Info query and returns the requested *RabInfo* in output, if available, or throws an exception. In our case, we need to filter with the UE IPv4 address, as follows:

```
List<String> ipV4Addresses = new ArrayList<>();
ipV4Addresses.add(ueIpv4Address);
RabInfo rabInfo = new RabInfo();
try{
    rabInfo = rniApi.rabInfoGET(null,
                      null,
                      ipV4Addresses,
                      null,
                      null,
                      null,
                      null,
                      null,
                      null,
                      null,
                         null
              );
} catch(Exception e){
    System.out.println(e.getMessage());
}
```

If the RAB information is available for the UE we want to track, we can proceed with the cell change subscription to receive notifications for the handovers of interest. We need to create a new *CellChangeSubscription* instance, setting the handover-related filtering criteria through the *CellChangeSubscriptionFilter CriteriaAssocHo* class. In our case, we are interested in the handover status "completed" (code 3), "rejected" (code 4) or "cancelled" (code 5), all of them related to the target UE. Finally, we need to set the callback URI that will be used to receive the notifications and send the subscription request through the *subscriptionPOST* method of the *RniApi* class.

```
if (rabInfo != null){
CellChangeSubscription subscription =
    new CellChangeSubscription();
CellChangeSubscriptionFilterCriteriaAssocHo filter =
    new CellChangeSubscriptionFilterCriteriaAssocHo();
List<Integer> hoStatus = new ArrayList<Integer>();
hoStatus.add(3);    //completed HO
hoStatus.add(4);    //rejected HO
hoStatus.add(5);    //cancelled HO
filter.addHoStatusItem(hoStatus);
AssociateId ueId = new AssociateId();
ueId.setType(TypeEnum.NUMBER_1);
ueId.setValue(ueIpv4Address);
filter.addAssociateIdItem(ueId);
subscription.setFilterCriteriaAssocHo(filter);
subscrip-
tion.setSubscriptionType("CellChangeSubscription");
subscription.setCallbackReference(
"http://10.30.8.10:8083/subscription/hoStatusChange");
try {
    rniApi.subscriptionsPOST(subscription);
} catch (ApiException e) {
    e.printStackTrace();
}
  [...]
```

Now the MEC application is ready to receive the cell change notifications. The last thing to do is to implement the logic to process them. For this, we need to override the *postNotification* method in our NotificationController, which offers as input parameter the *CellChangeNotification* received from the RNIS. The proposed implementation simply reads the *hoStatus* field, increments the counters of completed and failed handovers and computes the percentage of failed ones, writing the results on the standard output.

```
if (body.getHoStatus() == CellChangeNotifica-
tion.HoStatusEnum.NUMBER_3) {
  completedHandOver++;
  System.out.println("New HO completed successfully");
} else {
  failedHandover++;
  System.out.println("HO failed to complete with status " +
body.getHoStatus());
}
System.out.println("Total handovers: " + (failedHandover +
completedHandOver));
System.out.println("Completed handovers: "
  + completedHandOver);
System.out.println("Failed handovers: " + failedHandover);
System.out.println("Percentage of failed handovers: " +
  (failedHandover * 100.0) / (failedHandover +
    completedHandOver) + "%");
```

Exercise 9.2 Goal of the exercise: Write an emulated RNI MEC service that handles RAB queries for UEs with a given IPv4 address and issues Cell Change Notifications, thus implementing the MEC service side of the MEC application of Exercise 9.1.

NOTE This is just as an example to verify the correct working of the previous MEC application, the handovers should be emulated with a random time interval between 10 and 30 seconds, with a 80% probability of completed handovers. The development of the MEC service must make use of the OpenAPI representation of the RNI service.

Proposed solution

As in the previous Exercise 9.1, the first step involves the creation of the structure of our MEC service. Again, we can start from the RNIS OpenAPI available in the ETSI Forge repository and use the swagger-codegen-cli tool to autogenerate the classes of the application locally in our machine. Assuming the swagger-codegen-cli tool is available in our environment (*swagger-codegen-cli-3.0.22.jar* file in our case), we can launch the following command to generate the structure of the MEC service:

```
$ java -jar swagger-codegen-cli-3.0.22.jar \
  generate \
  -l spring \
  -i https://forge.etsi.org/gitlab/mec/gs012-rnis-
api/raw/stf593/RniAPI.yaml \
  -o MEC012service
```

The code will be generated for a REST server written in java using the Spring framework (command option *-l spring*), starting from the OpenAPI YAML specification defined with the -i option (*-i* https://forge.etsi.org/gitlab/mec/gs012-rnis-api/raw/stf593/RniAPI.yaml) and in the MEC012service folder (*-o MEC012service*). It should be noted that instead of using the option *-l java* as in the previous exercise, we are using the option *-l spring*. This option allows to create a REST server for the RNIS service, which acts as *provider* of the RNIS API.

As in the client case, the java code is located in the src folder. The information model classes are placed in the *io.swagger.model* package, while the implementation of the structure of the REST controllers for handling queries and subscriptions is available in the *io.swagger.api* package, in the *QueriesApiController* and *SubscriptionsApiController* classes, respectively. These are the two classes we need to modify to implement our MEC service.

In the autogenerated version of the software, the methods of the REST controllers are already filled with predefined code to be used as examples. In our case we need to modify (i) the method that provides a reply to the RAB Information query in the *QueriesApiController* class and (ii) the method that handles the subscriptions in the *SubscriptionsApiController* class.

```
public class QueriesApiController implements QueriesApi {

  [...]

  public ResponseEntity<RabInfo> rabInfoGET(
  [...]
```

```
public class SubscriptionsApiController
            implements SubscriptionsApi {

  [...]

  public ResponseEntity<CellChangeSubscription> subscrip-
tionsPOST(
      @ApiParam(value = "Subscription to be created" ,
                required=true )
      @Valid @RequestBody CellChangeSubscription body) {

  [...]
```

In the proposed implementation, the *QueriesApiController* class makes use of a utility factory class, called *RabInfoFactory*, which is used to build an emulated RAB information based on the UE IP address provided in the query filter. The ipAddress provided as input is translated in the *AssociateId* format expected in the query reply (*getIpV4AddressSample* private method) and then used to build the RabInfo (*getRabInfoSample* public method).

```
public class RabInfoFactory {

  private AssociateId getIpV4AddressSample(
                        String ipAddress){
      AssociateId associateId = new AssociateId();
      associateId.setType(AssociateId.TypeEnum.NUMBER_1);
      associateId.setValue(ipAddress);
      return associateId;
  }

  public RabInfo getRabInfoSample(String ipAddress){
    RabInfo rabInfo = new RabInfo();

    [...]

    RabInfoCellUserInfo rabInfoCellUserInfo =
            new RabInfoCellUserInfo();
    [...]
```

```
      List<AssociateId> associateIdList =
              new ArrayList<>();
      if(ipAddress!=null)
       associateIdList.add(getIpV4AddressSample(ipAddress));

      RabInfoUeInfo rabInfoUeInfo= new RabInfoUeInfo();
      rabInfoUeInfo.setAssociateId(associateIdList);
      List<RabInfoUeInfo> ueInfoList = new ArrayList<>();
      ueInfoList.add(rabInfoUeInfo);
      rabInfoCellUserInfo.setUeInfo(ueInfoList);

      List<RabInfoCellUserInfo> rabInfoCellUserInfoList =
              new ArrayList<>();
      rabInfoCellUserInfoList.add(rabInfoCellUserInfo);
      rabInfo.setCellUserInfo(rabInfoCellUserInfoList);
      return rabInfo;
    }
```

The *getRabInfoSample* method of the *RabInfoFactory* is then invoked in the *QueriesApiController* class, in the custom implementation of the *rabInfoGET* method.

```
public class QueriesApiController implements QueriesApi {

  [...]

  private RabInfoFactory rabInfoFactory;

  @org.springframework.beans.factory.annotation.Autowired
  public QueriesApiController(ObjectMapper objectMapper,
                  HttpServletRequest request) {
     [...]
     rabInfoFactory = new RabInfoFactory();
  }

  [...]
```

```
public ResponseEntity<RabInfo> rabInfoGET([...]) {
  String accept = request.getHeader("Accept");
  if (accept !- null &&
      accept.contains("application/json")) {
    RabInfo rabInfo = new RabInfo();
    if (ueIpv4Address != null &&
        ueIpv4Address.size() > 0) {
      rabInfo =
    rabInfoFactory.getRabInfoSample(ueIpv4Address.get(0));
    }
    else{
      rabInfo = rabInfoFactory.getRabInfoSample(null);
    }
    try {
      String rabInfoStringFormat =
        objectMapper.writeValueAsString(rabInfo);
      return new ResponseEntity<RabInfo>(
        objectMapper.readValue(rabInfoStringFormat,
                               RabInfo.class),
        HttpStatus.OK);
    } catch (IOException e) {
      log.error(e.getMessage());
      return new ResponseEntity<RabInfo>
              (HttpStatus.INTERNAL_SERVER_ERROR);
    }
  }
[...]
```

The cell change subscriptions are handled in the *SubscriptionsApiController* class, where the *subscriptionPOST* method has been modified to limit the processing to the *CellChangeSubscription* only. All the subscriptions of this type are managed through a new class called *SubscriptionManager*, which keeps trace of the active subscriptions, and starts or stops the threads (implemented in the *HandoverThread* class) that generates the cell change notifications related to the given UEs.

In detail, the subscriptionsPOST method of the *SubscriptionsApiController* class invokes the *SubscriptionManager* to add a new subscription:

```
public ResponseEntity<CellChangeSubscription>
   subscriptionsPOST(
    @ApiParam(value = "Subscription to be created" ,
               required=true )
    @Valid
    @RequestBody CellChangeSubscription body) {
  String accept = request.getHeader("Accept");
  if (accept != null &&
      accept.contains("application/json")) {

    CellChangeSubscription cellChangeSubscription =
        subscriptionManager.addSubscription(body);
    String cellChangeSubscriptionStrFormat = null;
    try {
      cellChangeSubscriptionStrFormat =
  objectMapper.writeValueAsString(cellChangeSubscription);
      return new ResponseEntity<CellChangeSubscription>(
   objectMapper.readValue(cellChangeSubscriptionStrFormat,
   CellChangeSubscription.class),
   HttpStatus.CREATED);
    } catch (IOException e) {
      log.error(e.getMessage());
    }
  }
  return new ResponseEntity<CellChangeSubscription>
      (HttpStatus.NOT_IMPLEMENTED);
}
```

In the *SubscriptionManager*, the *addSubscription* method generates a new identifier for the received subscription, instantiates and starts a new thread for the generation of the cell change notifications and updates its internal maps:

```
public class SubscriptionManager {

  private ExecutorService service =
    Executors.newFixedThreadPool(100);
  private int successfulHandoverProb = 80;
  private HashMap<String, Future<Boolean>> handoverMap;
  private HashMap<String,CellChangeSubscription> subscrip-
tionMap;

  [...]

  public CellChangeSubscription addSubscription(
          CellChangeSubscription body){
    String subscriptionId = UUID.randomUUID().toString();
    HandoverThread handoverThread =
      new HandoverThread(body,
                         subscriptionId,
                         successfulHandoverProb);
    Future<Boolean> handoverCompleted =
      service.submit(handoverThread);

    LinkType linkType = new LinkType();
    linkType.setHref(
        "http://10.30.8.54:8080/rni/v2/subscriptions/"
        + subscriptionId);

    CaReconfSubscriptionLinks caReconfSubscriptionLinks =
        new CaReconfSubscriptionLinks();
    caReconfSubscriptionLinks.setSelf(linkType);
    body.setLinks(caReconfSubscriptionLinks);

    handoverMap.put(subscriptionId, handoverCompleted);
    subscriptionMap.put(subscriptionId, body);
    return body;
  }

  [...]
```

The full implementation of the MEC service is available in the Github repository (https://github.com/nextworks-it/MEC-applications). We leave the development of the methods for subscriptions query and deletion to the reader.

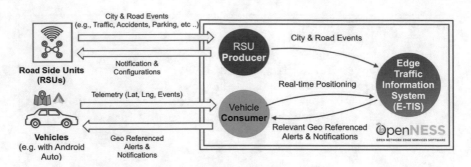

Fig. 9.18 Schematic architecture of the designed use-case to test the OpenNESS4J library

Exercise 9.3 Goal of the exercise: Write an exemplary MEC application that makes use of the OpenNESS4Java library described in Sect. 9.3.

More in detail: we want to build a smart mobility application where Road Side Units (RSU) deployed across a road infrastructure inform vehicles about relevant traffic events via a dedicated Edge Traffic Information System (e-TIS), exploiting OpenNESS services via the OpenNESS4Java library (schematically depicted in Fig. 9.18).

In such as system, the RSUs act as producers, as they register by OpenNESS to provide event notifications, the vehicles act as consumers, as they subscribe to relevant event notifications (e.g. based on proximity), the e-TIS acts as both as it subscribes to RSUs notifications, possibly filters and/or elaborates them, then publishes processed events to which vehicles are subscribed to. Each component is a separate Java application that should authenticate against OpenNESS via the OpenNESS4Java library. RSUs can be emulated by simply generating predefined events periodically. Vehicles can be emulated without any mobility-related features, or as following predefined GPX traces. The e-TIS may actually do some filtering and/or processing of RSU events before publishing its own notifications, or simply forward incoming notifications as is.

Proposed solution

The simplest solution considers (i) the RSU producing periodically the same event, (ii) the e-TIS simply forwarding RSU notifications as is, and (iii) the vehicle simply logging incoming notifications. Besides being reported below, the whole codebase is available on Github:

https://github.com/openness-4-java/openness-book.git

Here follows the commented RSU code:

```
   package it.unimore.dipi.book;

   import
it.unimore.dipi.iot.openness.config.AuthorizedApplicationConfiguration;
   import it.unimore.dipi.iot.openness.connector.EdgeApplicationAuthenticator;
   import it.unimore.dipi.iot.openness.connector.EdgeApplicationConnector;
   import
it.unimore.dipi.iot.openness.dto.service.EdgeApplicationServiceDescriptor;
   import
it.unimore.dipi.iot.openness.dto.service.EdgeApplicationServiceNotificationDes
criptor;
   import it.unimore.dipi.iot.openness.dto.service.EdgeApplicationServiceUrn;
   import it.unimore.dipi.iot.openness.dto.service.NotificationFromProducer;
   import
it.unimore.dipi.iot.openness.exception.EdgeApplicationAuthenticatorException;
   import
it.unimore.dipi.iot.openness.exception.EdgeApplicationConnectorException;
   import org.jetbrains.annotations.NotNull;

   import java.util.ArrayList;
   import java.util.List;

   /**
    * A simple emulated RSU providing parking occupancy notifications every
second for 10 times
    */
   public class RSU {

       public static final String OPENNESS_APP_ID = "openness4j-rsu";
       public static final String NAME = "parking";
       public static final String VERSION = "1.0";

       public static void main(String[] args) throws EdgeApplicationAuthentica-
torException, EdgeApplicationConnectorException, InterruptedException {
               // create authenticator providing URL of OpenNESS EAA
               final EdgeApplicationAuthenticator eaa = new EdgeApplicationAuthen-
ticator(Etis.OPENNESS_AUTH_URL);
               // request authentication providing data of app
               final        AuthorizedApplicationConfiguration        aac        =
eaa.authenticateApplication(Etis.OPENNESS_NAME_SPACE,        OPENNESS_APP_ID,
Etis.OPENNESS_ORG);
```

```
        // create connector providing authentication configuration just ob-
tained, URL of OpenNESS EA, and URL of OpenNESS websocket channel for notifi-
cations
        final EdgeApplicationConnector eac = new EdgeApplicationConnect-
or(Etis.OPENNESS_API_URL, aac, Etis.OPENNESS_WS);
        // create service descriptor
        final EdgeApplicationServiceDescriptor rsuService = createServiceDe-
scriptor();
        // registers
        eac.postService(rsuService);
        int occupancy = 0;
        // for 10 times...
        for (int i = 0; i < 10; i++) {
            // ...sends updated occupancy...
            final NotificationFromProducer notification = new Notification-
FromProducer(
                    NAME,
                    VERSION,
                    occupancy++
            );
            eac.postNotification(notification);
            // ...each second
            Thread.sleep(1000);
        }
    }


    @NotNull
    private   static   EdgeApplicationServiceDescriptor   createServiceDe-
scriptor() {
        // create notification descriptor
        final List<EdgeApplicationServiceNotificationDescriptor> notifica-
tions = new ArrayList<>();
        final EdgeApplicationServiceNotificationDescriptor parkingNot = new
EdgeApplicationServiceNotificationDescriptor(
                NAME,
                VERSION,
                "parking occupancy"
        );
        notifications.add(parkingNot);
        // create service descriptor
        return new EdgeApplicationServiceDescriptor(
                new                  EdgeApplicationServiceUrn(OPENNESS_APP_ID,
Etis.OPENNESS_NAME_SPACE),
                "parking occupancy",
```

```
                    "http://eaa.openness:7070/rsu/fake/uri",
                    "ready",
                    notifications,
                    ""
        );
    }

}
```

Following instead is the e-TIS code:

```
    package it.unimore.dipi.book;

    import
it.unimore.dipi.iot.openness.config.AuthorizedApplicationConfiguration;
    import it.unimore.dipi.iot.openness.connector.EdgeApplicationAuthenticator;
    import it.unimore.dipi.iot.openness.connector.EdgeApplicationConnector;
    import
it.unimore.dipi.iot.openness.dto.service.EdgeApplicationServiceDescriptor;
    import it.unimore.dipi.iot.openness.dto.service.EdgeApplicationServiceList;
    import
it.unimore.dipi.iot.openness.exception.EdgeApplicationAuthenticatorException;
    import
it.unimore.dipi.iot.openness.exception.EdgeApplicationConnectorException;

    /**
     * A simple e-TIS acting as a proxy between RSUs and vehicles:
     *  - scans OpenNESS services looking for the RSU
     *  - subscribes to its notifications
     *  - (continues in EtisHandle)
     */
    public class Etis {

        /*
         * Configuration data: change URLs as needed
         */
        public static final String OPENNESS_APP_ID = "openness4j-etis";
        public static final String OPENNESS_NAME_SPACE = "openness4j";
        public static final String OPENNESS_ORG = "it.unimore.dipi";
        public     static     final     String     OPENNESS_AUTH_URL     =
"http://eaa.openness:7080/";
        public     static     final     String     OPENNESS_API_URL      =
"https://eaa.openness:7443/";
```

```java
    public static final String OPENNESS_WS = "wss://eaa.openness:7443/";
    private static final String TARGET_SERVICE_URN_ID = "openness4j-rsu";


    public static void main(String[] args) throws EdgeApplicationAuthentica-
torException, EdgeApplicationConnectorException {
        // create authenticator providing URL of OpenNESS EAA
        final EdgeApplicationAuthenticator eaa = new EdgeApplicationAuthen-
ticator(OPENNESS_AUTH_URL);
        // request authentication providing data of app
        final        AuthorizedApplicationConfiguration        aac        =
eaa.authenticateApplication(OPENNESS_NAME_SPACE,             OPENNESS_APP_ID,
OPENNESS_ORG);
        // create connector providing authentication configuration just ob-
tained, URL of OpenNESS EA, and URL of OpenNESS websocket channel for notifi-
cations
        final EdgeApplicationConnector eac = new EdgeApplicationConnect-
or(OPENNESS_API_URL, aac, OPENNESS_WS);
        // request list of registered services
        final        EdgeApplicationServiceList        services        =
eac.getAvailableServices();
        if (services != null && services.getServiceList() != null) {
            for    (EdgeApplicationServiceDescriptor    service:    ser-
vices.getServiceList()) {
                // find relevant service to subscribe to (RSU in our case)
                if                                          (ser-
vice.getServiceUrn().getId().equals(TARGET_SERVICE_URN_ID)) {
                    // create handle for incoming notifications (in our case
will use same connector to forward notifications)
                    final EtisHandle handle = new EtisHandle(eac);
                    // setup the websocket notification channel (necessary
before subscription) providing data of app and just created handle
                    eac.setupNotificationChannel(OPENNESS_NAME_SPACE,
OPENNESS_APP_ID, handle);
                    // subscribe to the target service notifications (RSU in
our case) providing the notifications descriptor, and producer data (namespace
and app id)
                    eac.postSubscription(service.getNotificationDescriptorLi
st(),             service.getServiceUrn().getNamespace(),             ser-
vice.getServiceUrn().getId());
                    break;
                }
            }
        }
    }

}
```

The e-TIS uses a separate class to handle incoming notifications, whose commented code is below:

```
package it.unimore.dipi.book;

import com.fasterxml.jackson.databind.ObjectMapper;
import it.unimore.dipi.etis.process.EtisLiveCoding;
import it.unimore.dipi.iot.openness.connector.EdgeApplicationConnector;
import it.unimore.dipi.iot.openness.dto.service.*;
import
it.unimore.dipi.iot.openness.exception.EdgeApplicationConnectorException;
import it.unimore.dipi.iot.openness.notification.AbstractWebSocketHandler;

import java.io.IOException;
import java.util.ArrayList;
import java.util.List;

/**
 * The notification handle for a simple e-TIS acting as a proxy between RSUs
and vehicles:
 *   - (continues from Etis)
 *   - decodes notifications incoming from RSU
 *   - registers itself as a OpenNESS service delivering the decoded notifi-
cations
 *   - forwards the received notification
 */
// extend provided abstract class for convenience
public class EtisHandle extends AbstractWebSocketHandler {

    // required to register as a OpenNESS service and to forward notifica-
tions
    private final EdgeApplicationConnector eac;
    // required to decode incoming notifications
    private final ObjectMapper json;
    // required to avoid re-registering
    private static boolean IS_PUBLISHED = false;

    public EtisHandle(EdgeApplicationConnector eac) {
        this.eac = eac;
        this.json = new ObjectMapper();
```

```
        }

        /*
         * we care only about incoming notifications
         * other available methods are
         *  - onWebSocketConnect
         *  - onWebSocketError
         *  - onWebSocketClose
         *  - awaitClose (for graceful shutdown)
         */
        @Override
        public void onWebSocketText(String message) { // called for every incom-
ing notification
            try {
                // decode notification
                final NotificationToConsumer inbound = json.readValue(message,
NotificationToConsumer.class);
                if (!IS_PUBLISHED) {
                    // register as OpenNESS service once
                    selfPublish(inbound);
                }
                // create "copy" of received notification...
                final NotificationFromProducer outbound = new Notification-
FromProducer(
                        inbound.getName(),
                        inbound.getVersion(),
                        inbound.getPayload()
                );
                // ...to be forwarded
                this.eac.postNotification(outbound);
            } catch (IOException e) {
                e.printStackTrace();
            } catch (EdgeApplicationConnectorException e) {
                e.printStackTrace();
            }
        }


        private void selfPublish(NotificationToConsumer inbound) throws EdgeAp-
plicationConnectorException {
            final List<EdgeApplicationServiceNotificationDescriptor> notifica-
tions = new ArrayList<>();
            // create new notification descriptor from incoming notification
            final EdgeApplicationServiceNotificationDescriptor n1 = new EdgeAp-
plicationServiceNotificationDescriptor(
```

```
                    inbound.getName(),
                    inbound.getVersion(),
                    "description"
          );
          notifications.add(n1);
          // create own service descriptor (unnecessary data added just for
completeness)
          final EdgeApplicationServiceDescriptor etis = new EdgeApplication-
ServiceDescriptor(
                    new                               EdgeApplication-
ServiceUrn(EtisLiveCoding.OPENNESS_APP_ID,                EtisLiveCod-
ing.OPENNESS_NAME_SPACE),
                    "description",
                    EtisLiveCoding.OPENNESS_ETIS_URL,
                    "status",
                    notifications,
                    "info"
          );
          // register itself as OpenNESS service
          this.cac.postService(etis);
          IS_PUBLISHED = true;
    }


  }
```

Finally, here follows the vehicle code:

```
    package it.unimore.dipi.book;


    import
it.unimore.dipi.iot.openness.config.AuthorizedApplicationConfiguration;
    import it.unimore.dipi.iot.openness.connector.EdgeApplicationAuthenticator;
    import it.unimore.dipi.iot.openness.connector.EdgeApplicationConnector;
    import
it.unimore.dipi.iot.openness.dto.service.EdgeApplicationServiceDescriptor;
    import it.unimore.dipi.iot.openness.dto.service.EdgeApplicationServiceList;
    import
it.unimore.dipi.iot.openness.exception.EdgeApplicationAuthenticatorException;
    import
it.unimore.dipi.iot.openness.exception.EdgeApplicationConnectorException;


    import static it.unimore.dipi.book.Etis.OPENNESS_NAME_SPACE;
```

```
/**
 * A simple emulated vehicle subscribing to parking occupancy notifications
 */
public class Vehicle {

    public static final String OPENNESS_APP_ID = "openness4j-vehicle";
    private static final String TARGET_SERVICE_URN_ID = "openness4j-etis";

    public static void main(String[] args) throws EdgeApplicationAuthentica-
torException, EdgeApplicationConnectorException {
        // create authenticator providing URL of OpenNESS EAA
        final EdgeApplicationAuthenticator eaa = new EdgeApplicationAuthen-
ticator(Etis.OPENNESS_AUTH_URL);
        // request authentication providing data of app
        final       AuthorizedApplicationConfiguration       aac       =
eaa.authenticateApplication(OPENNESS_NAME_SPACE,              OPENNESS_APP_ID,
Etis.OPENNESS_ORG);
        // create connector providing authentication configuration just ob-
tained, URL of OpenNESS EA, and URL of OpenNESS websocket channel for notifi-
cations
        final EdgeApplicationConnector eac = new EdgeApplication Connect-
or(Etis.OPENNESS_API_URL, aac, Etis.OPENNESS_WS);
        // request list of registered services
        final       EdgeApplicationServiceList       services       =
eac.getAvailableServices();
        if (services != null && services.getServiceList() != null) {
            for    (EdgeApplicationServiceDescriptor    service:    ser-
vices.getServiceList()) {
                // find relevant service to subscribe to (RSU in our case)
                if                                       (ser-
vice.getServiceUrn().getId().equals(TARGET_SERVICE_URN_ID)) {
                    // create handle for incoming notifications
                    final VehicleHandle handle = new VehicleHandle();
                    // setup the websocket notification channel (necessary
before subscription) providing data of app and just created handle
                    eac.setupNotificationChannel(OPENNESS_NAME_SPACE,
OPENNESS_APP_ID, handle);
                    // subscribe to the target service notifications (e-TIS
in our case) providing the notifications descriptor, and producer data
(namespace and app id)
                    eac.postSubscription(service.getNotificationDescriptorLi
st(),           service.getServiceUrn().getNamespace(),           ser-
vice.getServiceUrn().getId());
                    break;
                }
            }
        }
    }
}
```

Also here, a separate class handles incoming notifications, as follows:

```
package it.unimore.dipi.book;

import com.fasterxml.jackson.databind.ObjectMapper;
import it.unimore.dipi.iot.openness.dto.service.NotificationToConsumer;
import it.unimore.dipi.iot.openness.notification.AbstractWebSocketHandler;

import java.io.IOException;

/**
 * The notification handle for a simple emulated vehicle receiving notifica-
tions from the e-TIS
 */
// extend provided abstract class for convenience
public class VehicleHandle extends AbstractWebSocketHandler {

    // required to decode incoming notifications
    private final ObjectMapper json;

    public VehicleHandle() {
        this.json = new ObjectMapper();
    }

    @Override
    public void onWebSocketText(String message) { // called for every incom-
ing notification
        try {
            // decode notification
            final NotificationToConsumer inbound = json.readValue(message,
NotificationToConsumer.class);
            // just log it in our case
            System.out.printf("Got notification < %s > from < %s > with con-
tent:    %s\n",    inbound.getName(),    inbound.getProducer().getId(),    in-
bound.getPayload());
        } catch (IOException e) {
            e.printStackTrace();
        }
    }

}
```

Exercise 9.4 Goal of the exercise: Write an exemplary MEC application that makes use of the UALCMP bundled with the ServerlessOnEdge framework described in Sect. 9.5 above to create an ETSI MEC application context.

Proposed solution

After compiling the ServerlessOnEdge framework, let's start the UALCMP:

```
./ueapplcmproxy &
```

This creates a default ETSI MEC UALCMP server listening to http://www.localhost:6500/ (for Mx2 messages –using v1 of the APIs at the time of writing) and a gRPC server listening to http://www.localhost:6477/ (for configuration).

Now add a new application called test:

```
echo "add-lambda test" | ./ueapplcmproxyclient
```

You can retrieve the list of applications using curl:

```
curl -H Content-Type:application/json \
    -XGET \
    http://127.0.0.1:6500/mx2/v1/app_list
```

The output is:

```
{
    "ApplicationList": {
        "appInfo": [
            {
                "appCharcs": {
                    "bandwidth": 0,
                    "latency": 0,
                    "memory": 0,
                    "serviceCont": 0,
                    "storage": 0
                },
                "appName": "test",
                "appProvider": "OpenLambdaMec",
                "appSoftVersion": "1.0"
            }
        ]
    }
}
```

Note that `OpenLambdaMec` is a fantasy application provider used in the example and assigned automatically by default to all applications by `ueap-plcmproxy`. Now, to associate all clients requesting the test application to a service responding (e.g.) to `localhost:10,000` run:

```
echo "associate * test localhost:10000"| \
    ./ueapplcmproxyclient
```

The current association table can be retrieved via:

```
echo table | ./ueapplcmproxyclient
```

whose output is:

```
* test localhost:10000
```

You can now create a JSON-formatted file containing the context creation request according to the Mx2 specifications:

```
{
    "appInfo": {
        "appName": "test",
        "appProvider": "OpenLambdaMec",
        "appSoftVersion": "1.0"
    },
    "associateUeAppId": "1234",
    "callbackReference": "http://localhost:6600/"
}
```

Let's assume the file is called request; then via the following `curl` command you can create the context on the UALCMP:

```
curl -H Content-Type:application/json \
    -XPOST \
    -d@request \
    http://127.0.0.1:6500/mx2/v1/app_contexts
```

which returns the following JSON object:

```
{
    "appInfo": {
        "appName": "test",
        "appProvider": "OpenLambdaMec",
        "appSoftVersion": "1.0",
        "referenceURI": "localhost:10000"
    },
    "associateUeAppId": "1234",
    "callbackReference": "http://localhost:6600/",
    "contextId": "dd241701-353e-4dd6-b4ae-aa5905d5abd2"
}
```

The context is now active and can be retrieved with the following command:

```
echo contexts | ./ueapplcmproxyclient
```

which returns:

```
127.0.0.1 test localhost:10000
```

Exercise 9.5 Goal of the exercise: Deploy a working interactive augmented reality edge application that demonstrates the value low-latency edge computing, based on the edge-native application OpenRTIST.

Proposed solution

The OpenRTIST augmented reality application, developed to demonstrate the value low-latency edge computing, can be easily deployed in many edge and cloud environments. Full deployment and use instructions at https://github.com/cmusatyalab/openrtist.

Step 1	**Procure and setup server environment**
	• Recommended: Local Ubuntu 18.04 LTS (Bionic) server; cloud-based servers also supported • Download OpenRTIST docker container • Start docker container • Open port 9099 on firewall • Identify server URL
Step 2	**Procure and setup client environment**
	• Android: – Install OpenRTIST app from Google Play Store – Configure server address in OpenRTIST home screen • Python/Docker (Windows, Linux and other systems supporting Python): – Configure client for Python > = 3.5 – Configure client for Docker – Download OpenRTIST docker container – Start docker container
Step 3	**Start OpenRTIST application**
	• On Client, open OpenRTIST application • Connect to OpenRTIST server URL
Step 4	**Experience application**
	• View application performance in a variety of environments (Wi-Fi, Mobile, local server, cloud server). • The experience will vary based on the network latency and compute performance of server and client.

Exercise 9.6   Goal of the exercise: Gain familiarity with edge network emulation to understand how an emulation environment can help analyze edge application performance or support edge network design.

NOTE This exercise introduces the AdvantEDGE emulator and outlines how use the platform to create and deploy an edge network emulation scenario. Full deployment and use instructions can be found at: https://github.com/InterDigitalInc/AdvantEDGE

Proposed solution

AdvantEDGE is an open mobile edge emulation platform that runs on Docker and Kubernetes, and it is an ideal environment for edge application developers or edge network designers to experiment with edge technologies, services and network configurations in agile iterations. [OEC009] is an excellent example of this platform utilized in cutting "edge" research.

Before creating an edge network scenario, it is beneficial to understand some AdvantEDGE platform concepts. To emulate an edge network, AdvantEDGE uses a hierarchical network model composed of domains (i.e. operator networks), logical zones (consistent with ETSI MEC), network points of access (4G, 5G or Wi-Fi), devices (edge nodes, cloud nodes, and terminals), as well as applications. Details can be found here: https://github.com/InterDigitalInc/AdvantEDGE/wiki/platform-concepts.

Step 1	**Procure and setup the AdvantEDGE platform**
	• Recommended: Machine with Ubuntu 18.04 LTS or 20.04 LTS • Setup prerequisites: Dockers, Kubernetes, Helm • Download the AdvantEDGE pre-built libraries from GitHub • Deploy AdvantEDGE following steps here: https://github.com/InterDigitalInc/AdvantEDGE/wiki#getting-started
Step 2	**Create an Edge Network Scenario** *Note: If you need help with the scenario, at any step, please click the (?) button, where you will find help and tips*
	• Open the AdvantEDGE GUI in a browser • Select *Configure* and click *New*. Name your scenario and save. • Define an operator domain (logical domain); click *new* under network elements; select a name and apply. • Add a zone to the operator domain (e.g. *zone1*). • Add a point of access (POA) within the zone; then add another POA. • Add an edge node at a zone level; this emulates an edge node at an aggregation site in the zone, such as a central office. • Add an edge node at one of the POA locations (i.e. edge at a cell site). • Add terminals with one connected to each of the POAs. • Add additional elements as desired. • When finished, save your scenario; it is ready to be deployed.
Step 3	**Deploy an Edge Network Scenario**
	• Click *Deploy* and select your scenario from the drop down. • AdvantEDGE will instantiate the network defined in the scenario. • Observe the edge network topology and verify that each element is placed as expected. • Issue a mobility event to transition a terminal across POAs (e.g. *poa1* to *poa2*); observe this change in the network topology. • Continue to send additional mobility events as desired. • Terminate the scenario.

This exercise used a simple example to introduce AdvantEDGE. However, the platform includes a wide set of capabilities not utilized in this exercise, including dynamic network performance emulation (latency, throughput, packet loss and jitter), deploying your own workload and edge applications, interworking with external nodes or apps (terminals and edge), visualizing key-performance indicators with integrated metric dashboards, a geospatial information service, and implementations of standardized edge services—such as ETSI MEC location service, RNIS, WLAN access info, app context transfer, etc.

For detailed information and addition exercise examples, please visit the AdvantEDGE Wiki at https://github.com/InterDigitalInc/AdvantEDGE/wiki

References

[OEC001]   Website Open Edge Computing Initiative - www.openedgecomputing.org
[OEC002]   Ha, K., Abe, Y., Chen, Z., Hu, W., Amos, B., Pillai, P., Satyanarayanan, M.;
           Adaptive VM Handoff Across Cloudlets, Technical Report CMU-CS-15–113,
           Computer Science Department, Carnegie Mellon University, June 2015 - http://
           elijah.cs.cmu.edu/DOCS/CMU-CS-15-113.pdf
[OEC003]   Kiryong Ha, Mahadev Satyanarayanan, OpenStack++ for Cloudlet Deployment,
           Technical Report CMU-CS-15–123 - http://elijah.cs.cmu.edu/DOCS/CMU-CS-15-
           123.pdf
[OEC004]   Wang, J, Feng, Z., George, S., Iyengar, R., Pillai, P., Satyanarayanan, M., Towards
           Scalable Edge-Native Applications; Proceedings of the Fourth IEEE/ACM
           Symposium on Edge Computing (SEC 2019), Washington, DC, November 2019 -
           http://elijah.cs.cmu.edu/DOCS/wang-sec2019.pdf
[OEC005]   S., Eiszler, T., Iyengar, R., Turki, H., Feng, Z., Wang J., Pillai, P., Satyanarayanan,
           M., OpenRTiST: End-to-End Benchmarking for Edge Computing IEEE Pervasive
           Computing, Volume 19, Issue 4, October-December 2020 - http://elijah.cs.cmu.edu/
           DOCS/george2020-openrtist.pdf
[OEC006]   Satyanarayanan, M., Eiszler, T., Harkes, J., Turki, H., Feng, Z., Edge Computing for
           Legacy Applications, IEEE Pervasive Computing 2020 - http://elijah.cs.cmu.edu/
           DOCS/satya2020-edgevdi.pdf
[OEC007]   OpenTPOD: Create DNN object detectors without any programming (February
           2020), Video - https://youtu.be/UHnNLrD6jTo
[OEC008]   OpenScout - ??
[OEC009]   Low Latency Interconnecct – technical report??

Chapter 10
MEC in Action: Performance, Testing and Ecosystem Activities

We finally arrived at the end of this book. This last chapter is very important as it provides an overview of the many ongoing activities devoted to put MEC in the actual market, e.g. from performance assessment, to conformance testing, proof-of-concepts, hackathons and deployment trials.

In fact, in order to kick off the MEC market and deploy MEC products in operations, all stakeholders must be involved and suitably convinced about the adoption of this technology. As we have clarified multiple times in this book, there are essentially two main categories of stakeholders: infrastructure owners and application developers. The first step is to convince infrastructure owners about the business need, and thus about MEC performance and related gains with respect to a traditional cloud system (by defining proper metrics and measurement methodologies and guidelines). Furthermore, proof of concepts and trials are great ways to start with even early implementations of MEC systems (not necessarily fully MEC complaint, of course), and represent another way for stakeholders to try MEC before the actual commercial launch. On the other side, engaging software developers by means of Hackathons or MEC Sandbox is critical, to help them in learning about edge applications development and service creation.

Finally, when deploying MEC in their infrastructures, operators and service providers need to verify the actual compliance of MEC products with consolidated standards (e.g. the MEC application enablement framework, the MEC APIs, etc.), where a testing framework and test suites are essential tools for this purpose.

All these aspects are described in this chapter, as an essential completion of the rest of the whole work in this book.

© Springer Nature Switzerland AG 2021   341
D. Sabella, *Multi-access Edge Computing: Software Development at the Network Edge*, Textbooks in Telecommunication Engineering, https://doi.org/10.1007/978-3-030-79618-1_10

10.1   Performance Assessment, Metrics, Best Practices and Guidelines

Performance assessment is essential to measure and quantify the actual implements enabled by MEC. In order to do that in a fair way (against traditional "cloud" implementations), a proper assessment methodology is needed, together with a consistent definition of measured KPIs (key performance indicators): these metrics are, in fact, intended to be used in order to demonstrate the benefits of deploying services and applications on a ME host compared to a centralized cloud.

The GS MEC-IEG 006 specification [IEG-006] published in the first phase of the ISG MEC introduced these aspects on MEC *metrics*, by including best practice and guidelines on *how* to evaluate the results.

10.1.1   MEC Metrics

Generally, MEC metrics are introduced with different purposes: evaluating the improvement given by MEC (as perceived by the end user) and assessing the benefits of different MEC deployment options (thus giving insights from a technologic point of view). With respect to these two goals, all metrics introduced in the MEC-IEG 006 deliverable can demonstrate the improvements of MEC solutions at least in the two following ways:

1. comparison between MEC and non-MEC solutions;
2. assessment of MEC deployments: comparison between different ME host positions within the network.

MEC metrics can be classified into two main groups: functional metrics and non-functional metrics (see below).

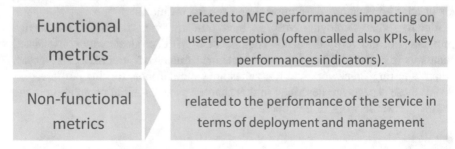

| Functional metrics | related to MEC performances impacting on user perception (often called also KPIs, key performances indicators). |
| Non-functional metrics | related to the performance of the service in terms of deployment and management |

Examples of **functional metrics**: Latency (both end-to-end and one-way), energy efficiency, throughput, goodput, loss rate (number of dropped packets), jitter, number of out-of-order delivery packets, QoS and MOS. Each of the functional metrics should be defined on per service basis. Note that the latency in localization (time to fix the position) is different from latency in content delivery.

Examples of **non-functional metrics**: Service lifecycle (instantiation, service deployment, service provisioning, service update (e.g. service scalability and elasticity), service disposal), service availability and fault tolerance (aka reliability), service processing/computational load, global MEC host load, number of API requests (more generally number of events) processed/second on MEC host, delay to process API request (north and south), number of failed API requests.

The reader should notice that in both cases, one could measure all the *statistics* over the above metrics (maximum value, mean and minimum value, standard deviation, the value of a given percentile, etc.).

As anticipated, several MEC metrics are possible, when it comes to performance evaluation. The table below provides more details on the most used categories of metrics for MEC systems.

Metric	Definition	Comments
Latency	*In communications, latency refers to a time interval whose measurement quantifies the delay elapsed between any event and a consequent target effect*	The concept of latency is wide and encompasses manifold (specific) metrics. Relevant latency examples are: Round-Trip Time, One-Way Delay (OWD), Setup Time, Service Processing Time, Context-update time **Note**: The latency definitions contained in MEC-IEG 006 are referring to latency measured on **application level**
Energy Efficiency	*In general, the ability of a mobile system to perform a certain work (e.g. transmit a volume of traffic or satisfying the QoS requirements for a certain service) by minimizing the power consumption*	As defined in standard in the ETSI TC EE specifications [EE-202706, EE-202336-12], it expresses the relationship between consumed power (or energy) and the production of a certain selected basic KPI of interest $\eta = \frac{power}{KPI}$ As an example, considering KPI = Traffic, it expresses the consumed watt of power per transferred traffic (in bps), or equivalently the energy per number of transferred bits
Network throughput	*Defined as measurement in terms bit rate units (e.g. kbps) at application level, in both upstream and downstream directions of the communication*	Since this is a metric at application level, it is categorized as **functional metric** **Note**: Throughput measurements could be performed both at transmitter side and at the receiver side
System resource footprint	*Referring to a cloud system, or a telecommunication system, in general, is referred to as the amount*	All the metrics considered here are **non-functional**. Current examples: • Computational load

(continued)

(continued)

Metric	Definition	Comments

	of system resources consumed, both in terms of a node's capacity and also in terms of communication requirements	○ The processing/computational time/load measures the amount of CPU processing time/cycles, and memory usage a service requires to operate
		• Non-user data volume exchange
		  ○ A service deployed with MEC requires the coordination of the modules running across different elements, this including the exchange of non-user data between entities to support application and user mobility
		  ○ Measurements: (1) between the MEC host and the radio network nodes (2) between MEC hosts (3) between the MEC host and the operational network management

Quality	*Traditionally, QoE (quality of experience) measures the global system performance using both subjective and objective measures of customer satisfaction*	The QoE metrics strongly depend on the service/application under analysis Since new services are implemented thanks to the flexibility of MEC, the definition of QoE metrics becomes a relevant aspect QoE metrics can be roughly classified into: • Objective and service-independent metrics about quality • Objective and service-dependent metrics about quality • Subjective and service-dependent metrics about quality • Objective metrics about user comfort

10.1.2 Performance Assessment of MEC

As anticipated, MEC metrics are introduced for different *comparison* purposes:

1. assessment of the performances of MEC with respect to non-MEC solutions;
2. evaluation of MEC deployments (comparison between different ME host positions within the network).

The figure below is showing an example of comparison between MEC and non-MEC systems, where in both cases a KPI (e.g. the latency) is measured between point (a) and (a') at the terminal side. In the non-MEC case, the KPI is

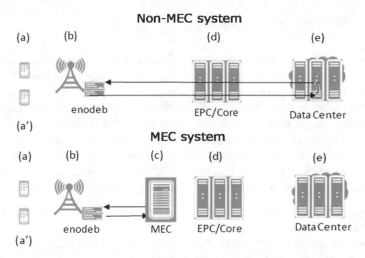

Fig. 10.1 End-to-end system comparison for MEC performance assessment. *Source* [IEG-006] ©
European Telecommunications Standards Institute 2017. Further use, modification, copy and/or
distribution are strictly prohibited

related to a situation where the remote data center (e) is involved, together with all
the other elements, e.g. eNodeB (b), EPC/Core (d). In the MEC case, only eNodeB
(b) and MEC (c) are involved in the measurement and should be taken into account.
This aspect will be clearer in the next sections, where we will provide more details
on the measurement methodology, with some practical examples (Fig. 10.1).

10.2 Measurement Methodology and Examples

Measurements can be obtained using different approaches:

- **Dedicated service monitoring tools** that measure the relevant functional metrics, either by the service itself or externally, and apply both to MEC and
  non-MEC solutions. Such metrics can be exposed to the MEC host as a mean to
  learn the resulted utilization and performance.
- **Common service monitoring** inside the MEC host that measures the
  non-functional metrics from each individual service.
- **Service measurement tools** inside the MEC host that measures the service
  non-functional metrics.

Additional tools are needed to generate workload and challenge the service in
terms of service scalability, availability/reliability.

In the following subsections, we provide some examples of measurement
applied to some well-known and widely used metrics.

10.2.1   *Evaluation of Latency*

The latency is the most commonly considered metrics, when it comes to MEC performance evaluations. We could even argue that latency improvement is one of the "MEC myths", in the sense that for the majority of the people, this is the *only* metric associated to the need for MEC. Obviously, this is an oversimplification, as other metrics can be relevant as well (depending on the context and the particular use case). Anyway, latency is certainly a critical metric for MEC.

MEC benefits on latency are expected to be particularly relevant in some of the use cases defined in ETSI GS MEC 002 [MEC-002]. For instance, in the following **use cases,** the evaluation of latency is pivotal for assessing MEC gains:

- Mobile video delivery optimization using throughput guidance for TCP (defined in clause A.2 of the ETSI GS MEC 002 spec).
- Local content caching at the mobile edge (defined clause A.3).
- Security, safety, data analytics (clause A.4).
- Augmented reality (AR), assisted reality, virtual reality (VR), cognitive assistance (clause A.5).
- Gaming and low-latency cloud applications (clause A.6).
- MEC edge video orchestration (clause A.10).
- Vehicle-to-infrastructure communication (clause A.14).

As a side note, depending on the particular use case considered (for example, from the above list of use cases), the related assessment methodology may be different.

An example of **passive measurement** of latency is provided in Fig. 10.2, where the traffic transmitted and received at the client side and received at server side (MEC or traditional server).

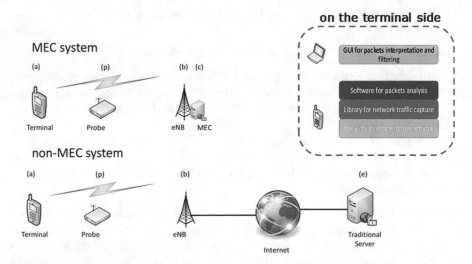

Fig. 10.2 End-to-end system comparison for MEC performance assessment. *Source* [IEG-006] © European Telecommunications Standards Institute 2017. Further use, modification, copy and/or distribution are strictly prohibited

In order to correctly parse this traffic, some packet analysis and filtering software and libraries are needed on the **terminal**, possibly with some GUI (graphical user interface) available for data logging and post-processing (see figure below). Nevertheless, it is worth mentioning that any application at the terminal side could be even source itself of a delay, as it may require computation capabilities that may slow down the actual process of reception at the terminal.

To overcome the performance problems related to the involvement of the mobile terminal in the measurements, an **external probe** can be used. This probe (p) is inserted in the radio interface and permits to acquire the signal transmitted/received at the terminal side, and the related timing (relevant in case of latency measurements). As an important side note, the probe is by definition an intermediate node between two communicating elements, and as such it cannot be perfectly aware of the actual reception of all the packets transferred between the two elements. Thus, in some cases, the passive measurement by means of probes can be still insufficient to understand the behavior of the system.

Active measurements

Differently from the pure observation of passive measurements, active measurement methods inject so-called probe traffic into the network at a traffic source and measure the outcomes at a probe traffic receiver. Hence, active measurement methods affect the network traffic.

In addition, simple PING tests using ICMP messages assess the average RTT latency of each communication link in the system.[1]

10.2.2 Evaluation of Energy Efficiency (EE)

As anticipated, different kind of EE metrics can be defined, depending on the scope of the assessment, and thus on the source of power consumption considered. In fact, EE can be defined as

- at component level (e.g. PA, chip, other components of network equipment or of a terminal, etc.);
- at node level (e.g. terminal, eNB, SGW, etc.);
- at network level (e.g. a set of nodes including or not terminals, only RAN equipment or including CN nodes, etc.).

There are several scenarios of measuring energy efficiencies. For all scenarios, the evaluation is done by comparing *baseline* and *frontline* systems.

The *baseline* is defined as a deployment without an MEC server in place and the methodology for the assessment foresees

---

[1]Active measurements can benefit from the adoption of specific protocols aimed at testing, for instance, the protocols defined by IETF IP Performance Metrics (IPPM) Working Group.

1. the injection of traffic from some users to a remote application server;
2. the measurement of energy consumed by the elements involved in the system;
3. the computation of EE metrics.

Compared to the baseline system, the *frontline* foresees an MEC server within the RAN network. In this case, the energy consumption is computed by taking into account not only this additional element, but also by considering the saving due to application and services residing on the MEC server.

In the following, we provide some examples of EE measurement setup, respectively, to assess energy efficiency at *network side* and at *terminal side*.

EE measurement setup 1: network side

This measurement setup can be applied to use case A.6 on "Gaming and low latency cloud applications" defined in ETSI GS MEC 002, where savings on infrastructure are beneficial for the operator (note: in this case, the EE is a non-functional metric). Power consumption and energy efficiency measurements of individual mobile network elements are described in several standards (for example, ETSI ES 202 706 specification, for radio base stations) [EE-202706].

The MEC-IEG 006 deliverable describes energy consumption and MN energy efficiency measurements in operational networks, and thus the measurement setup of the baseline system should be based on ETSI EE 203 228 specification (ref. [EE-203228]). In particular, the energy consumption of the MN can be measured

- by means of metering information provided by utility suppliers;
- by mobile network-integrated measurement systems;
- moreover, sensors can be used to measure site and equipment energy consumption.

Baseline: measurement without the MEC server.

Figure 10.3 provides an example of network cluster assessment in absence of MEC, where measurements of power consumption are done by means of external measurement tools (e.g. current clamps or more accurate tools).

In this case, power consumption of base stations should be assessed by using mobile network-integrated measurement systems, when available according to

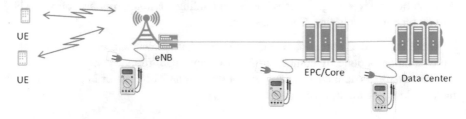

Fig. 10.3 End-to-end system comparison for MEC performance assessment. *Source* [IEG-006] © European Telecommunications Standards Institute 2017. Further use, modification, copy and/or distribution are strictly prohibited

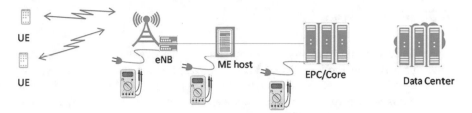

Fig. 10.4 End-to-end system comparison for MEC performance assessment. *Source* [IEG-006] © European Telecommunications Standards Institute 2017. Further use, modification, copy and/or distribution are strictly prohibited

ETSI ES 202 336–12 (ref. [EE-202336-12]). Similar methodology can be used for other equipment involved in the measurements, as shown in the above figure.

Frontline: measurement with the MEC server.

Figure 10.4 provides an example of network cluster assessment in presence of MEC, where measurements of power consumption are done by means of external measurement tools (e.g. current clamps or more accurate tools).

In this case, power consumption of base stations should be assessed by using mobile network-integrated measurement systems, when available according to ETSI ES 202 336–12 (ref. [EE-202336-12]). Similar methodology can be used for other equipment involved in the measurements, as shown in the above figure.

EE measurement setup 2: terminal side

This measurement setup can be applied to use case A.23 on "Application computation off-loading" defined in ETSI GS MEC 002, where savings on terminal are beneficial for the end user (note: in this case, the EE is a functional metric).

This "application computation offloading" use case provides the benefits to prolong the battery life of the mobile device (IoT, phone, vehicle, etc.) and can be more appropriate to run on the MEC server. In this case, the end-to-end system can be overall more efficient and can be a way for the operator to generate revenue. Assessment of end-to-end EE gains by offloading the UE of application should be computed by considering two competing aspects:

- % Battery life increase.
- % Increase in server power.

Moreover, the actual gain (and opportunity to actually perform the offloading decision) depends also on the application that is offloaded from the phone.

Note: As a deepening on this use case, the reader is suggested to have a look again at the case study in Sect. 8.4 and Exercise 8.1, covering the fundamental **trade-offs** between *energy consumption* and *latency* for the different offloading opportunities considering both computation and communication aspects of task offloading and execution.

10.3   MEC Testing

As emphasized in the previous sections, performance evaluation is an important aspect for assessing the benefits of MEC with respect to a cloud system without MEC. In addition to that, conformance of MEC products and their compliance to the standards is important to ensure interoperability in E2E systems, especially in presence of multiple stakeholders in the value chain (Fig. 10.5).

For this purpose, ETSI MEC has published a general testing framework for MEC Technologies (MEC-025). Based on the testing methodology guidelines and framework, ETSI MEC developed a multi-part conformance test specification (MEC-032) for the MEC service APIs and the MEC Application Enablement API. This multi-part document specifies the whole test suite to verify API conformance for server implementations.

The table below describes in more detail the purpose of the various published documents for this purpose.

Fig. 10.5 MEC workflow, from standardization to OpenAPI and testing. (© European Telecommunications Standards Institute 2020. Further use, modification, copy and/or distribution are strictly prohibited)

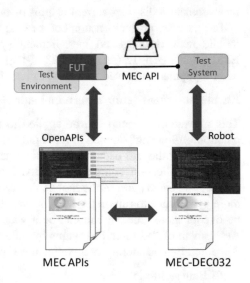

Document	Purpose
MEC-DEC 025 MEC testing framework	List the functionalities and capabilities required by an MEC compliant implementation Specify a testing framework defining a methodology for development of interoperability and/or conformance test strategies, test systems and the resulting test specifications for MEC standards
MEC-DEC 032–1 MEC API conformance test specification part I test requirements and implementation conformance statement (ICS)	Part 1 includes the test requirements and Implementation Conformance Statement (ICS)
MEC-DEC 032–2 MEC API conformance test specification part II test purposes (TP)	Part 2 includes a test suite structure (TSS) and test purposes (TPs) using the standardized notation TDL_TO: the test purposes are written in ETSI test description language (TDL), with the test objectives (TO) extension[a]
MEC-DEC 032–3 MEC API conformance test specification part III abstract test suite (ATS)	Part 3 includes an abstract test suite (ATS) written in a machine-readable specification language

[a]For more information on ETSI TDL and the TOP open source project visit https://tdl.etsi.org

In particular, **Test Purposes** (published in the ETSI forge website[2]) describe, for each individual functionality specified by the MEC API standards, the description of what is to be tested. A Test Purpose describes the components involved and the expected behavior of the Implementation Under Test (IUT), together with all the needed information on the requirements to be tested from the base specifications (Fig. 10.6).

```
7    *** Test Cases ***
8    Get list of locations of User Equipments
9        [Documentation]   Test ID: TP_MEC_PLAT_LOC_001_OK
10       ...
11       Set Headers   {"Accept":"application/json"}
12       Get    /location/v2/users/${zone_id}
13       ${output}=   Output   response
14       Set Suite Variable   ${response}   ${output}
15       Check HTTP Response Status Code Is   200
16       Check HTTP Response Body Json Schema Is   userInfo
17       Log   Check Location for userInfo element
18       Should be Equal   ${response['body']['userInfo']['zoneId']}   ${zone_id}
19       Log   Location OK
20
```

Fig. 10.6 Example of test case implementation in the robot framework language

---

[2]MEC API Test Purposes: https://forge.etsi.org/rep/mec/gs032p2-test-purposes.

Also **test implementations** of the document 032–3 are published in ETSI Forge website,[3] both in robot framework and TTCN-3 languages (openly available and released under BDS-3 license). In both cases, the ETSI Forge websites report also useful information on installation of the frameworks, usage and execution of the published test cases.

10.4   OpenAPI Representations of MEC Services APIs

DECODE WG provides OpenAPI representations for the MEC Services APIs on ETSI Forge (https://forge.etsi.org/). The rationale behind this initiative is that, as we explained at the beginning of Part 4, software developers are essentially not spending much time to read and double check the standards, as they need to minimize time to market for their applications. As a consequence, in order to better satisfy their needs, there is an effort in the DECODE WG to complement the definitions for each method and resource defined in the standard specifications, and to produce machine-readable representations of the MEC APIs (Fig. 10.7).

For this purpose, DECODE WG is working to provide for all APIs a supplementary description file compliant to the OpenAPI specification [OAS-spec], which is defining "*a standard, programming language-agnostic interface description for HTTP APIs, which allows both humans and computers to discover and understand the capabilities of a service without requiring access to source code, additional documentation, or inspection of network traffic*" (Fig. 10.8).

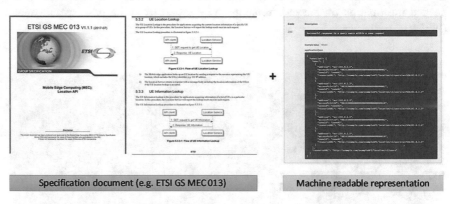

Specification document (e.g. ETSI GS MEC013)   Machine readable representation

Fig. 10.7 MEC API specifications and their OpenAPI representations. (© European Telecommunications Standards Institute 2020. Further use, modification, copy and/or distribution are strictly prohibited)

---

[3]MEC Robot Test Suite: https://forge.etsi.org/rep/mec/gs032p3-robot-test-suite.
   MEC TTCN-3 Test Suite: https://forge.etsi.org/rep/mec/gs032p3-ttcn-test-suite.

Fig. 10.8 MEC services APIs published on ETSI forge (https://forge.etsi.org/rep/mec)

The table below lists all the OpenAPI representations of MEC Services APIs currently available in the ETSI Forge website, with their respective implementation status (and the related version of the MEC specification).

API in ETSI forge	MEC Spec.	Implementation status
V2X information service API	ETSI GS MEC 030	Partial
MEC application support API and MEC service management API	ETSI MEC GS 011	Full (v2.1.1)
Application mobility service API	ETSI GS MEC 021	Partial
Application package, lifecycle and granting management API	ETSI GS MEC 010–2	Partial
WLAN information API	ETSI GS MEC 028	Partial
Fixed access information API	ETSI MEC GS 029	Partial
UE application interface API	ETSI GS MEC 016	Full (v2.1.1)
Bandwidth management API	ETSI GS MEC 015	Full (v1.1.1)
UE identity API	ETSI GS MEC 014	Full (v1.1.1)
Location API	ETSI GS MEC 013	Full (v1.1.1)
Radio network information API	ETSI GS MEC 012	Full (v1.1.1)

In addition to OpenAPI representations, the ETSI Forge contains a further supplementary file defining the data types in protocol buffers format, as defined in the Protocol Buffers Language Specification [ProtBuf-spec]. As a side note, the reader should keep in mind that in case of discrepancies between the supplementary files and the related data structure definitions in the specifications, the data structure definitions take precedence with respect to ETSI Forge implementations.

10.5 MEC Sandbox

Under its MEC enablement charter, DECODE offers a MEC Sandbox environment (https://try-mec.etsi.org/). The MEC Sandbox is an interactive online environment that enables developers to learn and experiment with ETSI MEC Service APIs.

The MEC Sandbox provides a set of experimentation scenarios combining different network access technologies (4G, 5G, Wi-Fi) and client devices in a macro-network configuration, set in Monaco. Sandbox users gain hands-on experience with the behavior and capabilities of MEC Service APIs. By providing a live and interactive environment, the MEC Sandbox exposes edge app developers with contextual information, highlighting how MEC services can offer significant performance gain for edge applications.

10.5.1 Sandbox Access and Configuration

Let us look at how a user can access the MEC Sandbox, configure a scenario within it and get started. First, the MEC Sandbox can be found online here: https://try-mec. etsi.org/. The MEC Sandbox home page presents information on the environment and some helpful MEC informational links. From the home page, a user needs to sign-in to the MEC Sandbox, with the presented instructions.

Once signed-in, an individual and isolated "sandbox" is instantiated, and the user is presented with the main MEC Sandbox dashboard. The top portion of the dashboard, as shown in Fig. 10.3, provides a map of Monaco, including placement of wireless points of access (PoA's) with their coverage area, color-coded network location zones, terminal devices and mobility paths. Zoom and drag controls on the Sandbox map allow a user to explore the Monaco network topology, terminal location and observe the MEC Sandbox behavior. A user can also control the level of detail by enabling different layers on the map (e.g. point of access coverage, terminal paths, etc.).

To the right of the map, the MEC Sandbox provides a set of configuration control points. Here, a user selects a network configuration from a set of predefined network topologies combining 5G, 4G and WLAN access technologies. Each configuration includes a varying number of macro-cells, small cells or WLAN hotspots, organized into color-coded network location zones, as shown in Fig. 10.9. The user may also choose the number and type of terminal devices to include in the Sandbox. Stationary terminals represent fixed wireless devices, such as smart city infrastructure. Low-velocity terminals represent pedestrians or perhaps slower moving autonomous devices. High-velocity terminals represent motor or electric

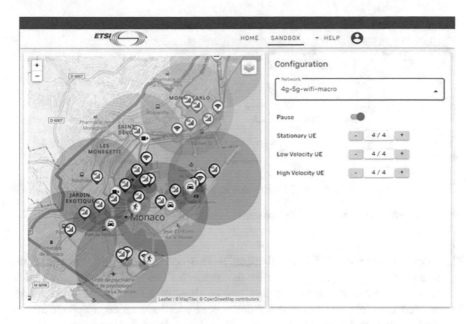

Fig. 10.9 ETSI MEC Sandbox dashboard

vehicles traveling on Monaco's streets and highways. Finally, the "pause" button provides a helpful control that stops the movement of the mobile terminals. However, while paused, all MEC Service APIs remain active and continue to respond to requests.

Clicking on any icon on the map opens a pop-up that provides details on the current state of a PoA or terminal, such as IP or MAC address, velocity, wireless attachment details (e.g. connected cell id or BSSID), location zone id, geo-location coordinates (latitude and longitude), etc.

10.5.2 Sandbox MEC Service API Interaction

The primary feature of the MEC Sandbox is live interaction with MEC Services APIs. The Sandbox provides two options for users to interact with its services:

1. **"Try-it in the browser"**—This option enables a user to interact with MEC Services directly from their browser via a SwaggerUI, just like the browsable MEC APIs on ETSI Forge (https://forge.etsi.org/), but with live, dynamic behavior. This option is the easiest in getting started and observing API behavior in a user-friendly, readable manner.

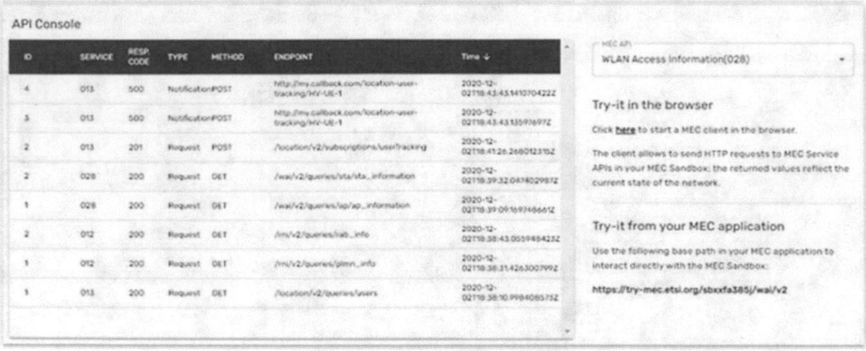

Fig. 10.10 ETSI MEC Sandbox API Console

2. **"Try-it from your MEC application"**—This option allows edge applications to access the MEC APIs served by the Sandbox. The MEC Sandbox provides user unique service endpoints that can be directly called by the user's edge application. This makes sense for developers who have an edge application or environment and would like to try MEC services from their software.

These "Try-it" options are available in the MEC Sandbox's API Console at the bottom of the dashboard, as shown in Fig. 10.10. Regardless of which "Try-it" option a user may utilize, the MEC Sandbox displays MEC service requests and subscription notifications in the API Console as they are received or issued by the MEC services. Clicking on a service request opens a pop-up containing detailed information such as the response code, endpoint path, timestamp and request/response body parameters.

For a deeper dive, let us explore using the MEC-013 Location Service. We can query the location service to learn a terminal's network location (i.e. its point of attachment to the network) and geo-location. Using a high-velocity terminal as an example, we find the terminal's id by clicking on its icon on the map (e.g. 10.100.0.1). With the terminal's id, we can then invoke the following location service query in the sandbox (e.g. within the SwaggerUI):

```
GET https://try-mec.etsi.org/<sandbox-id> /location/v2/que-
ries/users?address=10.100.0.1
```

The MEC Sandbox responds to this query with the following information: In this case, the terminal is attached to a 5G small cell in "zone02", at the indicated latitude and longitude.

```
  "userList": {
    "resourceURL": "https://try-
mec.etsi.org/sandbox/location/v2/queries/users",
    "user": [
      {
        "accessPointId": "5g-small-cell-7",
        "address": "10.100.0.1",
        "locationInfo": {
          "latitude": [43.734253],
          "longitude": [7.421442],
          "shape": 2,
          "timestamp": {
            "nanoSeconds": 0,
            "seconds": 1606942338
          }
        },
        "resourceURL": "https://try-
mec.etsi.org/sandbox/location/v2/queries/users?address=10.100.0.1",
        "timestamp": null,
        "zoneId": "zone02"
      }
    ]
  }
```

As of this writing, the MEC Sandbox includes the following: MEC Service
APIs: Location (MEC-013), Radio Network Information (MEC-012) and WLAN
Access Information (MEC-028). New configurations and MEC services may be
added to the sandbox in the future.

	Useful reference/source code
	More detailed and up-to-date information on the MEC Sandbox can be found on the MEC Wiki

Web link to the reference
https://mecwiki.etsi.org/index.php?title = https://mecwiki.etsi.org/index.php?title=MEC_
Sandbox_Help

10.6   MEC Ecosystem: Proof-of-Concepts, Trials, Hackathons

As described in the first chapter of the book, within ETSI MEC ISG, the
Deployment and ECOsystem DEvelopment (DECODE) Working Group is focused
on enabling and easing the MEC implementation path for mobile operators, edge
service providers and applications developers. We have also seen that
DECODE WG provides OpenAPI representations for the MEC Services APIs on
ETSI Forge, testing and compliance frameworks. Furthermore, DECODE WG
sponsors Proof-of-Concepts, Trials, Plugtests™ & Hackathons, in order to better
engage all MEC ecosystem stakeholders.

10.6.1 MEC PoCs (Proof-of-Concepts)

ETSI ISG MEC has created a MEC PoC framework (which can be found in the MEC Wiki page [MEC-Wiki]) to coordinate and promote multivendor Proof-of-Concept (PoC) projects illustrating key aspects of MEC technology. In fact, according to ETSI ISG MEC, *"Proofs of Concept are an important tool to demonstrate the viability of a new technology during its early days and or pre-standardisation phase"*. For that purpose, a number of PoC projects have been approved and promoted in multiple events and international venues.

A list of PoCs related to various use cases can be found in the MEC Wiki page, and summarized in the table below, together with the list of companies involved in each PoC project:

PoC #	Title	Team
1	Video User Experience Optimization via MEC—A Service-Aware RAN PoC	Intel—China Mobile—iQiYi
2	Edge Video Orchestration and Video Clip Replay via MEC	Nokia—EE—Smart Mobile Labs
3	Radio aware video optimization in a fully virtualized network	Telecom Italia—Intel UK Corporation—Eurecom—Politecnico di Torino
4	FLIPS—Flexible IP-based Services	InterDigital—Bristol is Open—Intracom—CVTC—Essex University
5	Enterprise Services	Saguna—Adva Optical Networking—Bezeq International
6	Healthcare—Dynamic Hospital User, IoT and Alert Status management	Quortus Ltd.—Argela—Turk Telecom
7	Multi-Service MEC Platform for Advanced Service Delivery	Brocade—Gigaspaces—Advantech—Saguna—Vasona—Vodafone
8	Video Analytics	Nokia—Vodafone Hutchison Australia—SeeTec
9	MEC platform to enable low-latency Industrial IoT	Vasona Networks—RIFT.io—Xaptum—Oberthur Technologies—Intel Corporation—Vodafone
10	Service-Aware MEC Platform to Enable Bandwidth Management of RAN	Industry Technology Research Institute—Linker Network—FarEasTone
11	Communication Traffic Management for V2X	KDDI Corporation—Saguna Networks Ltd.—Hewlett Packard Enterprise
12	MEC enabled Over-The-Top business	China Unicom—ZTE—Intel—Tencent—Wo video—UnitedStack
13	MEC infotainment for smart roads and city hot spots	TIM—Intel—Vivida—ISMB—City of Turin

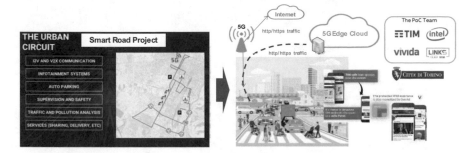

Fig. 10.11 PoC#13: MEC-based infotainment services in the framework of Turin smart road project [OVB-VTC20]

As an example of a recent proof-of-concept project, we describe here briefly the outcomes of the **PoC#13** on "*MEC infotainment for smart roads and city hot spots*" that was showcased at the Mobile World Congress 2019 in Barcelona. This project was a joint effort of multiple companies involved in the "Smart Road Project", a consortium of private and public subjects led by the City of Turin, working together with the interest on testing Cooperative, Connected and Automated Mobility (CCAM) in its territory [TRN-SmRd].

The goal of this PoC was to implement an MEC-based infotainment service for smart roads in 5G environments, using the OpenNESS platform and a commercial-ready client-server application, where the server application (called OverBrowser) provided a customized customer experience on the browser screen of any connected device (see Fig. 10.11).

The target of the implemented use case was to provide a clientless interactive communication/infotainment platform which can be executed over the MEC servers distributed in smart roads and city hot spots exploiting OverBrowser services by Vivida. OverBrowser is a service platform to create new, very visible and highly interactive, communication spaces, on the browser screen of any connected device, via clientless network activation. Main advantage of OverBrowser at the edge consists in accelerating frequently accessed contents by improving the content transfer speed through the 5G network reducing the latency and the backbone bandwidth usage. Data and contents are generated locally in the MEC platform and transferred back directly to the Client App well fit with the vehicular scenario where several users are accessing infotainment contexts in mobility asking for low-delay services (e.g. gaming) and, in the long run, for scalability.

The showcase of this PoC was also accompanied by empirical results on performance, as proof of the benefits of MEC in the experimented use case. Figure 10.12 shows the methodology and test setup of the performance measurements, where local sniffing probes have been installed, in order to capture the traffic of all the components of the network. The real-time capture enabled a measurable feedback of the intrinsic latency between the hosts and network services, e.g. such

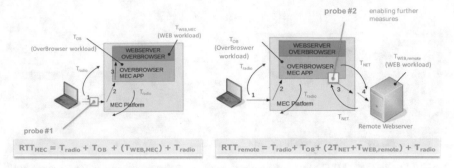

$$RTT_{MEC} = T_{radio} + T_{OB} + (T_{WEB,MEC}) + T_{radio}$$

$$RTT_{remote} = T_{radio} + T_{OB} + (2T_{NET} + T_{WEB,remote}) + T_{radio}$$

Fig. 10.12 PoC#13 testing methodology (comparison between MEC and non-MEC cases)

web server response time and latency. The performance gain was then evaluated by comparing RTT in the two cases (with and without MEC).

Test results showed a significant latency reduction of content load, indeed obtained by hosting the contents close to the user, to avoid data transfer across the Internet from the hosting provider. Figure below shows how delay gain between edge and remote is pretty constant for all packet sizes (more technical details can be found in the paper presented at the IEEE VTC2020-spring [OVB-VTC20]). Additional gains were obtained also in terms of transport network data saving, by supposing a video stream accessed from various users connected to the same access network. In fact, proxying of the video content in the MEC platform with OverBrowser can also provide a notable reduction of transport network usage, with evident benefits in terms of costs for operators and infrastructure owners (Fig. 10.13).

10.6.2 Trials and Plugtests™

After the successful campaign of MEC Proof of Concepts (PoC), ETSI ISG MEC has developed the MEC Deployment Trial Framework (MDT). The MDT framework (published as well in the MEC Wiki page) [MEC-Wiki] can be seen as the

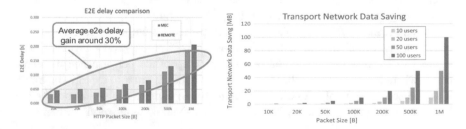

Fig. 10.13 MEC benefits in terms of average E2E delay and transport network data saving

next step of MEC PoC framework. The initiative is calling ETSI Members and nonmembers to form MDT teams and propose their solution to showcase their deployed system in a commercial network, relying on the principles of the MEC reference architecture and interfaces.

A list of PoCs related to various use cases can be found in the MEC Wiki page, and its current status (at the time of writing the present book) is summarized in the table below, together with the list of companies involved in each PoC project:

MDT #	Title	Companies
1	Deployment of CDN at the edge of the mobile network	China Mobile—Nokia
2	Introducing mobile edge computing in factory network	China Mobile—Huawei
3	Edge-Cloud VR cloud game scheme based on 5G network	China Unicom—Huawei—Tencent—Intel
4	ARVR navigation based on 5G MEC	China Telecom—Huawei—21CN

A more recent initiative, bringing implementers together since 2017, is the ETSI Plugtests Programme, consisting in a series of international events for testing multivendor interoperability and API conformance. These are free sessions, open for any organization providing an implementation to test or to support the testing. Operators and academia are also admitted as observers.

The ETSI NFV Plugtests Programme, while initially focused on interoperability of NFV implementations, from 2020 extended its focus to include participation of MEC implementations (this edition was also the first fully remote event of the series). The **NFV&MEC Plugtests 2020** offered NFV and MEC solution providers and open source projects an opportunity to meet and assess the level of interoperability of their NFV and edge solutions, while they validated and provided feedback on NFV and MEC specifications and APIs.

The 2020 event included a wide range of test sessions covering:

- NFV interoperability and API conformance.
- MEC and MEC-in-NFV interoperability and API conformance.

In particular, as far as MEC is concerned, the following tests have been conducted, according to the MEC interoperability test plan [PLG20-plan]:

1. **Multivendor MEC and MEC-in-NFV interoperability**, focused on MEC application lifecycle, traffic rules management and enforcement and MEC services and MEC APIs;
2. **MEC API conformance**, based on MEC-DEC 032 (robot framework and TTCN-3 Test Suites for MEC 010, MEC 011, MEC 012 and MEC 013 specifications).

The **MEC interoperability tests**, based on the test configurations described in the table below, involved multiple MEC platforms and applications, and included several elementary tests: MEC App on-boarding, start/stop, status query/change, or MEC service registration/update/deregistration/consumption, or again activation/ setup/update/deactivation of traffic and DNS rules from MEC applications. The results of these MEC interoperability tests revealed overall a high success rate (73.4% of passed tests), as a first sign of good level of interoperability.

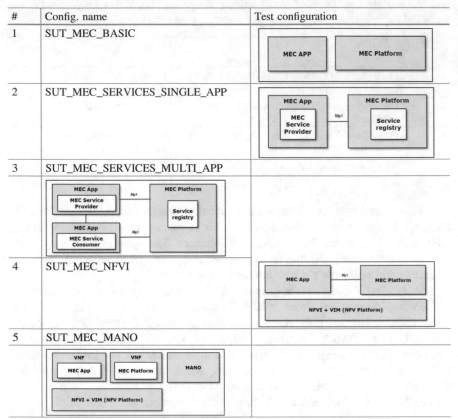

#	Config. name	Test configuration
1	SUT_MEC_BASIC	
2	SUT_MEC_SERVICES_SINGLE_APP	
3	SUT_MEC_SERVICES_MULTI_APP	
4	SUT_MEC_NFVI	
5	SUT_MEC_MANO	

The **MEC API Conformance Test Sessions** aimed at validating the conformance of the participants FUTs to the MEC011, MEC012 and MEC013 API specifications, while validating the API conformance robot and TTCN-3 test suites. More information on the test configuration and results of this 2020 edition of the NFV&MEC Plugtests can be found in the Plugtests Report [PLG20-report].

Next events of the ETSI Plugtests Programme are expected to continue promoting interoperability and validation of standards for both NFV and MEC technologies, since there is a growing interest and participation in the MEC testing, including multi-app interoperability scenarios and MEC APIs testing.

10.6.3 MEC Hackathons

Events like Hackathons are a great way to engage the developer community to get introduced, use and test a new technology, these events help in advertising great standards to future stakeholders. In MEC hackathons, where technology is introduced and offered in a controlled environment, implementation from developers can provide valuable feedback, which helps the MEC standard group to test the relevance of the standards both on the present status and future updates required.

Every year there are new proposals of MEC Hackathons, according to the **MEC Hackathon framework** published by ETSI ISG MEC (see MEC Wiki page here). These initiatives, coming from several organizing committees leveraging partnerships with key global companies, are often also supported by industrial associations or open source communities, or international bodies in the space. The goal of these initiatives is always to provide state-of-the-art technologies and support to hackathon participants, together with attractive prizes for the winners, in order to better attract application SW developers' communities, and encourage the adoption of edge computing technologies.

In the past, similar series of MEC Hackathons event has taken place:

- **September 18–19, 2018**[4]: Three parallel events: Berlin (co-located with Edge Computing Congress), Beijing (China), and Turin (Italy).
- **September 17–18, 2019**[5]: Two parallel events: London, UK (co-located with Edge Computing Congress) and Shenzhen (China).
- **November 18, 2019,**[6] in collab. with LF Edge and Akraino: San Diego (USA) (with KubeCon + CloudNativeCon North America).

In the follow-up, a similar MEC Hackathon competition was organized as co-located with the Droidcon Italy conference, and thus named **Droidcon MEC Hackathon 2020** (link here: https://it.droidcon.com/2020/hackathon/), primarily organized by Intel in collaboration with TIM, CISCO, Links Foundation, Equinix, endorsed by ETSI and supported by GSMA and the City of Turin. The main target of that 2020 Hackathon event was to provide an edge SW development environment for the developers' teams at the Hackathon to *develop mobile applications for advanced services in MEC-enabled 5G networks*, using ETSI MEC technologies and APIs.

For the successful organization of the hackathon, the organizing committee provided a development environment composed of several key components,

---

[4]https://www.etsi.org/newsroom/news/1345-2018-10-news-etsi-mec-hackathons-bring-developers-together-in-china-germany-and-italy-to-trial-edge-computing-solutions.

[5]https://www.etsi.org/newsroom/press-releases/1648-2019-09-developers-at-mec-hackathons-endorsed-by-etsi-challenged-to-trial-edge-computing-for-5g-in-uk-and-china.

[6]https://www.etsi.org/newsroom/press-releases/1648-2019-09-developers-at-mec-hackathons-endorsed-by-etsi-challenged-to-trial-edge-computing-for-5g-in-uk-and-china.

described here below.

Hardware setup at the Droidcon 2020

For the hackathon two types of end-to-end environment, setups were offered, i.e. allowing *F2F* (face-to-face) and *remote* participation (virtual setup).

1. Face-to-Face setup

In collaboration with Cisco, for the developers who wanted to participate in an on-site hackathon location in Italy, MEC pod set up was offered, constituted by important software and hardware components allowing the application MEC development. A sample pod setup for *F2F* participation looks like below, i.e. (Fig. 10.14).

Figure 10.15 shows how a client application (or another management device) from a developer team could access the edge node present in the F2F event infrastructure.

2. Virtual setup

In previous editions of hackathons, all developers used to apply for Face-to-Face participation. Nevertheless, because of the COVID pandemic in 2020, a new problem arose, i.e. uncertain situation about attendants, related to travel restrictions and changing rules overtime. As a consequence, there was the need to offer them the possibility to remotely participate, and to provide each admitted team remote access to a full E2E edge SW development environment. In this perspective, the

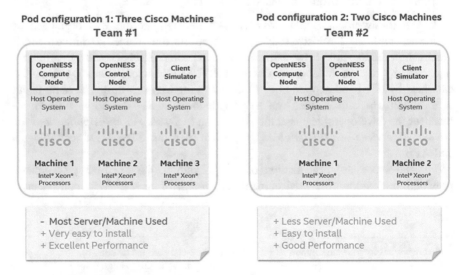

Fig. 10.14 Pod configurations for F2F participation (example for two developer team's participation at the Hackathon venue)

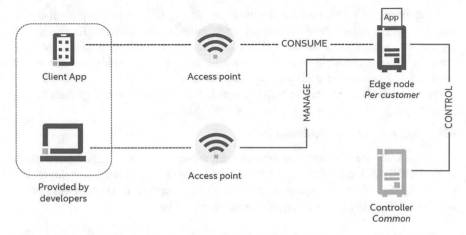

Fig. 10.15 The logical architecture of the F2F event infrastructure

Hackathon Organizing Committee (HOC) provided remotely accessible isolated environments to the participating teams, each with its own OS, on a single and powerful platform running software stacks like OpenNESS, oneAPI and OpenVINO. A sample pod setup for *virtual* participation looks like as shown in Fig. 10.16.

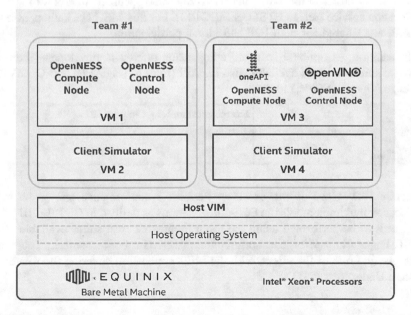

Fig. 10.16 Pod setup for remote participation (example for two developer team's participation at the Hackathon venue)

In addition, as a general consideration, creating virtual setup using open source virtualization technologies is also resulting in multiple advantages, like saving cost on the creation of new virtual pods, easier and fast replication of the pods in scenario of new multiple teams participating in the short notice taking care of isolation and privacy of each participating team.

More detail about this hackathon event and other information can be found in the white paper published for this event [INT-WP-hkt].

MEC software at the Droidcon 2020

The main software development environment provided to the admitted participants was OpenNESS toolkit (www.openness.org) as open source reference implementation of MEC platform, equipped with an ETSI compliant implementation of the MEC Location API (both briefly described here below):

- **OpenNESS**: As described earlier in this book, this is an edge computing software toolkit that enables highly optimized and performant edge platforms to onboard and manage applications and network functions with cloud-like agility across any type of network. For this MEC Hackathon, the OpenNESS Release 20.06 was used, including both controller node and compute node running on a single operating system.
- **MEC location API**: MEC location services are designed to provide information about the location of user equipment (UE), and are used by MEC applications to interact with the location service for retrieving the desired information. As described earlier in this book, the ETSI ISG MEC specified an API for the MEC location service (ref. to ETSI GS MEC 013). For this MEC Hackathon, also this API was offered for both F2F and virtual participation.

In addition, as optional software components offered to developers, the organizing committee made available also oneAPI (refer to [oneAPI]) and OpenVINO (refer to [OpenVINO™]) toolkits.

 | Exercises/examples—Chapter 10

Exercise 10.1 Goal of the exercise: Perform a conformance test on an MEC application that implements the DNS rule functions as defined in GS MEC 011. The test needs to be performed using the robot framework[7] test suites available in the ETSI Forge [https://forge.etsi.org/rep/mec/gs032p3-robot-test-suite]. The robot tests play the role of the client, with the MEC application covering the role of the System Under Test (SUT).

---
[7]https://robotframework.org/.

Proposed solution

Step 1: Environment preparation

As a first step, we need to download the robot framework tests from the ETSI Forge repository (branch *2.1.1-fix-plu*) on the local machine and the software required to run the test scripts, as documented in the test repository (see README.md file).

a. Download of robot framework tests

```
$ git clone https://forge.etsi.org/rep/mec/gs032p3-robot-
test-suite
$ git checkout 2.1.1-fix-plu
```

b. Installation of software dependencies

```
$ sudo apt install python3 python3-pip
$ cd gs032p3-robot-test-suite
$ sudo pip3 install -r requirements.txt
```

Step 2: Configuration of the environment to execute the test

The tests for the DNS rule functions are available in the SRV/DNS folder, which is structured as shown in Fig. 10.17.

The *environment* folder contains a file *variables.txt* with the configuration of the parameters to run the tests in the local environment, for the given application (e.g. the definition of the Function Under Test (FUT), the IP addresses, etc.).

The *jsons* folder contains the payload of the HTTP messages required by some of the MEC application API methods (usually for PUT and POST messages). Both these two folders contain data that the tester should provide before running the tests.

Name	Last commit	Last update
..		
environment	Updating from v2.0.9 to v2.1.1. Cause 7: MEC Application s...	7 months ago
jsons	Updating from v2.0.9 to v2.1.1. Cause 7: MEC Application s...	7 months ago
schemas	Updating from v2.0.9 to v2.1.1. Cause 7: MEC Application s...	7 months ago
PlatDnsRules.robot	Fix on check response contains keyword	7 months ago
README.md	Creating structure	1 year ago

Fig. 10.17 Structure of DNS rules test folder

The *schemas* folder contains the json schema for the MEC application API data models and they are used to validate the json payload provided by the MEC application in its responses.

The *XXXrobot* file(s) (*PlatDnsRules.robot* in our case) define the tests to be executed.

As mentioned before, the tester needs to configure the parameters in the *variables.txt* file. This file contains two blocks of variables. The generic ones include information related to the FUT, while the specific ones are related to the execution of the tests. Some of these variables are used to get specific data or to generate failure scenarios.

```
# Generic variables
${SCHEMA}            http                # http or https
${HOST}              10.192.2.172        # IPAddress of the FUT
${PORT}              8081                # port of the FUT
${response}          {}                  # no data is needed
${TOKEN}             Basic YWxhZGRpbjpvcGVuc2VzYW11 # Authorization
token, if it is needed
${apiRoot}                               # apiRoot of the FUT
${apiName}           mec_app_support # apiName of the FUT
${apiVersion}        v1                  # apiVersion of the FUT
```

```
# Specific variables
${APP_INSTANCE_ID}    5abe4782-2c70-4e47-9a4e-0ee3a1a0fd1f
        ## ID of the MEC instance
${DNS_RULE_ID}        e0deee2b-6e50-4f33-ab09-8bf0585025d3
        ## Existing Rule ID
${NON_ESISTENT_DNS_RULE_ID}   NON_ESISTENT_DNS_RULE_ID
        ## Not existing rule ID, generates 404 in GET method
${INVALID_ETAG}       INVALID_ETAG
        ## Invalid ETag: generated 412 in PUT method
${SOME_IP_ADDRESS}    146.241.7.3
        ## IP address associated with the FQDN resolved by the DNS rule
```

Moreover, the files in the jsons folder must be updated with valid data.

The information to be provided for our exercise is the payload related to the update of a DNS rule. Let's suppose we want to change the state, from *ACTIVE* to *INACTIVE*, for a given DNS rule with the following parameters:

```
{
        "dnsRuleId": "e0deee2b-6e50-4f33-ab09-8bf0585025d3",
        "domainName": "www.example.com",
        "ipAddressType": "IP_V4",
        "ipAddress": "146.241.7.3",
        "ttl": 300,
        "state": "ACTIVE"
}
```

To disable the DNS rule, we need to change the *state* parameter to *INACTIVE*. In this case, the new json payload (*DnsRuleUpdate.json*) should be as follows:

```
{
        "dnsRuleId": "e0deee2b-6e50-4f33-ab09-8bf0585025d3",
        "domainName": "www.example.com",
        "ipAddressType": "IP_V4",
        "ipAddress": "146.241.7.3",
        "ttl": 300,
        "state": "INACTIVE"
}
```

The file *jsons/DnsRulesUpdateError.json* is used to generate a bad request (400 error) during the update request, since it contains an error on the data model.

```
{
        "dnsRuleId": "e0deee2b-6e50-4f33-ab09-8bf0585025d3",
        "domainName": "www.example.com",
        "ipAddressType": "IP_V4",
        "ipAddress": "146.241.7.3",
        "ttl": 300,
        "state": "UNKNOWN_VALUE"
}
```

The *state* parameter cannot be set to "UNKNOWN_VALUE", so this will generate the error during the execution of the test.

Step 3: Tests execution.

Once all the required information is in place, we can start running the tests. Each test suite contains different test cases that can be executed all-in-one, or the tester can provide the list of the desired test cases (identified by name) within the robot command and execute single test cases.

To run the tests, the tester needs to use the "robot" CLI command:

```
$ robot  -T \    ## -T is used to generate timestamp on
output files
  -d outputDir \ ## -d provides the output directory for
output files
  FileName.robot ## Test suite file
```

In case of DNSRules, the command to launch is

```
$ robot -T -d /tmp PlatDnsRules.robot
```

In the above command, the whole test suite is executed. If we want to test part of the test suite, we can use the *–test* argument (one for each test case we want to execute):

```
$ robot -T -d /tmp \
   -test TC_MEC_SRV_DNS_003_BR --test TC_MEC_SRV_DNS_003_NF \
   PlatDnsRules.robot
```

Step 4: Check results

A first view of the execution result will be shown in the console, live, during the execution. Each test will be signed as "PASS" or "FAIL".

Figure 10.18 shows the result execution for each test case present in the test suite PlatDnsRules.robot and the overall results.

In the last part of the report, three different files are listed. These files can be used to better investigate the results.

The *report.html* file, depicted in Fig. 10.19, describes the status of the execution. A first visual impact on the result is given by the background of the page, which differs from green to red based on the overall result. If all tests have passed, the background is set to green. Otherwise, it is set to red.

The *log.html* file, on the other hand, contains a much more detailed information for each of the test case executed, as shown in Figs. 10.20 and 10.21.

The generic view of the *Log.html* is showing the list of the executed test cases. For each of them, structured in a tree format, it is possible to navigate the single keywords used in each of the test cases, as depicted in Fig. 10.21.

The last generated file, *output.xml*, is more appropriate to be read by a software (i.e. Jenkins), rather than being read by a human, since this is in an XML format.

Fig. 10.18 Execution report—console view

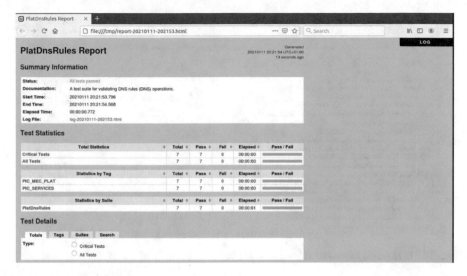

Fig. 10.19 Report.html page

Fig. 10.20 Log.html page

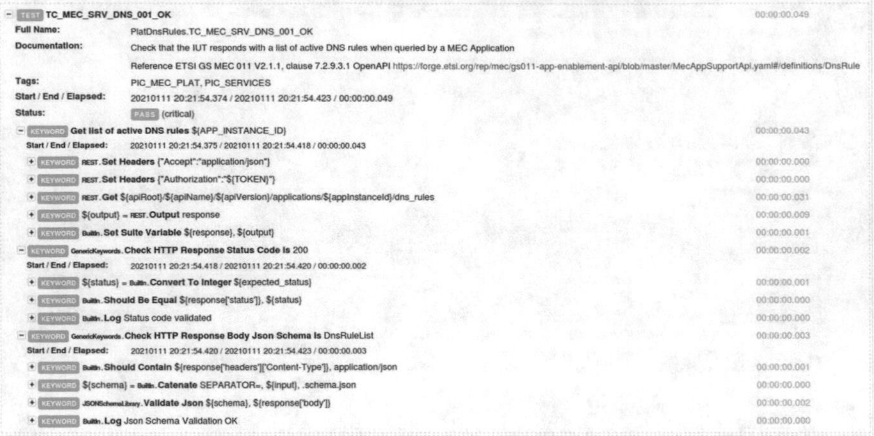

Fig. 10.21 Detailed keywords related to a test case

Exercise 10.2 Goal of the exercise: Get familiar with configuring an MEC Sandbox scenario and navigating the MEC Sandbox dashboard. With this knowledge, understand how to use the MEC-012 Location Service to discover the edge network topology and track the location of a terminal device.

In this exercise, we will need to use the MEC Sandbox. Sign-in to the MEC Sandbox at https://try-mec.etsi.org/ and follow the provided instructions.

The "getting started" and help guide on the ETSI MEC Wiki (https://mecwiki. etsi.org/index.php?title=MEC_Sandbox_Help) provide information on MEC Sandbox usage.

For this exercise, we will use "Try-it in the browser" option and the SwaggerUI interface for MEC services. However, the same exercise can be done from a user's edge application using the "Try-it from your MEC application" endpoints.

MEC Sandbox Configuration

The first action is to select what access network configuration to use (4G, 5G, Wi-Fi) and the number and type of terminal devices. For this exercise, we will use the "4 g-macro" network scenario with two low- and high-velocity terminals (or UEs) and no stationary terminals, as shown below. Observe the MEC Sandbox dashboard to gain an understanding of the network layout and how devices are moving in Monaco.

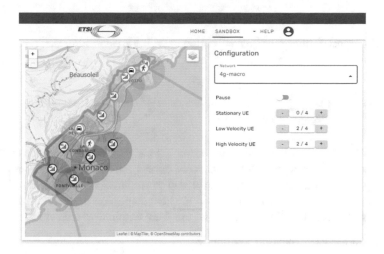

MEC-012 Location Service: Network Discovery

At times when an edge application is deployed and instantiated in the MEC system, the app may not be pre-configured with a network layout or terminal device information. The MEC Location Service can be utilized to discover how the network is organized, how many points of attachment are present, plus the number and location of terminal devices.

In the Sandbox dashboard, select "Location (013)" from the API Console drop down and, then open its SwaggerUI page (either in a new tab or window).

 Helpful hint
Opening the SwaggerUI in a new window allows side-by-side comparison between MEC API responses in the SwaggerUI and the MEC Sandbox dashboard

1. Learn how the network is organized

From SwaggerUI, query the "zones" endpoint to determine the network layout, the number of zones in the network, points of attachment and terminal devices:

```
GET https://try-mec.etsi.org/<sandbox-id>/location/v2/que-
ries/zones (no parameters)
```

The MEC Sandbox responds with a zone list that includes for each zone: number of access points, number of users, a resource URL to query the specific zone and its zone id. How many zones are in the network? Which zone has the most points of attachment and users?

2. Learn a terminal's id to start tracking its location

From SwaggerUI, query the "users" endpoint with a zone id filter (i.e. zone with the most terminals)

```
GET https://try-mec.etsi.org/<sandbox-id>/location/v2/que-
ries/users?zoneId=zone03
```

The MEC Sandbox responds with a user list for each terminal device connected in the requested zone, including connected access point, terminal address (IPv4), device geo-location, connected zone id and a resource URL to query that user terminal directly (this may be helpful in the next step). Are there any high-velocity terminals connected in this zone? What are their terminal addresses?

3. Track a device's location via queries

Periodically, query the "users" endpoint with an address filter (i.e. 10.100.0.1 == high-velocity terminal #1):

```
GET https://try-mec.etsi.org/<sandbox-id>/location/v2/que-
ries/users?address=10.100.0.1
```

For each query, the MEC Sandbox responds with the user terminal information for high-velocity terminal #1, including connected access point id, terminal address, device geo-location, connected zone id, etc. Monitor device 10.100.0.1 on the MEC Sandbox map and when it transitions across access points and zones. When does this terminal enter zone04? Can you also query the "users" endpoint to match this?

MEC-012 Location Service: User Tracking Subscription

MEC services include subscriptions for asynchronous report notifications to edge applications based on events of interest. The MEC location service includes subscriptions for user tracking, zonal tracking and zonal status. Here, we will use the user tracking subscription to track movements of a terminal in the network, as opposed to the previous query-based method.

1. Create a user tracking subscription

Subscriptions are created by using a POST to the appropriate resource URL (e.g. userTracking).

```
POST https://try-mec.etsi.org/<sandbox-
scriptions/userTracking
```

The following parameters define a subscription tracking high-velocity terminal #1 with notifications issued when this device enters, leaves or transfers between PoAs in all location zones.

```
"userTrackingSubscription": {
  "clientCorrelator": "0123",
  "callbackReference": {
    "notifyURL": "http://my.callback.com/location-user-
tracking/high-velocity-ue1"
  },
  "address": "10.100.0.1",
  "userEventCriteria": ["Entering", "Transferring", "Leaving"]
}
```

2. User Tracking Notifications

Once the subscription is created, the Sandbox Location Service will POST a notification whenever high-velocity terminal #1 does a handover in the network (within or between zones). Each notification is visible in the Sandbox API Console, including detailed notification body parameters: new connected PoA id, previously connected PoA id, connected zone id, and the user tracking event. Note the notification code is HTTP response code "500" (indicating a failure). However, this is normal since the provided callback notify URL is invalid in the "Try-it in the browser" option and the Sandbox is not capable of reaching that notify URL location. As an exercise, try setting up a publicly reachable server and use that URL instead.

How does the user tracking notifications is compared to the query-based method? Under what conditions are notifications more appropriate than queries? How could the user event criteria be optimized?

3. Cancel or remove the user tracking subscription –

The subscription can be removed or cancelled by issuing a DELETE to resource URL returned on the initial creation. For example:

```
DELETE https://try-mec.etsi.org/<sandbox-id>/location/v2/sub-
scriptions/userTracking/1
```

	Quiz–Part 4 (Chapters 9–10)

(1) **From an Application Developer perspective:**

    a. MEC application packaging and on-boarding is triggered by the Client App via Mx1

    b. MEC application packaging and on-boarding is triggered by the Device App via Mx1

    c. MEC application packaging and on-boarding is triggered by the Client App via Mx2

    d. MEC application packaging and on-boarding is triggered by the Device App via Mx2

(2) **According to ETSI MEC specifications, the MEC application instantiation:**

    a. Can be triggered only via MEC operator-internal channel

    b. Can be triggered only via request from the Customer Facing Service Portal

    c. Can be triggered only via request from the Device app via User app LCM proxy

    d. Can be triggered in all the three above ways

(3) **Open Network System Services Software (OpenNESS):**

    a. is an open source software

    b. is a standard defined by ETSI MEC

    c. is a MEC application example provided by ETSI MEC as open source implementation

    d. none of the above

(4) **The Akraino API portal:**

    a. is an Akraino portal where all ETSI MEC APIs are published

    b. is an Akraino portal where the API subcommittee has made it a mandatory requirement for each project to report APIs exposed and consumed as part of the project release review

    c. is an Akraino portal where the API subcommittee has made it a mandatory requirement for each project to refer to ETSI MEC APIs

    d. none of the above

(5) **The Open Edge Computing Initiative:**

    a. was created to provide open standards and specifications

    b. is an industry consortium created with the aim to provide guidance to ETSI MEC standard

    c. is an industry consortium created with the aim to address the fundamental challenge of bringing together IT and telecom industry to shape edge computing

    d. was created with the aim to provide guidance to other industry consortia in the edge computing space

(6) **When it comes to MEC performance evaluation:**

    a. functional metrics are related to MEC performance impacting on user perception (e.g. latency, number of API requests)

    b. functional metrics are related to performance of the service in terms of deployment and management (e.g. latency, throughput)

    c. functional metrics are related to MEC performance impacting on user perception (e.g. latency, throughput)

    d. functional metrics can be measured only in terms of average values (e.g. latency, throughput), while other statistics (e.g. maximum value, minimum value) are not meaningful

(7) **MEC metrics are introduced for different comparison purposes:**

    a. assessment of the performances of MEC with respect to non-MEC solutions and evaluation of MEC deployment options

    b. assessment of the performances of MEC with respect to non-MEC solutions and evaluation of MEC solutions from various vendors

    c. assessment of the performances of MEC with respect to non-MEC solutions, with the purpose of stimulating continuous improvement of MEC products

    d. assessment of the performances of MEC with respect to non-MEC solutions, with the purpose of ETSI compliance of MEC products

(8) **MEC testing is intended for:**

    a. conformance of MEC products and their compliancy; this is important to ensure interoperability in E2E systems, especially in presence of multiple stakeholders in the value chain

    b. assessment of the performances of MEC with respect to non-MEC solutions; this is important to ensure interoperability in E2E systems, especially in presence of multiple stakeholders in the value chain

    c. assessment of the performances of MEC in various deployment options; this is important to ensure interoperability in E2E systems, especially in presence of multiple stakeholders in the value chain

    d. conformance of MEC products and their compliancy; this is important for stimulating continuous improvement of MEC products

(9) **DECODE WG provides OpenAPI representations for the MEC Services APIs on ETSI Forge** (https://forge.etsi.org/)

    a. This is an effort in the OpenAPI alliance to complement the definitions for each method and resource defined in the standard specifications, and to produce some machine-readable representation of the MEC APIs

b. This is an effort in the DECODE WG to complement the definitions for each method and resource defined in the standard specifications, and to produce some machine-readable representation of the MEC APIs

c. This is an effort in the OpenAPI alliance to provide reference implementations coming from MEC products that are aligned with the definitions for each method and resource defined in the standard specifications, and to produce some machine-readable representation of the MEC APIs

d. This is an effort in the DECODE WG to provide reference implementations coming from MEC products that are aligned with the definitions for each method and resource defined in the standard specifications, and to produce some machine-readable representation of the MEC APIs

(10) **Among the MEC metrics defined in MEC-IEG 006, Network Energy Efficiency (EE) is defined as**

a. Ratio between a KPI (data rate) and transmitted power
b. Ratio between a KPI (data volume) and consumed power
c. Ratio between a KPI (data rate) and consumed energy
d. Ratio between a KPI (data volume) and consumed energy
e. Ratio between a KPI (data rate) and radiated power
f. Ratio between a KPI (data volume) and transmitted energy
g. None of the above

References

[IEG-006]   ETSI GS MEC-IEG 006 V1.1.1 (2017–01), "Mobile Edge Computing; Market Acceleration; MEC Metrics Best Practice and Guidelines". Link: https://www. etsi.org/deliver/etsi_gs/mec-ieg/001_099/006/01.01.01_60/gs_mec-ieg006v010101p.pdf

[EE-202706]   ETSI ES 202 706–1 V1.6.0 (2020–11), "Environmental Engineering (EE); Metrics and measurement method for energy efficiency of wireless access network equipment; Part 1: Power consumption - static measurement method". Link: https://www.etsi.org/deliver/etsi_es/202700_202799/20270601/01.06.00_50/es_20270601v010600m.pdf

[EE-202336-12]   ETSI ES 202 336–12 V1.2.1 (2018–12): "Environmental Engineering (EE); Monitoring and control interface for infrastructure equipment (power, cooling and building environment systems used in telecommunication networks); Part 12: ICT equipment power, energy and environmental parameters monitoring information model". Link: https://www.etsi.org/deliver/etsi_es/202300_202399/20233612/01.02.01_50/es_20233612v010201m.pdf

[EE-203228]   ETSI ES 203 228 V1.3.1 (2020–10): "Environmental Engineering (EE); Assessment of mobile network energy efficiency". Link: https://www.etsi.org/deliver/etsi_es/203200_203299/203228/01.03.01_60/es_203228v010301p.pdf

[OAS-spec]   OpenAPI Specification. Link: https://github.com/OAI/OpenAPI-Specification
[ProtBuf-spec]   Protocol Buffers Language Specification. Link: https://developers.google.com/protocol-buffers/

[MEC-Wiki]   ETSI MEC Wiki page: https://mecwiki.etsi.org/
[OVB-VTC20]   D. Sabella, D. Brevi, E. Bonetto, A. Ranjan, A. Manzalini and D. Salerno,
                "MEC-based infotainment services for smart roads in 5G environments," 2020
                IEEE 91st Vehicular Technology Conference (VTC2020-Spring), Antwerp,
                Belgium, 2020, pp. 1–6. https://doi.org/10.1109/VTC2020-Spring48590.2020.
                9128807. Link: https://ieeexplore.ieee.org/document/9128807
[TRN-SmRd]   Torino Smart Road project: https://www.torinocitylab.it/en/experiment-to/
                smart-road-track
[SmRd-Decree]   Italian decree for autonomous vehicles testing (in Italian): https://www.
                gazzettaufficiale.it/eli/id/2018/04/18/18A02619/sg
[PLG20-report] ETSI CTI, "ETSI NFV&MEC Plugtests (Remote), 15–19 June 2020—Plugtests
                Report V1.0.0 (2020–08)". Link: https://portal.etsi.org/Portals/0/TBpages/CTI/
                Docs/ETSI_NFV&MEC_2020_Plugtests_Report_v1_0_0.pdf
[PLG20-plan]   ETSI CTI, "ETSI NFV&MEC Plugtests (Remote), 15–19 June 2020—MEC
                interoperability Test Plan V1.0 (2020–06)". Link: https://portal.etsi.org/Portals/
                0/TBpages/CTI/Docs/ETSI_NFV_MEC_Plugtests_2020_-_MEC_IOP_Test_
                Plan_v1_0_0.pdf
[INT-WP-hkt]   https://networkbuilders.intel.com/solutionslibrary/providing-a-hackathon-edge-
                platform-for-application-developers

Annex A
MEC Terminology (Phase 1 and Phase 2)

As described in Chap. 2, the MEC work is divided in phases. Initially, the standardization was focused on the introduction of MEC technology, where the acronym was intended as "Mobile Edge Computing", to promote and accelerate the advancement of edge cloud computing in mobile networks. Only in a second phase, the group decided to change the meaning of MEC acronym, by widening the scope of the work to "Multi-access Edge Computing", thus not only by considering mobile networks but also other accesses (e.g. fixed networks) and non-3GPP networks (e.g. Wi-Fi).

In this annex, we will quickly clarify the differences between MEC terminology used in Phase 1 (ref. [MEC001p1]) and Phase 2 (ref. [MEC001p2]), also by describing the new terms/acronyms introduced by ETSI ISG MEC from Phase 2 (Tables A.1 and A.2).

Table A.1 MEC terminology comparison (Ph. 1 vs. Ph. 2)

Term/Acronym in phase 1	Term/Acronym in phase 2	Comments
MEC (Mobile Edge Computing)	MEC (Multi-access Edge Computing)	Expanding to non-cellular networks (e.g. Wi-Fi, fixed access, etc.)
Mobile edge application	MEC application	Updated definition: "application that can be instantiated on a MEC host within the MEC system and can potentially provide or consume MEC services"
Mobile edge host	MEC host	Updated definition: "entity that contains a MEC platform and a virtualization infrastructure which provides compute, storage and network resources to MEC applications"

(continued)

© Springer Nature Switzerland AG 2021
D. Sabella, *Multi-access Edge Computing: Software Development at the Network Edge*, Textbooks in Telecommunication Engineering, https://doi.org/10.1007/978-3-030-79618-1

Table A.1 (continued)

Term/Acronym in phase 1	Term/Acronym in phase 2	Comments
Mobile edge host level management	MEC host level management	
Mobile edge management	MEC management	
Mobile edge platform	MEC platform	
Mobile edge service	MEC service	
Mobile edge system	MEC system	
Mobile edge system level management	MEC system level management	
UE application	Device application	Application running in the device that has the capability to interact with the MEC system via the user application lifecycle management proxy

Table A.2 New MEC terminology in phase 2

New Term/Acronym in phase 2	Definition	Comments
Application context	Set of reference data about an application instance that is used to identify it, enable lifecycle management operations and associate it with its device application	
Application rules and requirements	Rules and requirements associated to MEC applications, such as required resources, maximum latency, required or useful services, traffic rules, DNS rules, mobility support, etc	
Client application	Application software running on a device (e.g. UE, laptop with internet connectivity) in order to utilize functionality provided by one or more specific MEC application(s)	Note: The definition of device is not limited to UEs
Device application	Application running in the device that has the capability to interact with the MEC system via the user application lifecycle management proxy	
Lifecycle management	Set of functions required to manage the instantiation, maintenance and termination of a MEC application instance	
MEO	MEC Orchestrator	
MEP	MEC Platform	
MEPM	MEC Platform Manager	

References

[MEC001p1] ETSI GS MEC 001 V1.1.1 (2016-03) Mobile Edge Computing
            (MEC) Terminology http://www.etsi.org/deliver/etsi_gs/MEC/001_
            099/001/01.01.01_60/gs_MEC001v010101p.pdf
[MEC001p2] ETSI GS MEC 001 V2.1.1 (2019-01) Multi-access Edge Computing
            (MEC) Terminology https://www.etsi.org/deliver/etsi_gs/mec/001_
            099/001/02.01.01_60/gs_mec001v020101p.pdf

Annex B
Functional Blocks and Reference Points in the MEC System

The present annex is describing all functional blocks and reference points defined by the ETSI MEC architecture [MEC003v2], and according to the MEC terminology [MEC001v2].

Functional blocks

MEC host	Entity that contains a MEC platform and a virtualization infrastructure which provides compute, storage and network resources to MEC applications

(continued)

© Springer Nature Switzerland AG 2021
D. Sabella, *Multi-access Edge Computing: Software Development at the Network Edge*, Textbooks in Telecommunication Engineering,
https://doi.org/10.1007/978-3-030-79618-1

(continued)

MEC platform	Collection of functionality that is required to run MEC applications on a specific MEC host virtualization infrastructure and to enable them to provide and consume MEC services, and that can provide itself a number of MEC services

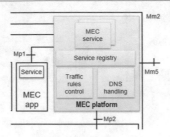

MEC apps	Applications that can be instantiated on a MEC host within the MEC system and can potentially provide or consume MEC services

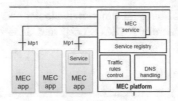

User App LCM proxy	The UALCMP allows device applications to request on-boarding, instantiation, termination of user applications and when supported, relocation of user applications in and out of the MEC system. It also allows informing the device applications about the state of the user applications *Note: A user application is a* MEC application *that is instantiated in the MEC system in response to a request of a user* via *an application running in the device (device application)* The UALCMP authorizes requests from device applications in the device (e.g. UE, laptop with internet connectivity) and interacts with the OSS and the multi-access edge orchestrator for further processing of these requests

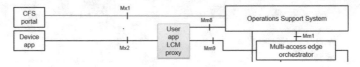

(continued)

(continued)

Device App	Applications in the device (e.g. UE, laptop with internet connectivity) that have the capability to interact with the MEC system via a user application lifecycle management proxy (UALCMP)

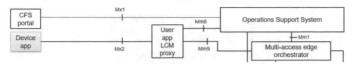

Operations Support System (OSS)	It refers to the OSS of an operator

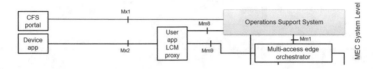

Multi-access edge orchestrator	The Multi-access Edge Orchestrator (MEO)

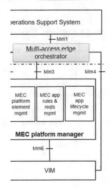

(continued)

(continued)

MEC platform manager	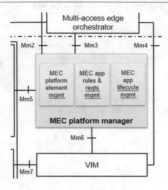
Virtualization Infrastructure Manager (VIM)	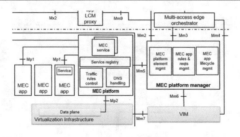

Reference points

Mp1, Mp2, Mp3

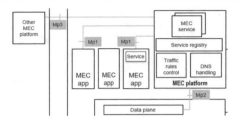

Mx1, Mx2

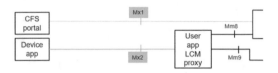

Mm1, Mm2,
Mm3, Mm4

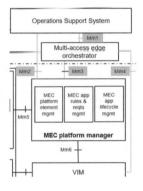

Mm5, Mm6, Mm7

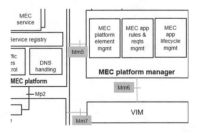

(continued)

(continued)

Mm8, Mm9

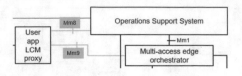

Annex C
MEC Software Resources

C.1—Location API Simulator

Currently, the deployment of edge hardware and software is having a strong acceleration due to the market interest in the potential benefits of this technology. Additionally, the COVID pandemic has further accelerated the need for high bandwidth and low latency services to be provided anywhere. Unfortunately, it is not yet possible, for developers, to interact with production edge servers: notably, the edge servers of an MNO network have a direct link to the BTS that provides the location information and are even more critical to be involved in the development and testing of experimental services.

For these reasons, it could be useful to have available an environment, as similar as possible to a production server. From this idea, LINKS Foundation started the development of a Location Service simulator that will permit developers to create, test and debug applications that retrieve position information of the UE through the Location API. The high-level architecture of the simulator is depicted in Fig. C.1.

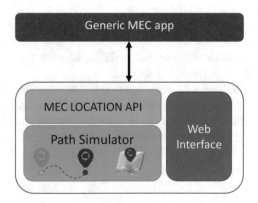

Fig. C.1 High-level view of the Location Service simulator

© Springer Nature Switzerland AG 2021
D. Sabella, *Multi-access Edge Computing: Software Development at the Network Edge*, Textbooks in Telecommunication Engineering, https://doi.org/10.1007/978-3-030-79618-1

Fig. C.2 Web interface to interact with the Location Service simulator

To allow the testing and debugging of services, the simulator can provide the position of different users that are moving on a certain path. The path is a collection of points in GPX format. The file with the different paths can be easily uploaded thanks to the integrated web interface as shown in Fig. C.2.

When the simulation is started, all the information about the users is made available by the Location API block through an interface fully compliant with the API, as defined by the ETSI MEC group.

In this way, developers can create their application and test it in a friendly environment, defining the behavior of the simulated users. The application can then be installed on an edge server (like OpenNESS) keeping the same code also for interacting with a real Location API service.

The exercises at the end of Chap. 5 use this Location Service Simulator as a convenient exemplary tool for edge developers, in order to consume Location APIs from their applications. Here below some practical instructions are presented, from those needed to install and run the suite, in order to be able to implement some concrete examples on the usage of the MEC Location API (e.g. tracking of multiple nodes, displaying them in a map).

How to retrieve the software

To retrieve the software, an email has to be sent to https://mec-services@linksfoundation.com, getting back a link from which it is possible to download a docker container, enveloped in a tarball file, named as *locationapi.tar. gz.*

How to install the SW

The only requirement to install the software is to have *docker* installed on your system (Linux, Windows, MacOS). Thus, the container can be easily loaded and run.

Once the compressed file has been downloaded and *docker* installed, the software can be uploaded to the local docker repository and added to the other docker images with the command below:

$ docker load < locationapi.tar.gz.

Then, its availability can be checked through the list printed out with the following command:

$ docker images.

Finally, the software can be executed:

$ docker run – –network host location-api:latest.

which allocates a web server as localhost and uses port 8080.

Annex D
All Quiz Results

QUIZ PART 1—(Chaps. 1 and 2).

(1) **The lowest level of Cloud computing service model is:**

    a. PaaS
    b. **IaaS**
    c. SaaS
    d. Depends on the OS

(2) **According to Mathis formula, the QoE performance of a TCP network is limited by:**

    a. Latency and bandwidth
    b. Latency and throughput
    c. **Packet loss and round trip time**
    d. Packet Loss and bandwidth
    e. None of the above

(3) **Let's consider the tables below, showing Latency and Bandwidth distributions. We can say that:**

    a. home C and A are probably connected to the same ISP
    b. **the link between home A and B is the best in terms of CDF of Latency and Bandwidth**
    c. the link between home A and B is the worst in terms of CDF of Latency and Bandwidth
    d. **home A and B are connected to ISP1 and home C with ISP2**
    e. considering the median value, the link A and B is equivalent to B and C
    f. None of the above

© Springer Nature Switzerland AG 2021
D. Sabella, *Multi-access Edge Computing: Software Development at the Network Edge*, Textbooks in Telecommunication Engineering,
https://doi.org/10.1007/978-3-030-79618-1

	5%	10%	50%	90%	95%
Home A & B	18.5	19.2	26.4	77.8	133.6
Home B & C	36.4	37.2	44.9	87.2	98.0
Home C & A	38.8	39.3	44.9	75.1	92.6

(a) Latency Distribution (milliseconds)

	5%	10%	50%	90%	95%
Home A & B	0.5	0.6	1.9	2.3	2.3
Home B & C	0.5	0.7	0.8	0.9	0.9
Home C & A	0.5	0.5	0.8	0.9	0.9

(b) Upload Bandwidth Distribution (Mbps)

(4) **In the NIST's Cloud Computing Reference Architecture (CCRA), a Cloud Broker is:**

    a. **an entity that manages the use, performance and delivery of cloud services, and negotiates relationships between Cloud Providers and Cloud Consumers**

    b. an intermediary that provides connectivity and transport of cloud services from Cloud Providers to Cloud Consumers

    c. a party that can conduct independent assessment of cloud services, information system operations, performance and security of the cloud implementation

    d. a person or organization that maintains a business relationship with, and uses service from, Cloud Providers

(5) **MEC means "multi-access" edge computing because:**

    a. MEC users must use multiple access technologies

    b. **ETSI is defining an access agnostic technology, thus applicable in principle to 4G/5G, Wi-Fi or fixed access**

    c. ETSI is defining an access agnostic technology, thus applicable to 5G networks connected contemporarily to Wi-Fi and fixed access

    d. ETSI MEC system can be connected to multiple Wi-Fi access points

(6) **MEC is based on virtualized infrastructure. Please select the correct sentence below:**

    a. MEC is only intended to be deployed in Virtual Machines (VMs)

    b. MEC 027 is clarifying that support that needs to be provided by MEC when containers are used for MEC platforms

c. **MEC 027 is identifying the additional support that needs to be provided by MEC when MEC applications run as containers**

d. as ETSI NFV is also working on containers, the MEC work is aligned with this group, where applicable. Thus, MEC is not studying containers support.

(7) **MEC in NFV is:**

a. **a MEC architecture variant, applicable when MEC is not standalone but deployed in a network based on the ETSI NFV framework**

b. a MEC architecture flavor, applicable when MEC standalone is deployed in a NFV environment

c. a MEC architecture flavor, applicable when MEC is not standalone but deployed in a network based on the ETSI NFV framework

d. a MEC architecture variant, applicable when MEC standalone is deployed in a NFV environment

(8) **5GAA (5G Automotive Association) consider MEC as a key technology for C-V2X use cases. In particular:**

a. an example of highly relevant 5GAA use-case is given by "Software Updates", because they are frequently and continuously updating the whole in-car software package

b. a relevant example is provided by VRU (Vulnerable Road User) discovery use-case, where the VRU detection can be updated by means of frequent Software Updates in the car

c. "Cooperative Lane Change (CLC) of Automated Vehicles" can be done by switching frequently MEC platforms in different lanes, which are serving the same MEC application

d. **an example of MEC relevance is given by the use-case on "High Definition (Local) Maps", due to the real-time and local nature of the data transferred**

(9) **Akraino blueprints are:**

a. Research projects leveraging open source implementations from LF Edge

b. **A definition from Akraino Edge Stack community of "declarative configurations of the entire stack i.e., edge platform that can support edge workloads and edge APIs"**

c. A commonly recognized definition of "reproductions of a technical drawing, an architectural plan, or an engineering design, using a contact print process on light-sensitive sheets"

d. A way, defined by LF Edge and properly covered by copyright, to print the word "Akraino" with a special blue painting and well-specified style

(10) **5G-ACIA:**

    a. is an SDO, similar to 5GAA but focused on connected vehicles and automated driving

    b. **is not an SDO, similar to 5GAA but focused on connected industries and automation**

    c. is not an SDO, similar to 5GAA but focused on connected vehicles and automated driving

    d. is an SDO, similar to 5GAA but focused on connected industries and automation

QUIZ PART 2—(Chaps. 3, 4 and 5).

(1) **According to the definition of Multi-access Edge Computing, the "core business" of ETSI MEC is:**

    a. Providing cloud computing and IT service environment to operators at the edge of the network

    b. Providing cloud computing to application developers at the edge of the network

    c. **Providing to application developers cloud computing and IT service environment at the edge of the network**

    d. All the previous are correct

(2) **The MEC architecture, as described in the ETSI GS MEC-003 specification:**

    a. Is fully compatible with the ETSI NFV framework

    b. Isn't based on a virtualization infrastructure, and cannot be implemented in a NFV environment

    c. Is based on a virtualization infrastructure, and can be implemented in a NFV environment

    d. **Is built upon and consistent with NFV principles**

    e. None of the above

(3) **The MEC reference point Mp3 is defined:**

    a. Between MEC app and MEC platform

    b. Between MEO and ME platform manager

    c. **Between different MEC platforms**

    d. Between MEC platform and the Data Plane

    e. None of the above

(4) **MEC applications:**

    a. **Can be consumers and/or producers of services**

    b. Can be only consumers of services

    c. Can be only producers of services

    d. Cannot consume nor produce any services

(5) **According to the ETSI MEC architecture:**

    a. **The MEC host is an entity that contains a MEC platform and a virtualization infrastructure which provides compute, storage and network resources, for the purpose of running MEC applications**

    b. The MEC platform is the collection of essential functionality required to run MEC hosts on a particular virtualization infrastructure and enable them to provide and consume MEC services

    c. The MEC application is an entity that contains a MEC platform and a virtualization infrastructure which provides compute, storage and network resources, for the purpose of running MEC hosts

    d. The MEC host is the collection of multiple MEC platforms, which are all running on the same virtualization infrastructure which provides compute, storage and network resources, for the purpose of running MEC applications

(6) **A MEC system is composed by:**

    a. A MEC host level with a single MEC host and multiple MEC platforms; a MEC system level, with a MEC orchestrator and a single MEC platform manager for the whole system

    b. **A MEC host level with one or more MEC hosts and MEC platforms; a MEC system level, with a MEC orchestrator and one or more MEC platform managers**

    c. A MEC system level with a single MEC host and multiple MEC platforms; a MEC host level, with a MEC orchestrator and one or more MEC platform managers

    d. A MEC host level with one or more MEC hosts and MEC platforms; a MEC system level, with a MEC orchestrator and a single MEC platform manager for the whole system

(7) **Mp1 reference point is:**

    a. Mandatory for all MEC applications (which can run on the MEC host without connection to the MEC platform)

    b. **Not mandatory for a MEC App (which can run on the MEC host without connection to the MEC platform)**

    c. Mandatory for all MEC applications (but without Mp1, the MEC app cannot run on the MEC host because there is no connection to the MEC platform)

    d. Not mandatory for a MEC App (but without Mp1, the MEC app cannot run on the MEC host because there is no connection to the MEC platform)

    e. Not even a reference point of ETSI MEC

(8) **Mp2 reference point is:**

    a. Connecting together all MEC applications with a single MEC platform
    b. Connecting together MEC applications running in different MEC hosts
    c. **Connecting together the MEC platform and the Data Plane**
    d. Connecting together the MEC application and the Data Plane
    e. Connecting together Data Planes running in different MEC hosts
    f. Not even a reference point of ETSI MEC

(9) **Mp3 reference point is:**

    a. Only defined for audio files and not in ETSI MEC
    b. Connecting together the MEC platform and the Data Plane
    c. Connecting together the MEC application and the Data Plane
    d. Connecting together Data Planes running in different MEC hosts
    e. **Connecting together MEC platforms running in different MEC hosts**

(10) **Mx2 reference point is:**

    f. Connecting together a MEC App and the MEC system via a proxy called UALCMP
    g. **Connecting together a Device App and the MEC system via a proxy called UALCMP**
    h. Connecting together a MEC App and the MEC system via a proxy called CFS
    i. Connecting together a Device App and the MEC system via a proxy called CFS

QUIZ PART 3—(Chaps. 6, 7 and 8).

(1) **CUPS is:**

    a. present since Rel.8 networks, but adopted only in Rel. 14, as a CU/DU split in the eNB
    b. a concept only present in 5G core networks
    c. consisting in control and user plane separation, introduced only from Rel.15 onward
    d. **part of the Rel. 14 standard**

(2) **5G is an important technology because:**

    a. 5G will satisfy contemporarily the needs of different scenarios (eMBB, URLLC, mMTC)
    b. 5G will try to satisfy the needs of different scenarios (eMBB, URLLC, mMTC), and not contemporarily
    c. 5G will introduce a lot of innovations, mainly related to new radio design, to support low latency and high throughput services
    d. **5G will support low latency and high throughput services, expanding the set of 4G addressable services**

(3) **MEC is compatible with 5G system because:**

   a. Edge computing is part of the 5G system architecture

   b. MEC is part of the 5G system architecture

   c. **Edge computing is part of the 5G system architecture, and MEC is leading standard for Edge Computing**

   d. MEC supports 5G system architecture

   e. Edge computing supports 5G system architecture, and MEC is leading standard for Edge Computing

   f. None of the above

(4) **According to the below diagram on technical 5G performance requirements set by the ITU-R:**

   a. 5G system should provide 20 Gbit/s of peak data rate, and always with less than 1 ms of latency

   b. **5G system should provide 20 Gbit/s of peak data rate, i.e. about 20 times of LTE systems**

   c. LTE system can provide up to 20 Gbit/s of peak data rate, but not less than 1 Gbit/s

   d. 5G system can provide up to 20 Gbit/s of peak data rate, or less than 1 Gbit/s

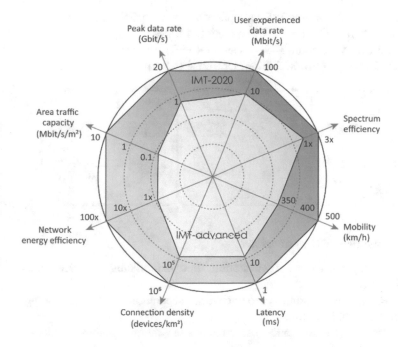

(5) **The 5G deployment option depicted in the figure below is:**

   a. Non-standalone, NR assisted, 5GC connected
   b. Standalone, NR assisted, 5GC connected
   c. **Non-standalone, LTE assisted, EPC connected**
   d. Standalone, LTE assisted, 5GC connected

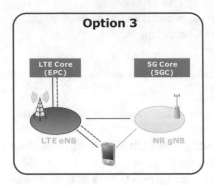

(6) **The 5G deployment depicted in the following three figures are, respectively:**

   a. Option 3 (NSA), Option 2 (SA), Option 4 (SA with EN-DC)
   b. Option 3 (NSA), Option 2 (NSA), Option 4 (NSA)
   c. Option 3 (NSA), Option 2 (SA), Option 4 (SA)
   d. **Option 3 (NSA), Option 2 (SA), Option 4 (NSA)**

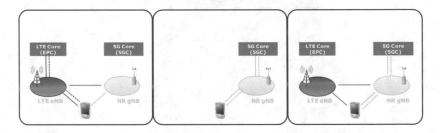

(7) **Key components of 5G systems are:**

   a. **New Radio, network softwarization, network slicing, Service-based architecture**
   b. New Radio, radio and core network separation, network slicing, service-based architecture
   c. New Radio, control and user plane separation, network slicing, service-based architecture
   d. New Radio, network services exposure, access network softwarization, service-based architecture

(8) **Consider the SSC modes below:**

    a. SSC mode 2 is also known as "make-before-break"

    b. SSC mode 3 is also known as "make-before-brake"

    c. SSC mode 3 is also known as "break-before-make"

    d. SSC mode 2 is also known as "make-before-fake"

    e. **None of the above**

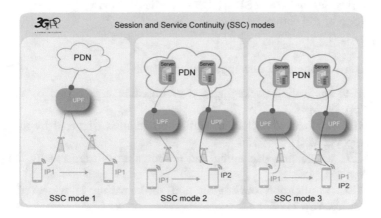

(9) **In the 5G Core, the Network Exposure Function (NEF) is in charge of:**

    a. **Exposure of capabilities and events**

    b. Exposure of entire network slices

    c. Mobility and session management control

    d. Connecting control plane with user plane functions

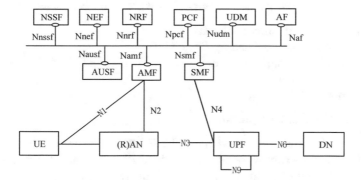

(10) **Considering MEC in 5G systems:**

    a. The AF is a particular implementation of MEC platform

    b. The UPF in the 3GPP architecture may correspond to some functionalities defined in ETSI MEC for a ME application

    c. The MEC application can be mapped with a UPF

    d. **The AF in the 3GPP architecture may correspond to some functionalities defined in ETSI MEC for a ME platform**

    e. None of the above

QUIZ PART 4—(Chaps. 9 and 10).

(1) **From an application developer perspective:**

    a. MEC application packaging and on-boarding is triggered by the Client App via Mx1

    b. MEC application packaging and on-boarding is triggered by the Device App via Mx1

    c. MEC application packaging and on-boarding is triggered by the Client App via Mx2

    d. **MEC application packaging and on-boarding is triggered by the Device App via Mx2**

(2) **According to ETSI MEC, the MEC application instantiation:**

    a. Can be triggered only via MEC operator-internal channel

    b. Can be triggered only via request from the Customer Facing Service Portal

    c. Can be triggered only via request from the Device app via User app LCM proxy

    d. **Can be triggered in all the three above ways**

(3) **Open Network System Services Software (OpenNESS):**

    a. **is an open source software**

    b. is a standard defined by ETSI MEC

    c. is a MEC application example provided by ETSI MEC as open source implementation

    d. none of the above

(4) **The Akraino API portal:**

    a. is an Akraino portal where all ETSI MEC APIs are published

    b. **is an Akraino portal where the API subcommittee has made it a mandatory requirement for each project to report APIs exposed and consumed as part of the project release review**

    c. is an Akraino portal where the API subcommittee has made it a mandatory requirement for each project to refer to ETSI MEC APIs

    d. none of the above

(5) **The Open Edge Computing Initiative:**

   a. Was created to provide open standards and specifications
   b. Is an industry consortium created with the aim to provide guidance to ETSI MEC standard
   c. **Is an industry consortium created with the aim to address the fundamental challenge of bringing together IT and telecom industry to shape edge computing**
   d. Was created with the aim to provide guidance to other industry consortia in the edge computing space

(6) **When it comes to MEC performance evaluation:**

   a. Functional metrics are related to MEC performance impacting on user perception (e.g. latency, number of API requests)
   b. Functional metrics are related to performance of the service in terms of deployment and management (e.g. latency, throughput)
   c. **Functional metrics are related to MEC performance impacting on user perception (e.g. latency, throughput)**
   d. Functional metrics can be measured only in terms of average values (e.g. latency, throughput), while other statistics (e.g. maximum value, minimum value) are not meaningful

(7) **MEC metrics are introduced for different comparison purposes:**

   a. **assessment the performances of MEC with respect to non-MEC solutions and evaluation of MEC deployment options**
   b. assessment the performances of MEC with respect to non-MEC solutions and evaluation of MEC solutions from various vendors
   c. assessment the performances of MEC with respect to non-MEC solutions, with the purpose of stimulating continuous improvement of MEC products
   d. assessment the performances of MEC with respect to non-MEC solutions, with the purpose of ETSI compliance of MEC products

(8) **MEC Testing is intended for:**

   a. **conformance of MEC products and their compliancy; this is important to ensure interoperability in E2E systems, especially in presence of multiple stakeholders in the value chain**
   b. assessment the performances of MEC with respect to non-MEC solutions; this is important to ensure interoperability in E2E systems, especially in presence of multiple stakeholders in the value chain
   c. assessment the performances of MEC in various deployment options; this is important to ensure interoperability in E2E systems, especially in presence of multiple stakeholders in the value chain
   d. conformance of MEC products and their compliancy; this is important for stimulating continuous improvement of MEC products

(9) **DECODE WG provides OpenAPI representations for the MEC Services APIs on ETSI Forge** (https://forge.etsi.org/)

    a. This is an effort in the OpenAPI alliance to complement the definitions for each method and resource defined in the standard specifications, and to produce some machine-readable representation of the MEC APIs

    b. **This is an effort in the DECODE WG to complement the definitions for each method and resource defined in the standard specifications, and to produce some machine-readable representation of the MEC APIs**

    c. This is an effort in the OpenAPI alliance to provide reference implementations coming from MEC products, that are aligned with the definitions for each method and resource defined in the standard specifications, and to produce some machine-readable representation of the MEC APIs

    d. This is an effort in the DECODE WG to provide reference implementations coming from MEC products, that are aligned with the definitions for each method and resource defined in the standard specifications, and to produce some machine-readable representation of the MEC APIs

(10) **Among the MEC metrics defined in MEC-IEG 006, Network Energy Efficiency (EE) is defined as:**

    a. Ratio between a KPI (data rate) and transmitted power

    b. Ratio between a KPI (data volume) and consumed power

    c. Ratio between a KPI (data rate) and consumed energy

    d. **Ratio between a KPI (data volume) and consumed energy**

    e. Ratio between a KPI (data rate) and radiated power

    f. Ratio between a KPI (data volume) and transmitted energy

    g. None of the above

Annex E
Acknowledgements

This book is a tremendous effort of many years. I wrote most of the content, but definitely its finalization and whole publication would not have been possible without the help and support of many friends and colleagues.

Here below is the list of all contributors and reviewers (chapter by chapter):

Chapter 1

Contributors: All sections—Dario Sabella (Intel, ETSI ISG MEC Chairman); Reviewed by Alex Reznik (HPE, former ETSI ISG MEC Chairman).

Chapter 2

Contributors: All sections—Dario Sabella (Intel, ETSI ISG MEC Chairman); Reviewed by Dario Sabella.

Chapter 3

Contributors: All sections—Dario Sabella (Intel, ETSI ISG MEC Chairman); Reviewed by Alice Li (Huawei, ETSI ISG MEC Vice-Chair).

Chapter 4

Contributors: All sections—Dario Sabella (Intel, ETSI ISG MEC Chairman); Sect. 4.4—Luca Cominardi, Angelo Corsaro, Olivier Hecart, Julien Enoch (Adlink), Exercise 4.1—Giada Landi, Elian Kraja, Pietro Piscione (Nextworks), Exercise 4.2—Luca Cominardi, Angelo Corsaro, Olivier Hecart, Julien Enoch (Adlink).
Reviewed by Alice Li (Huawei, ETSI ISG MEC Vice-Chair).

D. Sabella, *Multi-access Edge Computing: Software Development at the Network Edge*, Textbooks in Telecommunication Engineering, https://doi.org/10.1007/978-3-030-79618-1

Chapter 5

Contributors: Sect. 5.1—Dario Sabella (Intel, ETSI ISG MEC Chairman); Sect. 5.2 —Edoardo Bonetto, Daniele Brevi, Maurizio Floridia and Riccardo Scopigno (LINKS Foundation), Sect. 5.3—Jing Zhu (Intel), Sect. 5.4—Miltiadis Filippou (Intel), Exercise 5.2—Edoardo Bonetto, Daniele Brevi, Maurizio Floridia and Riccardo Scopigno (LINKS Foundation), Exercise 5.3 and Annex C—Jing Zhu (Intel), Exercise 5.4—Miltiadis Filippou (Intel).
    Reviewed by Dario Sabella (Intel, ETSI ISG MEC Chairman).

Chapter 6

Contributors: All sections—Dario Sabella (Intel, ETSI ISG MEC Chairman); Exercise 6.1 (Manuel Femia), Exercise 6.2 (Kishen Maloor).
    Reviewed by Uwe Rauschenbach (Nokia, ETSI ISG MEC delegate).

Chapter 7

Contributors: All sections—Dario Sabella (Intel, ETSI ISG MEC Chairman); Sect. 7.4.1—Changhong Shan (Intel), Sect. 7.5.1—Dario Sabella, Danny Moses (Intel), Sect. 7.5.2—Joey Chou, Yizhi Yao (Intel), Exercise 7.1—Giovanni Nardini, Giovanni Stea, Antonio Virdis (University of Pisa).
    Reviewed by Dario Sabella (Intel, ETSI ISG MEC Chairman).

Chapter 8

Contributors: All sections—Dario Sabella (Intel, ETSI ISG MEC Chairman); Sect. 8.3 (Masaki Suzuki, KDDI), Exercise 8.1—Miltiadis Filippou (Intel), Exercise 8.2—Giovanni Nardini, Giovanni Stea, Antonio Virdis (University of Pisa).
    Reviewed by Dario Sabella (Intel, ETSI ISG MEC Chairman).

Chapter 9

Contributors: Section 9.1—Dario Sabella (Intel, ETSI ISG MEC Chairman); Sect. 9.2—Neal Oliver, Purvi Thakkar, Amr Mokhtar, Sindhura Gaddam, Greg Allison (Intel); Sect. 9.3—Marco Picone, Stefano Mariani (UniMore); Sect. 9.4— Walter Featherstone (Samsung R&D Institute UK, ETSI MEC DECODE WG Chair), Jane Shen (Akraino API subcommittee Co-Chair); Sect. 9.5—Claudio Cicconetti (CNR); Sect. 9.6—Rolf Schuster (Director, Open Edge Computing Initiative), Jim Blakley (Living Edge Lab Associate Director, Carnegie Mellon University); Exercises 9.1 and 9.2—Giada Landi, Elian Kraja, Pietro Piscione (Nextworks); Exercise 9.3—Marco Picone, Stefano Mariani (UniMore); Exercise 9.4—Claudio Cicconetti (CNR); Exercise 9.5—Jim Blakley (Living Edge Lab Associate Director, Carnegie Mellon University); Exercise 9.5—Jim Blakley (Living Edge Lab Associate Director, Carnegie Mellon University), Bob Gazda (InterDigital), Debashish Purkayastha (InterDigital);
    Reviewed by Neal Oliver (Intel).

Chapter 10

Contributors: All sections—Dario Sabella (Intel, ETSI ISG MEC Chairman); Sect. 10.5—Bob Gazda, Debashish Purkayastha (InterDigital); Sect. 10.6.3— Anish Rawat (Intel); Exercise 10.1—Giada Landi (Nextworks), Elian Kraja (Nextworks), Pietro Piscione (Nextworks); Exercise 10.2—Bob Gazda, Debashish Purkayastha (InterDigital);

Reviewed by Michele Carignani (ETSI CTI).

Pri
by

Chapter 10

Contributors: All sections—Dario Sabella (Intel, ETSI ISG MEC Chairman); Sect. 10.5—Bob Gazda, Debashish Purkayastha (InterDigital); Sect. 10.6.3— Anish Rawat (Intel); Exercise 10.1—Giada Landi (Nextworks), Elian Kraja (Nextworks), Pietro Piscione (Nextworks); Exercise 10.2—Bob Gazda, Debashish Purkayastha (InterDigital);
   Reviewed by Michele Carignani (ETSI CTI).